T0292702

CAMBRIDGE LIBRARY COLLECTION

Books of enduring scholarly value

Life Sciences

Until the nineteenth century, the various subjects now known as the life sciences were regarded either as arcane studies which had little impact on ordinary daily life, or as a genteel hobby for the leisured classes. The increasing academic rigour and systematisation brought to the study of botany, zoology and other disciplines, and their adoption in university curricula, are reflected in the books reissued in this series.

Memoir and Correspondence of the Late Sir James Edward Smith, M.D.

Originally published in 1832, this two-volume account of the life of Sir James Edward Smith (1759–1828) was posthumously compiled by his wife, Pleasance (1773–1877). Smith trained originally as a doctor, but his independent wealth enabled him to pursue botany. Hugely influenced by the work of Linnaeus, he benefited greatly from the purchase of the latter's library and herbarium in 1783, upon the advice of his friend, Sir Joseph Banks. He was highly regarded throughout Europe as a botanist, and in 1788 founded the Linnean Society. He published various botanical works, of which the most important was *The English Flora* (1824–8), and assisted in the publication of many more. His wife recounts his character as well as his achievements, using both narrative and 'various familiar and domestic letters' to do so. Volume 1 includes letters from Banks and Samuel Goodenough, bishop of Carlisle and Smith's close botanical friend.

Cambridge University Press has long been a pioneer in the reissuing of out-of-print titles from its own backlist, producing digital reprints of books that are still sought after by scholars and students but could not be reprinted economically using traditional technology. The Cambridge Library Collection extends this activity to a wider range of books which are still of importance to researchers and professionals, either for the source material they contain, or as landmarks in the history of their academic discipline.

Drawing from the world-renowned collections in the Cambridge University Library, and guided by the advice of experts in each subject area, Cambridge University Press is using state-of-the-art scanning machines in its own Printing House to capture the content of each book selected for inclusion. The files are processed to give a consistently clear, crisp image, and the books finished to the high quality standard for which the Press is recognised around the world. The latest print-on-demand technology ensures that the books will remain available indefinitely, and that orders for single or multiple copies can quickly be supplied.

The Cambridge Library Collection will bring back to life books of enduring scholarly value (including out-of-copyright works originally issued by other publishers) across a wide range of disciplines in the humanities and social sciences and in science and technology.

Memoir and Correspondence of the Late Sir James Edward Smith, M.D.

Volume 1

Edited by Pleasance Smith

CAMBRIDGE UNIVERSITY PRESS

Cambridge, New York, Melbourne, Madrid, Cape Town,
Singapore, São Paolo, Delhi, Tokyo, Mexico City

Published in the United States of America by Cambridge University Press, New York

www.cambridge.org
Information on this title: www.cambridge.org/9781108037075

This edition first published 1832
This digitally printed version 2011

ISBN 978-1-108-03707-5 Paperback

Bust by F Chantrey. R.A. H.B.Love. delin. Engraved by W.Say.

SIR JAMES EDWARD SMITH,

President of the Linnæan Society

AGED 68.

Pub^d by Longman & Co London, Augt 1832.

MEMOIR

AND

CORRESPONDENCE

OF

THE LATE

SIR JAMES EDWARD SMITH, M.D.

FELLOW OF THE ROYAL SOCIETY OF LONDON;
MEMBER OF THE ACADEMIES OF
STOCKHOLM, UPSAL, TURIN, LISBON, PHILADELPHIA, NEW YORK, ETC. ETC.
THE IMPERIAL ACAD. NATURÆ CURIOSORUM,
AND
THE ROYAL ACADEMY OF SCIENCES AT PARIS;
HONORARY MEMBER OF THE HORTICULTURAL SOCIETY OF LONDON;
AND
PRESIDENT OF THE LINNÆAN SOCIETY.

EDITED
BY LADY SMITH.

"How delightful and how consolatory it is, among the disappointments and anxieties of life, to observe Science, like Virtue, retaining its relish to the last!"
Sketch of a Tour on the Continent, vol. ii. p. 60.

IN TWO VOLUMES.
VOL. I.

LONDON:
PRINTED FOR
LONGMAN, REES, ORME, BROWN, GREEN, AND LONGMAN,
PATERNOSTER ROW.

1832.

PREFACE.

IN the following pages, which contain the principal events in the life of Sir James Edward Smith, the writer has been tempted to insert various domestic and familiar letters, even from an early period, as it appeared to her that they mark the progress of his character, his predilection for botanical science, and other facts, more faithfully than a narrative composed of different materials would be likely to do.

From the vast accumulation of letters, preserved during a period of more than fifty years, the limits of this work would admit only of a selection; and many therefore are omitted, equally worthy of publication.

The letters of the Bishop of Carlisle, following the chapter upon Sir James's works, being chiefly critical, and relating to those publications, are kept together, that the subject may not be interrupted.

The correspondence of Mr. Davall and Sir James is kept separate from the general current of letters, as it relates principally to the botany of Switzerland, and their peculiar regard for each other. That of the Marchioness of Rockingham, their mutual friend, is, for similar reasons, mingled with theirs.

The letters of Mr. Caldwell, of Dr. Wade, &c. are inserted together, being descriptive of Ireland chiefly.

Those of the Abbé Corrêa De Serra and a few other friends have a reference to some political occurrences in Portugal which affected the Abbé's safety; and among them are a few concerning a magnificent plant, the *Cyamus Nelumbo*, whose history has been a subject of interest and speculation with classical botanists.

The correspondence of Mr. Roscoe and

Sir James is given apart from others, for the same reason as Mr. Davall's.

To her friend Dr. Boott, of Gower Street, the editor is under great obligation for his repeated acts of kindness in the course of the work, which she is happy thus to acknowledge: and for assistance in the selection of foreign correspondence she is indebted to Mr. Dawson Turner's friendship. For every thing else the compiler alone is answerable; —conscious as she is of the imperfections of the work, and unconscious probably of many that may have escaped her observation, it might seem unjust not to make this avowal.

Should it be inquired why no portrait of an earlier age is given, in preference to that which is prefixed to the work, the answer is contained in the fact, that of several delineations by different artists at various periods, none have been esteemed as likenesses, and that the bust of Sir James, by the hand of Chantrey, in the library of the Linnæan Society, conveys the only representation of him which retains the expression of his mind, through the features; and if the copy of it at the opening of this volume is less perfect than could be wished, it must be ascribed to the difficulty of fixing in a copy the expression which was often found impossible to catch from the original. Sir James Edward Smith's mental lineaments are less difficult to trace; they will be seen in various lights, sketched by a variety of hands, in the succeeding pages.

PLEASANCE SMITH.

Norwich, 1832.

CONTENTS.

VOL. I.

CHAPTER I.

CHAPTER II.

CHAPTER III.

D Portland

Saml: Carlisle –

M Rockingham

W: Roscoe

N Wallich

E. Davall.

J Sibthorp.

J E Smith. 1784.

Alex Humboldt

7 juil 1818 Bonpland

Bn Cuvier

B. Mirbel

C. Kunth.

Henry Muhlenberg

Jacob Bigelow

John Bradbury

B S Barton

Ben Waterhouse

John Brickell

DeWitt Clinton

Wm D Peck

David Hosack

Wm Darlington

Giovanni Scopoli

Grégoire Fontana

Scarpa.

Allion.

Barnaba Oriani.

Esprit Giorna

Ippolito Durazzo

W Roxburgh

Francis Buchanan

Dugald Stewart

J a Beattie.

Robt Brown

James-Brodie.

Alexr McLeay

A Menzies

Arthur Bruce

Obedient Humble Servt

Buchan

John Hope

John Walker

Henry Bryant.

H. Rose

J Pitchford

James Dickson

C Linnæus 1774.

A L De Jussieu

Gerard

Augusto Broussonet

Thouin

Desfontaines

a P. de Candolle

Joannes de Bragança

Joseph Corrêa de Serra

O d. Rodrigo de Souza Coutinho

Antoine J. Cavanilles

Joseph Antoine Pavon

MEMOIR

AND

CORRESPONDENCE

OF

SIR JAMES EDWARD SMITH.

CHAPTER I.

Introductory notice.—Birth of Sir J. E. Smith.—Pedigree of the Kinderley family.—Mr. Smith, the father of Sir James.—Early education.—His domestic amusements.—Fondness for history.—Began the study of systematic botany the day Linnæus died.—Trained by his father for merchandise.—Acquaintance with James Crowe, Esq., Mr. Pitchford, Mr. Rose, Rev. H. Bryant, T. J. Woodward, Esq.—Goes to Edinburgh.—Dr. John Hope.—Lord Monboddo.—Mr. Engelhart a young student from Sweden.—Mr. Batty.—Mr. Broussonet.—Forms a Society for the study of natural history at Edinburgh.—Tour in the Highlands.—Makes a collection of the native plants of Scotland.—Presented with a gold medal by Dr. Hope.—Made first President of the Natural History Society.—Letter of the Earl of Buchan upon that occasion.—Letter of Mr. Kindersley from Tinnenelly.

THE Correspondence of the late President of the Linnæan Society falling under the care of her who is the natural guardian of all the confidential as well as scientific communications it contains, it appeared to her improper and scarcely possible that any other person should overlook, or could have leisure to select from several thousand letters, those most fit for the public eye. This consideration had sufficient weight with her to attempt what otherwise she would have left in abler hands; but, as the letters form the principal and most engaging portion of the following Memoir, she has given her attention to put them in their proper places, and to burthen the reader with few that are trifling, although many will be interesting chiefly to persons who love to trace the doubts and progress of those learned men, who by degrees and with much mental labour, but more mental enjoyment, have raised the botanical department of natural science to its present high station and importance; and why may she not add another if not a stronger inducement to the work?—the delight of renewing some shadow of that choice society in which she has lived over again while preparing these Letters for the press.—Would that the ability to appreciate the virtues and talents of the lamented subject of this Memorial imparted equal ability to record them!

Sir James Edward Smith has been so long known as the possessor of the Linnæan Library and Herbarium, and as the original founder of the Society which bears the name of the illustrious Swedish naturalist, that some account may be expected here of his early years, and of the circumstances that led him to the choice of a profession offering few lucrative rewards to an aspiring and not independent man. "The last infirmity of noble minds" can hardly be said to have spurred him to the effort; for an effort it surely is, to choose a path through which we see but darkly where it leads.

Though enthusiasm and a love of fame had perhaps some influence, a love of science and of truth had greater still. He said to others, "The fairest flower in the garden of creation is a young mind, offering and unfolding itself to the influence of divine wisdom, as the heliotrope turns its sweet blossoms to the sun;" and may it not be said of him that taste and virtue fixed his choice?

He was born at Norwich the second of December 1759, was the eldest child of his parents, and for almost five years continued the only one. From infancy a delicacy of constitution marked his bodily frame; and an extreme susceptibility was no less obvious in his mental temperament. He was consequently more under the immediate care and direction of his mother than most children require to be, and it was from her, that at a very early period he imbibed a taste for flowers, which she had pleasure in cultivating. He seldom in after-life saw the delicate blue flowers of the wild succory, without recalling to mind, that, when, in infancy, their beauty caught his eye and attracted his admiration above most others, he tried in vain to pluck them from the stalk.

Probably the charm of this quiet amusement was greatly enhanced by a natural timidity, a diffidence amounting to a degree often painfully embarrassing, and which was never so obliterated from his remembrance, but that at times he would recur to events in his childhood, when for a word or almost a thought he feared was wrong, he experienced the pangs of a broken and a contrite spirit, and in his later years has felt that pity for his former sufferings, which he would have done for those of a different individual.

It was impossible for a mother not to be tenderly attached to such a disposition in her child; and accordingly a more than common affection subsisted between them, and he at all times spoke of her as his guardian and his friend.

The family of Kinderley, from whom Sir James is descended on the maternal side, is an old and opulent one in the north of England. His great-great-grandfather, Geoffery Kinderlee of Spalding in Lincolnshire, was the intimate friend of Daniel de Foe, the well-known author of Robinson Crusoe; who in one of the persecutions which he suffered, was sheltered from its effects in the house of Geoffery Kinderlee. He died in 1714, and is buried in Spalding churchyard under a stone bearing this character of him:

> "He was a very charitable and merciful man."
>
> "The pleasure which from virtuous deeds we have,
> Affords the sweetest slumber in the grave."*

Nathaniel Kinderley, Sir James's great-grandfather, lived at Saltmarsh, between Stockton and

* Of this ancestor many anecdotes are preserved. He was notorious for having had six wives, and it was whispered that he sought them in healthier counties, and that the change to the fens of Lincolnshire soon gave him an opportunity of trying his fortune again. He drove four horses in his coach, and had an appropriate

Durham, and inherited a considerable fortune from his father Geoffery before mentioned. He married Mary, grandaughter of the honourable Francis Pierpoint, uncle to Evelyn Earl of Kingston, by whom he had issue John, Nathaniel, Audrey, and Mary. He endowed a school for poor boys at Dundee, and attempted to civilize the inhabitants of the northern part of England near which he resided.

Being induced to speculate in a project for the improvement of the navigation of the river Dee, Mr. Kinderley lost a great portion of his property in that adventure: he also engaged in the Eaubrink drainage, and was the original projector of the measure which has within a few years been carried into effect and completed near Lynn in Norfolk. A canal in that district is still known by the name of the Kinderley Cut. But these schemes proved unsuccessful at the time, and ruinous to Mr. Kinderley's fortune. He left an estate at Setch within four miles of Lynn, which thirty years ago was valued at about 12,000*l.*, and was the residue of the property saved from the wreck of his unfortunate speculations.

airing for every morning in the week, in which his grandson John was his frequent companion. In these excursions he heard many anecdotes, which he transmitted, probably with some embellishments, to his successors.

His sixth wife survived him, and is supposed to have avenged her predecessors by her excessive care; for she made a flannel cap to his gold-headed cane, lest the coldness of the metal should affect his health, and this treatment he did not long survive.—But these have the air rather of jocular tales than of serious accusations, and we may believe "he was a very charitable and merciful man."

His eldest son John, the grandfather of Sir James, being obliged to turn his attention to some means of living, beyond the remnant of his paternal fortune, chose the clerical profession, and was sent to St. Andrew's, where he graduated. The corporation of Norwich presented him with the perpetual curacy of St. Helen's church in that city, and the vicarage of South Walsham in Norfolk; and the Countess of Leicester appointed him her domestic chaplain at Holkham.

The generosity of his nature was a check upon his enriching himself or his family, although he preserved his independence; but he seldom could take his dues from a necessitous parishioner, and it was his invariable custom, at the wedding of a poor couple, to receive the fee from the husband and slip it into the bride's hand. He was remarkable for the sweetness of his temper and for several little eccentricities of character; was a great pedestrian, and not unfrequently walked from Norwich to Holkham before breakfast, a distance of at least forty miles, always beginning his journey at one or two in the morning. His daughter on one occasion received the severest reproof he ever gave her for altering the clock to retard his hour of setting off.

He married Sarah, daughter of Mr. John Raining *,

* Mr. Raining gave the service of communion-plate still in use at the Octagon Chapel in Norwich; 500*l.* for charitable uses to his native town of Dumfries; 1200*l.* to found a school in North Britain; 1000*l.* among several schools and congregations about Norwich, or, if times of persecution should arise, to their ministers; besides other charitable legacies.

Mr. Raining was the intimate friend of the Rev. Robert Flea wealthy Dutch merchant, a woman of graceful and refined manners, and considerably more accomplished than ladies of her day usually were: she survived her husband many years. The Rev. John Kinderley died the 10th of April 1775, aged 69: his widow lived to the great age of 91, and died in 1799. They had two children: Frances, the mother of Sir James Smith, whose exemplary conduct through life requires no eulogium here, for her memory is still cherished by a numerous circle of friends and relatives; her life was protracted to the great extent of 88 years, when, without bodily in-

ming, V.D.M., who had been minister of the English church at Leyden, and was patronized by the Prince of Orange. He published a poetical essay on the death of King William; but Mr. Fleming is more remarkable as the author of a prophecy deduced from the obscure and highly poetic visions of the Apocalypse, contained in a "Discourse concerning the Rise and Fall of the Papacy," printed in 1701. The following passage was noticed soon after the commencement of the French Revolution by the editors of some London newspapers; and being received with suspicion by many readers, Sir James having in his possession a copy of Mr. Fleming's work, which descended from his predecessor Mr. Raining, he took it with him to London, for the purpose of showing it to his incredulous friends.

"There is ground to hope that about the beginning of another such century, things may alter again for the better; for I cannot but hope that some new mortification of the *chief supporters* of *antichrist* will then happen; and perhaps the *French monarchy* may begin to be considerably humbled about that time: that whereas the present *French King* takes the *sun* for his *emblem* and this for his *motto, Nec pluribus impar*, he may at length, or rather his successors, and the monarchy itself (at least before the year 1794), be forced to acknowledge that (in respect to neighbouring potentates) he is ever *singulis impar*."

firmity or any mental failing, she expired in February 1820;—and Nathaniel, lieutenant-colonel of the honourable East India Company's artillery service, whose only son was the friend through life of his cousin, the subject of this memoir.

Mr. Smith, the father of Sir James, was a man of strong understanding and of a cultivated mind. Having in his early youth occasion to reside some time at Clifton hot-wells for the recovery of his health, he was induced by the kindness of a lady there, who conceived a partiality for him as a clever and superior boy, to employ his leisure hours in learning French, and with the language he acquired a taste for reading the best authors of a country whose history more especially engaged his attention, and perhaps few men in his own or any station read more, or enjoyed in a greater degree the charm of good composition. He left remarks upon the style and character of most of the works belonging to the book-club of which he was a member; and they are indicative of the acute and sound judgement he possessed*.

* The following is a specimen, though a very short one, of Mr. Smith's notice of the books he read:

"*Some Thoughts concerning Education*, 1693.

"This little useful treatise was written by the celebrated Mr. Locke, and is truly valuable, although the luxury and effeminacy of the present times will not, cannot conform to the rules he delivers. He recommends a private education, because he says the first, the greatest object of education is virtue and goodness. In reading his observations, 'tis impossible not to reflect how very different is the mode of education in the great schools from that he thinks proper."—*Common-place Book.*

A habit of thinking for himself strengthened his understanding; and his son has often expressed himself deeply indebted to his father's, and it ought also to be added his mother's, encouragement not to follow any received opinion blindly and implicitly, but dare to think for himself and stand alone. For the free action this gave his mind, he to the last year of his existence expressed his obligation to both parents.

The education which Sir James received was entirely domestic: he never was at a boarding-school, and had even as a child, a dread of being sent to one. The best masters, however, which the city of Norwich afforded, attended him at home, and he acquired the knowledge of grammar only through the Latin tongue: an English grammar, he has frequently observed, he never had; nor did his proficiency in Latin extend beyond the rudiments of the language, till he had passed the usual period of a school-boy's age.

The French and Italian languages he acquired correctly, and made some progress in mathematics; and in the society of well-informed, sensible parents, those hours which in a public school are frequently grievous, or unavoidably wasted, those domestic evenings which expand the heart with the understanding, and "leave us leisure to be good," were devoted to reading, or lessons rendered pleasing by the associations connected with them.

His timidity has just been mentioned; but it was the timidity of a tender infant mind, fearful of doing wrong. As he grew older, mental courage was perhaps the most prominent feature of his character, and its supporting effect was felt in his presence, and diffused confidence. "He feared God, and knew no other fear."

It seems natural to suppose that his father's love of historical works led his son to take an interest in similar researches; for at the early period of eleven or twelve he composed an imaginary history of Scotland, very fairly and correctly written and expressed, in which two races of kings are accurately described. The writer of these pages has in her possession the drawing-room or saloon of these illustrious personages, and their place of interment under the chapel floor.

If invention be the proof of genius, it must be accorded to him even in this youthful occupation; for nothing is borrowed: though the dresses, language, and furniture strictly resembled realities he had seen, yet the materials he used and the manner of applying them to his minute population, show very uncommon originality of design.

The writer is not ashamed to acknowledge, that reading the history of this ideal court, its ladies, servants, and dependents, and the satirical verses and pasquinades upon some members belonging to it, has occasionally beguiled a winter's evening very agreeably, when the company of some young friend has been the occasion of introducing the "*Paper People*," as they were called, upon the tea-table: and at the same time his own playful recurrence to the scenes of his youthful happiness produced an enjoyment which will never return. "Man was made for relaxation as truly as for labour; and by a law of his nature, he finds perhaps no relaxation so restorative as that in which he reverts to his childhood; seems to forget his wisdom; leaves the imagination to exhilarate itself by sportive inventions; talks of amusing incongruities in conduct and events; smiles at the innocent eccentricities and odd mistakes of those whom he most esteems; allows himself in arch allusions or kind-hearted satire, and transports himself into a world of ludicrous combinations. It may be said that on these occasions the mind seems to put off its wisdom; but the truth is, that in a pure mind wisdom retreats, if we may so say, to its centre, and there unseen keeps guard over this transient folly, draws delicate lines, which are never to be passed in the freest moments, and, like a judicious parent watching the sports of childhood, preserves a stainless innocence of soul in the very exuberance of gaiety."

Whoever recollects his love of fun and drollery may perceive that it was an original part of his disposition; but at all times joined to such a sweetness of temper and true humanity as left no sting behind: a quick perception of the ridiculous, and especially the grave affectation of wisdom, was always irresistible.

Enough, and perhaps it may be said more than enough, has been bestowed upon this juvenile game; yet it may be considered as the embryo appearance of a taste, or rather passion, for historical records, which was at all times the relaxation most salutary to his spirits, and to which he daily looked forward in the pages either of a real or fictitious representation of human life, with a zest that almost identified himself with the characters he read of. Whether he saw a delineation of human passions and events in the pages of Robertson, Froissart, Brantome, or Roscoe, or met with them in those of Lady Morgan, Mrs. Ratcliffe, Richardson, Fielding, or Sir Walter Scott, his sympathies were called forth, and he would weep or rejoice as the master pen of the writer touched his heart and charmed his imagination.

It was his knowledge of history that made his company so delightful in travelling; for never did he pass a spot, marked by an event in our national history, without reminding his companion, and thus furnishing topics of discourse from place to place, and peopling a desert with forms long since at rest, and ideas that were as amusing as the prospect before them; while, to enrich the scene, and fill each step with interest, the *habitats* of plants were always noticed, and their sure recurrence gave a delight well known to those of similar pursuits.

The writer can never forget some occurrences of this kind, and will mention one, because it happened in a road generally thought dull and wearisome, except to men of the turf, and that is Newmarket heath, a tract he always passed with particular pleasure, from the exhilarating effect of the pure air.

Here it was, one "incense-breathing morn," he pointed out to her notice the *Carduus acaulis,* whose close and stunted growth his companion had attributed to the barren soil and frequent treading upon. Soon after a wide field covered with the rich blossoms of the red poppy and the viper's bugloss, which attracted their mutual admiration, reminded him that when Mr. Kindersley's East Indian servant first travelled into Norfolk and passed over this heath, he exclaimed in rapture, "Yonder are flowers worthy to adorn the gardens of *the gods*, and *here* they *grow wild!*"

Nor was this a solitary instance of the pleasure afforded to the admirer of nature in an unpromising tract of country. Sir James's friend the late Andrew Caldwell, Esq. of Dublin, describing his return out of Norfolk, observes, " It was dark before we reached Newmarket, and the night misty; I could only perceive this part of the road led over vast extended heaths. The favourite *Verbascum*, I could not help observing, gradually took its leave, and disappeared entirely some time before day declined. The *Centaurea Cyanus*, *Cichorium Intybus*, and *Echium* in profusion on the road sides—whose beautiful blue colours attracted *even the notice of my servant*. The sun rose in the utmost splendour just before we came to what I believe was Epping forest. The landscape was wild and enchanting enough for the warmest fancy of the poet and painter, yet this pleasure was lost to every one but me. Not a door or window in any house but was closed, and the inhabitants asleep. How continually we give up the most delightful hours! Yet, sensible of this as I am, I shall persevere in error, following the example that is always surrounding one. An opportunity was soon afforded of perfect contrast. The sky became clouded before eleven, and heavy showers were repeated the whole day afterwards. What a change from the brilliant light and the beautiful objects of the forest, to the gloom and confinement of streets and houses!

"The travelling observer of nature," Sir James remarks, "has, as it were, the enjoyment of a new sense in addition to those common to the rest of mankind. He can find amusement and instruction, where they bemoan themselves as in a wilderness; he can relieve his attention and refresh his spirits when wearied by common objects of observation, or troubled with disagreeable ones, and is stimulated with ardour to undertakings, prolific of pleasure in various ways, which the incurious, half-occupied mind would not think worth the pains of attempting. A still higher advantage is attached to the pursuit of natural history in a journey through an enlightened country, as well as in the journey of life itself. It is an unerring clue to an intercourse with the best minds. It brings those together who are connected by a most commendable, disinterested and delightful tie; it brings forth the best parts of every character."

Happy as he was in an excellent mother, he was no less so in *another maternal character*, to whom the writer owes an infinite debt of gratitude and love. Sir James has often reminded the person who records the circumstance, of the pleasure it gave him, when, in early life, and then almost a stranger, he first paid a visit to that inestimable parent, and found the apartment destined for himself, decorated by her hand with wreaths of the *Calluna* and *Erica*

Tetralix. The appearance of these beautiful flowers served to open an immediate communication of minds and taste, led to harmony of ideas on other subjects, and proved the beginning of a friendship which was never impaired by time or a nearer affinity with each other. These are the graces which compose the poetry of life; which require neither equipages nor liveries; which, instead of waste, create abundance; and best of all, unlock the treasures of a well-stored mind,

"Rich in the pure and precious pearls of splendid thought!"

Those only who have witnessed the effect which the tranquillity of the country, the sight of gardens, the unpacking even of dried specimens, had upon him, can form an idea of the serenity and charm such situations and objects produced;—a flow of happy spirits, never overbearing, a ready wit, an enjoyment which communicated its happiness to all about him, not a solitary pleasure, exclusive of society, but which made society itself more welcome.

In his "Biographical Memoirs of Norwich Botanists," published in the seventh volume of the Linnæan Society's Transactions, in 1804, Sir James has recorded an event worthy of remembrance.

"I became," he says, "at the age of eighteen, desirous to study botany as a science. The only book I could then procure was Berkenhout, Hudson's Flora having become extremely scarce. I received Berkenhout on the 9th of January 1778, and on the 11th began, with infinite delight, to examine the *Ulex europæus* (common furze), the only plant then in flower.

"I then first comprehended the nature of systematic arrangement, and the Linnæan principles; little aware, that, *at that instant,* the world was losing the great genius who was to be my future guide; for Linnæus died on the night of January the 11th, 1778."

In an age of astrologic faith, such a coincidence would have excited superstitious reflections, and the polar star of the great northern philosopher might have been supposed to shed its dying influence on his young disciple.

It was his father's intention to train his son to merchandise, with a design of his engaging in the importation of raw silk; but the thoughts of trade never satisfied him, and he passed some years in doubts and wishes that kept him from repose and enjoyment.

As an inclination for science unfolded itself, he formed connections more congenial to him; and a few of these, who knew his wishes, strongly urged his father to offer him a more suitable object.

Among the friends whom a love of botany procured him, must be mentioned the late James Crowe, Esq. of Lakenham, to whose constant attachment and friendship Sir James was indebted for much of the choicest social happiness he enjoyed in his subsequent residence in his native town. Mr. Crowe died in 1807, and on that occasion his friend drew up a short memoir of him, of which the following is a part. "He had," says his biographer, "for nearly thirty years past studied the botany of Britain with indefatigable zeal, and with peculiar success. A penetrating eye, and no less sagacious and discriminating mind, fitted him in an eminent degree for the study of nature. No man ever loved that science more, nor derived more satisfaction from the goodness and wisdom of the Creator as displayed in his works, to which he was constantly recurring. To the more difficult parts of British botany he had given peculiar attention, especially to the Mosses and Fungi, but above all to the Willows, a tribe of plants, which, however important in an œconomical point of view, may be said, before his time, to have been almost entirely unknown to botanists, so confused were their ideas concerning them.

"In public life Mr. Crowe was a warm and strenuous assertor of the genuine old English Whig principles; to which he was attached by early education, extensive reading and experience, but especially by his uncommon acuteness of judgement and manliness of sentiment, for

'Never Briton more disdain'd a slave.'"

Mr. John Pitchford, another of his early associates, "was one of a school of botanists in Norwich among whom the writings and merits of Linnæus were perhaps more early, or at least more philosophically, studied and appreciated, than in any part of Britain."

Of this school was Mr. Hugh Rose, "who to much classical learning added a systematic and physiological turn of mind." In 1780 a *gutta serena* deprived him of sight. This affliction he bore with exemplary patience; for though with the loss of his external visual organs he lost his darling amusement, "no one," observes Sir James, "could ever derive more consolation from looking within."

To these must be added the Rev. Henry Bryant, "a man of singular acuteness, well skilled in the mathematics, and sufficiently master of his time to devote a considerable portion of it to natural history." And, Thomas Jenkinson Woodward, Esq., of Bungay in Suffolk, whom he terms "his intimate and much-loved friend, the most candid, liberal and honourable of men." The Fungi and Sea-weeds were the vegetable tribes he more particularly studied.

To the late Dr. Manning, a physician of eminence, and to Robert Alderson, Esq., the present Recorder of Norwich, he considered himself more especially indebted for the accomplishment of his views towards going to Edinburgh to complete his education, and acquire what was necessary for the practice of physic, that being the profession towards which he at this time directed his attention. On the 14th of October, 1781, he began his journey to Edinburgh. This was his first separation from an affectionate family and home, and it was not therefore with unmixed feelings of delight that he attained the object of his sanguine wishes. He has sometimes recurred with emotion to his distressing sensations, when he turned from his father and a beloved brother, who accompanied him part of the way, to pursue the remainder of his journey alone.

As a man of inflexible integrity in all his mercantile transactions, and of moderation blended with generosity in his domestic regulations, Mr. Smith had been distinguished among his townsmen; but in the correspondence which follows, the reader may view his character nearer still, and will sympathize in the tenderest parental feelings and the deep sense of piety and virtue which influenced and pervaded his mind.

Mr. Smith to Mr. James Edward Smith.

My dear Son, Norwich, Oct. 27, 1781.

We received your letter with all the joy that our concern and anxiety for your welfare had prepared for us in the gratification of our wishes; and I cannot refrain from giving you the pleasure of receiving a letter from me before you may expect it. I am afraid you had an unpleasant journey to Sheffield, though you make the best you can of it; but 'tis over safe, and the remembrance of it will be pleasant hereafter. I can never regret the journey I took with you, as it has left some of the tenderest ideas my mind is possessed of, and although anxious and serious were the minutes, they were precious indeed. What would I not give for such another morning as we passed at Wansford! It is true the separation and the rest of that day was cutting, and as much as I could well bear; but every reflection on the prospect that attends you is a balm to heal the wounds that absence gives the mind. It is obvious, that in proportion to the difficulties we encounter in the pursuit of laudable acquirements, whether of riches, honours, or knowledge, in adequate proportion is the pleasure of overcoming them, and the enjoyment of the rewards we have obtained.

Our return home would have been very pleasant indeed, if our separation had not been so recent. We had delightful weather, fine roads, and very capital towns to pass. We lodged on Tuesday at Alconbury Hill, a most excellent house, the neatest I ever saw; on Wednesday we breakfasted at Fen Stanton, and dined at Cambridge, where I showed John, King's College chapel only, the walks, and fronts of the colleges, and the streets: we had time for no more; and came to Newmarket to sleep. It was one of the meetings, so next day we went upon the course, heard a deal of betting, and saw three matches run. There was very little company, but some great men; the Duke of Grafton, Marquis of Rockingham, Earl of Egremont, and, above, all Mr. Charles Fox. It was amusing and agreeable enough to see this picture of human life. God be with you, and bless all your undertakings!

Your ever affectionate Father,

JAMES SMITH.

Writing home to his father on the second of November 1781, he acquaints him with the success of his introductory letters, and informs him that the person from whom he expects to derive most comfort and advantage is Dr. Hope*. "He has the highest character for abilities and real goodness of heart, and is a man of the first consequence in this place: his behaviour was at first (as it generally is) a little reserved; but botanical subjects opening the way, he became perfectly affable, and treats me with almost paternal tenderness. Having found that I was quite a novice in the study of medicine, he talked the whole

* Dr. John Hope, professor of botany.

over with me, and recommends me, above all things, first to make myself master of Latin, for which purpose he has recommended me a master, who taught all his children, who is to come for an hour every day: the usual terms are a guinea a month, but I am to give after the rate of eight guineas a year, and expect six or eight months will do. I hope you will not disapprove of this expense, as it is quite necessary, and you may depend on my frugality in every case where I can save money without missing anything of real importance. Dr. Hope thinks that, with the utmost œconomy, I cannot spend less than 120*l*. a year; but I don't see how it can amount to near that.

"I am quite pleased with my lodgings and companions. My only fellow-lodger, besides Mr. Lubbock, is Mr. Engelhart, a most accomplished and agreeable young gentleman, whose father is physician to the King of Sweden.

"At Dr. Hope's I have seen Lord Monboddo*: he is a plain-dressing elderly man; he had on an ordi-

* The following notes concerning this nobleman's works are extracted from Mr. Smith's common-place book.

"*Ancient Metaphysics, or the Science of Universals*. J. Balfour, Edinburgh, 1783: 2 vols. 4to.—Lord Monboddo is the author of this very extraordinary work, and some other whimsical ones. It is amusing to see what great lengths the imaginations of some contemplative men will carry them in fanciful hypotheses, which the Abbé Buffier aptly calls philosophical romances. Indeed metaphysicians are a sort of knights errant in literature, who sally out in search of adventures in fancy's region; and their wildness and absurdity, like that of the knights historians, are more or less shocking to reason and probability, as they are more or less ingenious and penetrating; but they are always absurd in something,

nary gray coat, leather breeches, and coarse worsted stockings; he conversed with me with great affability about various matters, spoke of the great decline of classical learning at Edinburgh, and mentioned the Norfolk husbandry, which he said he had adopted.

"I often think of you, and imagine what is passing in the scenes which my friends render so dear to me. Pray give my most affectionate duty to my dear mother."

Mr. Smith to Mr. James Edward Smith.

My dear Son, Norwich, Nov. 12, 1781.

We received your letter with a pleasure equal to transport, for the satisfaction it gave us to hear how well you go on so far, which although I did never doubt, yet the confirmation of the hopes I had enter-

when they affect to discover what is out of the reach of the faculties of man to know, or even to comprehend."

"*Ancient Metaphysics, containing the History and Philosophy of Man, &c.* 378 pp.—This is the 3rd volume, and it appears in it that his lordship, Lord Monboddo, the author, proposed to continue the work by an inquiry into the state and condition of man to be expected after death, where I suppose his lordship will get to his furthest.

"I read only a part of the 1st and 2nd volumes;—this was so amusing I read it through. The wildness of the author's imagination and his credulity go beyond all bounds. There are some acute and sensible passages; but when his lordship tells you he believes there have been a race of men born with tails, another with only one leg, others twenty or thirty feet high; that the orang outang is *really* a man, and the *true standard* of our species in a natural state, and a great deal more such extravagant stuff, the sober reader must conclude his lordship's 'learning hath made him mad.'"

tained is their fruition, to which I am sensible in proportion to that fond and eager desire for your welfare and prosperity, from which they owe their existence. It cannot be doubted but you will recommend yourself wherever you are by those amiable qualities that gained and fixed you so many friends in your native place; and I trust to your discretion and knowledge of the world to distinguish and choose such among those that present themselves to your acquaintance, that you will be able to reap most advantage from in point of knowledge, true politeness, and sincere friendship. As for morals, you are too well grounded in virtue, and sound, unaffected piety, to make it at all necessary for me to mention them, as I am sure you will shun the immoral and profane, both from taste and principle.

We are happy to hear you are so satisfied with your lodgings, and the company you find in them, especially the young Swede, whose acquaintance must be both entertaining and useful, and his friendship may be of service when and where you don't expect it*; so it may be right to cultivate it, when you have sounded the heart; you know if that be not good, no reliance is to be had upon even warm professions. It is a great comfort to us that you are received so kindly by Dr. Hope, and that you have fallen into such hands as may supply in a great measure the place of a father to you; and I have no

* There appears something prophetic in this observation. It was Dr. Engelhart who recommended Sir James to Dr. Acrel, and was mainly instrumental in the acquisition of the Linnæan collections, as will be noticed hereafter.

doubt, my dear James, you will meet with many such friends during your stay in Edinburgh.

My dear, I cannot disapprove of any expense that is useful to your pursuit, therefore have no objection to a Latin master. Latin and Greek are necessary to your profession in more respects than being keys to the doors of science, into any of which you may enter if you have those keys; and I would wish you should have as good ones in your possession as any body else has; they should have no advantage over me in that respect, and I believe, between ourselves, there is a great deal in the parade of it, besides the use. The men of learning have agreed to stamp a high value upon classical learning: it sets them out of the reach of the vulgar, and of those who are their superiors in every other worldly advantage; yet I do not think it is all sterling worth, but a great deal of it imposition. I would not be without any of it that will be really useful to you, nor without enough of it to be creditable, but I would not sacrifice anything substantial to attain more; the knowledge of *things* is your proper study, and an acquisition of that knowledge will be the solid and profitable advantage of your attention *; that will be your grand aim; and as the study you have so delighted in, not only requires a mind formed for nice observa-

* The opinions on the subject of education in the above passage very much coincide with those of Milton.

" Though a linguist should pride himself to have all the tongues that Babel cleft the world into, yet if he have not studied the *solid things* in them, as well as the words and lexicons, he were nothing so much to be esteemed a learned man, as any yeoman or tradesman competently wise in his mother tongue only. Hence

tion, as the foundation of a genius successful in that branch, but also a patient discriminating judgement, joined to constant industry and close observation to seize the varieties of nature in her productions, I persuade myself that genius and taste, joined to those qualities which you possess in an eminent degree, will enable you to discern wherein the essential of medicine lies, and to discover not only as far into the nature of diseases, and what relates to them, as human knowledge has yet gone, but I go further, and flatter myself you will be distinguished for a judgement and penetration that surpasses most others, and such as will procure you the esteem and veneration of the world, as I am without doubt your conduct and behaviour in your practice will secure you the love of it.—You say I may depend upon your frugality in every case. I know I may, my dear; but I would not have you cramp yourself, nor deny yourself either any advantage or enjoyment upon that account. I am perfectly easy; satisfied that you would not wish for what I ought to refuse.

appear the many mistakes which have made learning generally so unpleasing and so unsuccessful: first, we do amiss to spend seven or eight years merely in scraping together so much miserable Latin and Greek as might be learned otherwise easily and delightfully in one year.

"I shall detain you now no longer in the demonstration of what we should *not* do, but strait conduct you to a hill-side, where I will point you out the right path of a virtuous and noble education: laborious indeed at the first ascent, but else so smooth, so green, so full of goodly prospect and melodious sounds on every side, that the harp of Orpheus was not more charming."—*Milton: Letter to Master Samuel Hartlib.*

Mr. D. showed us the Maccaroni rooms at Newmarket, where the ordinary for dinner is 28*s.* besides wine, and for supper 14*s.*, where every luxury is procured, and every vice, I am afraid, indulged. How much happier are the innocent, who know nothing of these excesses of the great and affluent!

I hope you have seen Dr. Hutton. Your meeting with Lord Monboddo, and at Dr. Hope's, pleases me much. A superficial view of singular and great characters is entertaining, and an acquaintance with them is honourable and useful.

I would not have you make a burthen of writing letters to any of your friends, because you will have enough of sedentary application without that.

Perhaps I may be your physician as long as I live, if you prove as great as Boerhaave or any other: but you will make allowance for a father's tenderness, for indeed, James, I love you as much as any father ever did a son, and I have the joy of supposing you will allow me to show it to the last. I will take care it shall not be ridiculous fondness, if I can; but fondness I must be indulged in.

Your affectionate Father,

James Smith.

The following is from Mr. James Dickson, author of "Four Fasciculi of Cryptogamic Plants."

To Mr. James Edward Smith.

Dear Sir, October 22, 1781.

I received yours. I am very glad to hear from you, and return my best thanks for the *Bryum rigidum.*

As such a rare plant has been found with you, I have no doubt but more will be found if looked after. You must not depend upon my judgement on Fungi, it being but a new study to me as yet. I find no author that is to be depended upon, and none worse than Mr. H.; and Mr. Lightfoot has so very few. Schæffer's figures, some good and some very bad; I know about an half of what is in Hudson to be sure of. I mean Agarici, and many he has not got. If you go to the Highlands of Scotland, I have not any doubt but you will find some new plants. I have received some from thence since Mr. Lightfoot's publication. I am now on the top study of Fungi; should be glad of all you can send me, and I will return you what I make of them. I received some from Mr. Crowe: I am not able to make them all out as yet. You are wrong in your doubts about the *Elymus arenarius*—I have seen the *Elymus philadelphicus* at Sir Joseph Banks's. I have a specimen of *Elatine Alsinastrum*, but did not find it myself. I shall ask Sir Joseph Banks about it: he is the only man that can inform me of it. The *Arbutus Andrachne* that sold at Dr. Fothergill's for fifty-one guineas was not half so large as that at Chelsea. I found a few days ago the *Boletus perennis*, which I had never seen before. Have you ever seen the *Lycoperdon pedunculatum?* I want it much.

I am, Sir, your very humble servant,

JAMES DICKSON.

Mr. Smith to Mr. James Edward Smith.

My dear Son, Norwich, Dec. 1781.

The manner in which you speak of your situation, the pleasure you take in the objects of your studies, the satisfaction the prospect gives you, the company you have got acquainted with, and, above all, the friendly manner in which Dr. Hope treats you, quite transports us; and as we have so much confidence in your prudence and virtue as to be quite satisfied that neither the examples of ——— will draw you into vice, nor the blandishments of beauty mixed with coquetry will steal you from yourself and us, we have no uneasy reflections on those considerations: but as it will give you more enlarged knowledge of the world, I doubt not you will be more confirmed in your principles of the excellence of virtue, and will receive a polish and ease of deportment from the other, which, if it does not enhance the intrinsic value of your mind, will set off your more valuable qualities, and altogether will recommend you to the esteem of the penetrating and the superficial, to people truly valuable and the world in general. The one is obtained by true merit; the other by external show of it: and there is nothing either vicious or base in courting the approbation of both by the talents they are adapted to admire, and both will be found useful if properly estimated. As for ———, you draw an amiable picture of him, and you may do him much good if he does you no harm; you also may gather knowledge, curious at least, from him: but beware of strict friendships. I don't mean, avoid them, but be cautious how you engage; they very frequently influence a man's prosperity, and oftener his happiness through life. But honey is to be gathered from the flowers of poisonous plants, with submission to you botanists;—I repeat submission, in every sense; for I delight to think you will soon be above the reach of my feeble pen in every branch of knowledge, ethics, and moral philosophy, as well as physic and the *belles lettres*, and I shall be proud of taking from, instead of giving lessons to you; so you will not be troubled with so much sermonizing; yet probably I may not leave it off at once, and ever throw in an observation which appears to me may be useful. 'Tis an effect of the imbecillity of age to hobble in their advice, as in their gait; and they fancy people think them wise, when they undertake to instruct others; but nothing is more fallacious, and nothing so common, as to see an old prating fool, a Polonius, think himself an overmatch for Hamlet.

I have declined the cottage; and as your absence makes a greater impression there than any other place, and nobody has the taste to take care of it, it gives your mother and me too much pain to see the spot so neglected as it is, and will be; and those plants you used to nurse and view with delight in their progress and perfection, those plots you used to ornament, vacant, and no hand able to cherish them. We cast a mournful eye upon them, and a sigh; frequently a pang is excited: thus a gloom hangs about us, instead of the former cheerful disposition when we looked with joy on what had been the effects of your beloved hands, and had given you health and amusement. So we are almost in the mind to let it.

There is a talk of some removes in the ministry, and change of measures; and indeed things wear a very serious appearance. It is time to think how to find means to save the nation, which is crumbling fast to pieces, and must soon be ruined if our affairs be no better conducted than they have been ever since Lord North has been at the helm; though I presume people where you are entertain different sentiments: but you are not at Edinburgh to learn politics. I esteem the Scotch much for their zeal for the protestant religion, yet I think two sermons at a time rather too much. I hope their kirks are warmer than our churches.

I am, and ever shall be,

Your most affectionate Father,

JAMES SMITH.

Mr. James Edward Smith to his Father.

Honoured Sir, Edinburgh, Dec. 31, 1781.

I was much entertained on Friday se'nnight at a *mourning concert* in honour of the Earl of Kelly, a member of the musical society here; this compliment is always paid to every member's memory soon after his death: the music is all of the sacred kind, and as fine as can possibly be, and the company and performers are all in mourning. It is very difficult to get admittance to these concerts; ladies have the preference, of course, and gentlemen are balloted for by the members. I was so fortunate as to be proposed, together with Mr. Engelhart, by his friend Sir William Forbes, and we were both admitted; but great numbers were rejected: and above a hundred people who had got tickets could not get in. The room is most elegant, of an oval form.

I am very happy in the Miss Riddels acquaintance. I have dined there twice with some great people, and their brother Sir James; he behaved very politely to me: his lady is my aunt Kindersley's friend, a literary lady. Sir James knows much of Norwich; is acquainted with Counsellor Chambers, Aldermen Harvey, Thurlow, Ives, the Morse family, and Dr. Manning, of whom he spoke much.

We have had twelve days vacation; on Wednesday next the college meets again. I have been very much taken up with invitations, as this is a time of feasting; though not so much so as in England. I was much surprised to see all the shops open on Christmas-day, which is not observed here at all, except by the church-of-England members, who have a most beautiful chapel here, where they perform cathedral service.

I do not perceive that the better sort of people are less neat here than elsewhere. I am sure in many places I visit, the most exquisite neatness is apparent.

My friend Engelhart introduced me one night as a visiter to the Medical Society, of which he is president.

My warmest prayers are offered up for a continuation of your health and happiness in the approaching new year, and many more after it. I am quite well and continue very happy.

I am, honoured Sir,

Your ever dutiful Son,

J. E. Smith.

Mr. Smith to Mr. James Edward Smith.

Dear James, Norwich, Jan. 17, 1782.

We are all much pleased that you pass your time so agreeably. You must begin to experience the advantage of travelling, of which I doubt not you will also reap the benefit by forming your own manners and carriage upon the best models that fall in your way: I mean, you will choose the best from each individual, for few are altogether perfect; and that after weighing well in your reflections what strikes you in the manners and behaviour or conversation of any person as polite, amiable and sensible, you would endeavour to trace the principle upon which such a manner was in general approved, and if it had its foundation in reason. If it was alloyed with any affectation, it will not stand that test; but if it had intrinsic beauty, and will bear examination, such a scrutiny will fix it so in your mind, that you will have it operate upon yourself without being too sensible of it; for when a man feels himself really acting after a pattern, he must be stiff and unnatural at best; 'tis very likely he may appear affected, which is equally disgusting and despicable to people of judgement and taste. I am satisfied, my dear, your own good sense makes this and many other things I write, and have written, unnecessary; but I am satisfied too, when it happens so, I have too great a share in your affection and esteem to apprehend any disdain.

You were much obliged to Mr. Engelhart for the concert. I like much you should gain an acquaintance with valuable foreigners; 'tis impossible to say how useful they may happen to be in every walk in life.

Yours ever,

JAMES SMITH.

The Rev. Henry Bryant to Mr. James Edward Smith.

Dear Sir, Heydon, Jan. 18, 1782.

I rejoice to hear that you are seated so much to your satisfaction, and trust you will find your account at last in the walk of life you have chosen, and which nature herself seems to have chosen for you. I am in no fear but you will eagerly catch hold of every favourable advantage you can meet with for improvement, and which in your present situation I know must be very many. I rejoice also to hear that you have already begun to make some accession to the Flora Scotica; which I am sure must be very imperfect. Lightfoot's situation in life, whatever advantages he might boast of, could never qualify him for giving a botanical account of all the plants of North Britain. To do this, he ought to have been stationary in it for some years, and to have had free communication with the chief botanists in that country.

This has been the mildest winter in Norfolk I ever remember, and consequently a fine season for cryptogamic botany. Crowe and I have made good use of our time, and have found a multitude of new things; many of which are not described, and I want you much to delineate some of them for me. I have found a new species of *Jungermannia*. I have found one good specimen of *Phallus caninus* Schæff., which agrees exactly with his figure, *t.* 330. I have found many specimens of Hudson's *Helvella planis*, which as it is a creature of his own, and consequently there can be no figure of it in any author, I will beg the favour of you to delineate one for me: and as it must be new to Dr. Hope, pray give him two specimens, with my compliments.

Pray remember your sincere Friend,

H. BRYANT.

Mr. James Edward Smith to his Father.

Honoured Sir, Edinburgh, Feb. 12, 1782.

I now sit down to give you the information you desire concerning my studies. In the first place, my progress in Latin satisfies myself, and my master too, as Dr. Hope tells me: indeed I find it very easy. I am at present reading Dr. Home's *Principia Medicinæ* (an excellent book), after which shall begin Celsus. I think one quarter more with my tutor will be sufficient; and in the summer shall continue my application to it regularly by myself, so that next winter I hope to be able to attempt a little Greek (a very little will do; but I would not be entirely ignorant of it.) Few physicians go further than the works of Hippocrates, which are very easy I am told. Indeed I am far from being singular in my deficiency in the knowledge of Latin; but I assure you no application shall be wanting to complete me in it as much as possible. I am really very fond of the language, and have got over the worst part: I am before most students in the knowledge of French and Italian, the former of which is of the greatest use; and both have assisted me wonderfully in acquiring Latin.

I have learned to pronounce Latin like Italian; as it is pronounced so here, as well as by all foreigners; I mean the vowels*.

I know no entertainment equal to Dr. Monro's physiological lectures; his remarks are so ingenious, satisfactory, and curious, that we could never be tired with hearing them. He cannot forbear throwing out remarks now and then, when he finds either of his great rivals Haller or Hunter in a mistake.

Your affectionate Son,

JAMES EDWARD SMITH.

* "Their speech is to be fashioned to a distinct and clear pronunciation, as near as may be to the Italian, especially in the vowels. For we Englishmen, being far northerly, do not open our mouths in the cold air wide enough to grace a southern tongue; but are observed by all other nations to speak exceedingly close and inward; so that to smatter Latin with an English mouth, is as ill hearing as law French."—*Milton on Education; in a Letter to Master Samuel Hartlib.*

Mr. Smith to Mr. James Edward Smith.

My dear Son, Norwich, Feb. 25, 1782.

We are all much pleased that you spend your time so agreeably, and hope nothing I have said will convey the idea that I think you too profuse in your amusements: on the contrary, as you rightly say, it is a part of your education *de vous apprivoiser à la grande foule;* besides, I look upon diversions as useful, nay necessary, to relieve your mind and renew its vigour, to exhilarate the spirits and give a zest to life, for which end the beneficent Author of our nature has given us the capacity of an almost innumerable variety of enjoyments, which are all lawful when they are expedient, when they promote our happiness and that of our friends and connections. I look upon the promotion or production of genuine true happiness to be the surest mark of virtue, if it is not *virtue itself.* Some philosophers call a mediocrity in all things, virtue: however that be, *medio tutissimus ibis* is an excellent maxim, and I am in no fear you should transgress; on the contrary, I would rather urge you forward to take pleasure than restrain you, for I am not in the least afraid you should go beyond what will do you real good. So, my dear, go to as many diversions as you like, see everything you can, and push forward your acquaintance with genteel, valuable people; and be not under any concern whether you spend a few pounds more or less in the year. I would not have you neglect any advantages, nor deny yourself any proper gratification for fear of swelling your expenses. Solomon says, "There is a time to scatter, and a time to gather:" do you scatter wisely, and I will endeavour to gather carefully, and hope I shall so far succeed as to leave a comfortable subsistence to every one that depends upon me for support. I think you had better not fix a time to leave off your tutor: 'tis impossible to tell where you may be situated, or how; and scholarship will recommend in all parts of the world. And as you have the elegancies of French and Italian, the useful Latin, with a little Greek, will be desirable. God be with you and bless you, my ever dear child!

Your affectionate Father,

JAMES SMITH.

Mr. James Edward Smith to his Father.

Honoured Sir, Edinburgh, March 11, 1782.

I want words to answer as it deserves that part of your letter concerning my expenses: can only say, your kind indulgence will have a most powerful influence in making me as œconomical as real prudence will allow of. With respect to diversions interrupting my application, I dare say you have no apprehension of that: as you know my inclination for the pursuit of science, you will easily believe that to be the highest pleasure I can enjoy, as I am at present circumstanced with every advantage for that pursuit, and at full liberty to explore the vast stores of knowledge that are presented to me on every side.

The diversions here will be over in a few days; as

the session (the courts of justice) rises then. I have not been to Archer's-hall again, nor to the concert, but have been to three assemblies in the new town, in a room opened only this season, where there is a subscription-ball every Friday. I have a very useful friend in Lady Gordon, with whom I became acquainted at the Queen's assembly, and who always finds me a partner when I am at a loss for one. It is a common complaint among the students, that the Edinburgh people are very proud, and that students are despised by them: I have not experienced any neglect on account of my profession, nor do I believe that any one who behaves decently will experience it.

The Miss Riddels are excellent acquaintances for me; they have very good connections, and are perfectly conversant with genteel life: they are my privy counsellors in all matters of etiquette, and are quite unreserved and familiar with me.

Mr. Martineau advises me to get into the Medical Society; but there is a law made, that no more can be admitted this year. The Earl of Buchan was made an honorary member last Saturday; but that is an extraordinary thing. I always find means to get in as a visitor: I was there last Saturday fortnight, and *spoke twice*, from which I hope you will think I have got rid of some of my *mauvaise honte*. The members were disputing on the analogy of the diseases of brutes with those of men, and how far the method of cure, which succeeds in theirs, might be applied to ours. I ventured to represent the danger which might happen from trusting too much to this analogy, considering that many plants are poisonous to some animals and wholesome to others, of which I gave several instances. The president very politely thanked me for my observations. I find they are wonderfully ignorant of natural history: and even my little knowledge of the subject gives me an importance which I hope will be of great advantage, and may perhaps in some measure atone for my deficiency in classical learning.

James Edward Smith.

To his mother, on the following day, he expresses himself thus:—"My happiness, honoured madam, in my present situation is completed by your expressing so much happiness in my prospects, as well as my father. I cannot help considering it, as you say, peculiarly directed by the Almighty, and therefore I recur immediately to him when any gloomy ideas present themselves; as I hope I have the most perfect confidence in him, and trust he will preserve us all to be a blessing to each other. But if he thinks fit to separate us, I hope we could acquiesce; and we know that not a single kind thought can ever be lost, or lose its reward. I have met with a number of young play-fellows, as you said I should. The children of Dr. Duncan are very pretty, and remarkably sensible; and here are a sweet little boy and girl, the children of Dr. Adam, whom I often play with. Mrs. Adam is a very beautiful polite woman, and the children in perfect order; the little lass told her mamma I was 'a bonny man.' 'Ay,' says her brother, 'and a good man too!'"

Mr. Smith to Mr. James Edward Smith.

Dear Son, Norwich, March 1782.

When a man really takes delight in his business, be it what it will, it is hardly ever seen that he is unsuccessful. You cannot conceive the joy it gave your affectionate mother and me to hear you have spoken twice at the Medical Society: you have broke the ice, and have good ground to expect you will be distinguished from the common herd. I would have you proceed with firmness and due confidence: one of the most certain prognostics of victory in every conflict, is a dependence upon one's self, so that a man does not quite miscalculate his powers; but that includes the idea of so much vanity or ignorance, neither of which I am sure will dupe you, that I don't take them into the account when I consider your attempts.

I have seen your brother since I wrote before. What gave me much pleasure was, to see that he read the dedication to Dryden's Virgil, and tasted the beauties of it besides; for you know his dedications are looked upon as the best in our language, and masterly performances;—and how few boys read dedications and prefaces!

I am, dear James, with the strongest affection,

Your loving Father,

JAMES SMITH.

N.B. 'Loving' is an out-of-fashion term, and has not been in use since our grand- or great-grand-fathers; but they were as honest, sincere, and virtuous as the present age; and as I am not ashamed to be related to them, though many are, I do not blush to use their phrase. To say true, I could find no better; and I know, my dear boy, you will not despise it.

Mr. J. Pitchford to Mr. James Edward Smith.

Dear Sir, Norwich, April 1782.

It is now high time to answer your obliging letter. I am very sure you have but little time to botanize; and am far from being jealous, as an admirer of Flora, that you should neglect her for the more useful parts of science, which I make no doubt at this time have taken possession of your heart. This is as it should be; nor can I see that the competent knowledge of botany, which you say is considered at Edinburgh as an essential part of medical education, can really be so very necessary; unless no more is meant than a knowledge of the species employed in medicine. This is so very necessary, that Linnæus, you know, makes one of the obstacles to the improvement of physic to arise from an ignorance of the species intended. Botany, to be sure, ought to be pursued as an amusement *only* (except by those who write upon it); and as such this present letter, and I'm afraid my future ones, will contain scarce anything else, unless you will improve me by informing me what new discoveries are making in physic, and what are the principal studies in which you are at present engaged.

Let us, however, take a little ride together upon our favourite north and south British steeds, and communicate our discoveries to each other. In the first place let me thank you for your discovery of *Lichen miniatus*, in which you are perfectly right; the plant brought last summer by Messrs. Crowe and Woodward being no other, notwithstanding Mr. Browne is confident that Sir Joseph Banks and Dr. Solander named it *deustus*. Many discoveries have been made in Norfolk indeed, owing to Mr. Crowe's industry, and Mr. Dickson's having been down here for ten days: he left Norwich last Friday night. They found little less than thirty new species in the Cryptogamia; among which, two very pretty *Pezizas*, a foreign *Jungermannia*, now named *hypnoides*; and by Mr. Bryant, a plant figured in Dillenius under the name of *Sphærocephalus terrestris minimus &c.* What think you of the finding *Hydnum imbricatum* and *Lycoperdon coliforme*, or *Fungus pulverulentus coli instar perforatus cum volva stellata, R. Syn.* 28. 12? It is twice as large as the common *stellatum*, and the *vesica* at least six times. It is an elegant *Fungus*, and much deserves a place in the Flora; I counted fifteen rays, and about ten perforations about the size of small peas; not with their *oribus acuminatis*, but *laceratis*, as in the *Bovista*. As I know you are an admirer of Ray, I thought this information would give you pleasure, as it is probable the plant may not have been found (at least known by botanists) since his time. *Hydnum imbricatum* was scarce less valuable to Mr. Dickson; he said the sight of it alone was worth five guineas: but the other is in my opinion the greater discovery, and Mr. Dickson was not a little rich in carrying them home. They were both found by Mr. Stone* of Bungay. For my own part, I dreamt of *Lycoperdon* the night I saw it, and thought I had found four or five; but, alas! it was only a dream. *Lichen pullus* I have not yet seen: *horizontalis* is one of Mr. Crowe's,—at least so marked. *Fucus palmatus* I have not seen except in Gmelin. *Lichen parellus* grows very common here; I did not know the plant 'till lately, but took it for *pertusus*, which I now have. The London botanists know very little of *Agaricus quercinus* and *betulinus*, nor do I believe writers are clear about them. We are certain of your *quercinus* from Batarra's figure, which I'm afraid does not accord with Schæffer's: the *betulinus* we are in the dark about. Young Linnæus is at London, and turns out better than they expected, showing no want of genius; but he has put a stop to his publication, since Sir Joseph Banks's discoveries, which Mr. Crowe was told amount with the *Spe. Pl.* to the number of 40,000. I will not say a word more upon botany, except that I beg my respects to Dr. Hope; and whatever is in my power to procure for you and your friends you may command. And I know you will take pity on a poor botanist, who must depend upon his friends for any thing he gets new.

My wife desires her best respects to you. As I have

* Robert Stone, Esq., late of Bedingham Hall, Norfolk, died 5th January, 1829.

filled this letter, she is determined to fill the next herself.

I remain your affectionate humble Servant,

J. Pitchford.

P.S. This afternoon I saw a letter of yours to your sister Fanny: I respect you for such a mark of your affection to her. Messrs. Rose and Humphrey desire their compliments.

Mr. James Edward Smith to his Father.

Honoured Sir, Edinburgh, 15th April, 1782.

——, myself, and four or five friends, who have a turn for natural history, have lately formed a society for the prosecution of that study. Dr. Walker the new professor, who is a most amiable, worthy and ingenious man, no sooner heard of it than he offered us his museum to meet in, with the use of his books and specimens; and he begged to be admitted an ordinary member, which he accordingly was, and about seven young men besides. Dr. Hope was made an honorary member, as he cannot often attend us; but Dr. Walker, who has no business to follow but natural history, foresees the consequence this society may be of to him, and is resolved to support it as much as possible. Several men of genius and rank have petitioned to be admitted as ordinary members, among whom are the Earls of Glasgow and Ancram, and Lord Dacre, son to the Earl of Selkirk,—three young noblemen of fine parts and great fortunes We have had two public meetings:

at the first Dr. Walker was president, and at the last I had that honour; and the other members are to take it in turn: four visitors are admitted every night. We meet every Friday evening, from six to nine o'clock; and two papers are to be produced and discussed at every meeting, the members taking it in turn to write them. I did not accept the office of president without great anxiety; but I went through it with credit, as I knew the power I held, which is absolute for the time in all societies. I have great hopes that this will be a most respectable and useful institution, and am very proud of having been one of its first founders. As I told Dr. Walker at his first coming I could not attend him this year, but should the next, if his hour suited me; he was so generous as to give me a ticket for his present course, saying I might perhaps find some opportunities of attending him. He also told me I had studied more of natural history than anybody he had before met with in this country; but in this I doubt he was a little premature in his decision, as I doubt not but he will find many more learned than myself, upon examining. It is accidental my not having mentioned Dr. Hutton; he is one of my best and most agreeable acquaintances, a man of the most astonishing penetration and remarkable clearness of intellects, with the greatest good humour and frankness; in short, I cannot discover in what his oddity (of which I heard so much) consists. He is a bachelor, and lives with three maiden sisters; so you may be sure the house and every thing about it is in the nicest order. I step in when

I like, and drink tea with them; and the Doctor and I sometimes walk together. He is an excellent mineralogist, and is very communicative, very clear, and of a candid though quick temper; in short, I am quite charmed with him. He has a noble collection of fossils, which he likes to show:—by the way, I do not mean to prosecute this study any further than is necessary and proper for me to be acquainted with; it requires infinite attention and labour, and there are few certain conclusions to be found. I shall endeavour to get a general knowledge of every branch of literature as it falls in my way; but believe I shall find enough to employ me in the strict line of my profession, with the two first kingdoms of nature by way of relaxation; for I am fully persuaded that an intimate acquaintance with these is not only peculiarly ornamental, but highly necessary, to form an accomplished physician, as literature now stands: and am sure the benefit I have derived, wherever I have been, and am continually deriving, from the little knowledge of this kind which I am possessed of, is greater than could have been imagined,—I mean with respect to introducing me to the literary world; for if I had been without such an introduction, I might have drudged here perhaps a couple of years before I could have done anything to have signalized myself, or have been taken half the notice of which I now am.

I promised to give you some account of my young acquaintances. The name of the one I have contracted most intimacy with is Batty; he comes from Kirby Lonsdale in Westmoreland; is about twenty, and has had an excellent education*. He is a good Greek scholar; but what principally endears him to me are his refined feelings and great sensibility, joined with a strong judgement, and a mind whose native simplicity and purity have been preserved by an education in a sequestered and virtuous part of the world, where luxury and vice have made very little progress indeed, compared with ours. Mr. Batty has a fine ear for music, plays on the German flute, and sings well; and there is something in his appearance that pleased me at first sight. There are a few others about taking their final leave of Edinburgh: this I consider as a very great alloy to the happiness which a scientific man enjoys in a seat of learning like this. I have a numerous acquaintance, with whom I visit or walk with occasionally. I have written Mr. Rose an account of our new society; as I thought it would please him.

Your ever dutiful and affectionate Son,

JAMES EDWARD SMITH.

Mr. James Edward Smith to his Mother.

Honoured Madam, Edinburgh, May 16, 1782.

I have a plan in agitation to take a little tour

* Robert Batty, M.D. received his classical education under the Rev. Mr. Wilson, a very celebrated schoolmaster of Kirby Lonsdale. He was early patronized by the late Sir Richard Jebb, who sent him into Italy with one of his patients. He is a member of the Royal College of Physicians, F.L.S., and senior physician of the Brownlow-street Hospital.

to the Highlands on foot, at the instigation of Dr. Hope, who is desirous his son (a fine youth of about sixteen), and a few others, should be of the party. Dr. Hope thinks we might be out a month for about three pounds each; but I should be for taking some kind of horse to carry baggage, which would make the expense more. Perhaps my father will allow me to lay out five or six pounds in a scheme of this kind, as I may not have another opportunity of seeing the country; however, there is time enough to think of it, as we would not go till August.

The winter classes all finished the end of last month. I was quite melancholy at the conclusion; for besides being really sorry that the lectures were over, it made me vapourish to see so many students going away, and all the places which used to be so cheerful and busy quite vacant and gloomy. Our Natural History Society goes on gloriously. Dr. Black, professor of chemistry, is become an honorary member, and spoke there last Friday. Dr. Walker is there constantly, and generally speaks.

Dr. Hope means to give a medal this year for the best collection of the native plants of Scotland and plants of the materia medica, and will extend the benefits of it not only to his pupils, but to all the members of our Society. I think it will be worth my trying for, and have but little doubt of getting it, if I try. Such an honour is surely worth taking some pains for, and ought not to be neglected by a young man, as such things are generally thought more of at a distance than on the spot. I believe I have never given you an account of my friend Lady Reay, to whom I was introduced by the Miss Riddels, and who honours me with her particular attention:—she is the widow of Lord Reay, and has two daughters, the elder of whom only I have yet seen, who is a very pleasing unaffected young lady, about sixteen. Lady Reay never goes to public places: she has read a good deal, and is highly polished in her manners. I have a general invitation to go when I please to see her.

A few weeks ago I read a paper before the Natural History Society on collecting and preserving plants, which was debated on for three hours, and procured me much commendation from Dr. Walker and Dr. Hope.

When I was at Sheffield, my cousin T. Smith introduced me to an intimate friend of his, a son of Dr. Younge*, a physician there, who is to come and study here next winter. I have engaged him a room in Mrs. Beveridge's house.

I am, my dear Madam,

Your dutiful, affectionate Son,

J. E. SMITH.

Mr. Smith to Mr. James Edward Smith.

My dear James, Norwich, May 2, 1782.

O my dear son! how much gratitude we think is due to your great Creator, from us, who are so inexpressibly concerned for your temporal and eternal

* Afterwards his companion in a tour on the continent.

happiness, that your mind is so naturally formed for virtue, that our minds are entirely at ease on that account, which is the most important of our concerns about you nevertheless, and that they are only the subordinate cares for your health and prosperity that give us much anxiety. Thanks to you for the account you give us of your intimates! it delights us that you find such congenial to your own mind. The establishment of your new society for natural history, in which you appear so conspicuously, fills us with pleasure which need not be described; we flatter ourselves it is a presage that you will arrive at eminence in your profession, and reap a good harvest of honour and profit. How happy will that make both your most affectionate parents in the decline of life! What joy is it to us to contemplate, that as you must now be the fabricator of your own fortune, you begin your career so happily! You are certain you have our daily prayers for the continuance of the protection and favour of our universal Parent and Benefactor, whose blessing I have no doubt you omit not to solicit yourself, with a piety which I trust never will forsake you.

I would by no means oppose your journey to the Highlands, if it would answer any good purpose, of which you are to be the judge; and Dr. Hope's sending his son along with you is a pledge of its safety: but by no means go without one baggage horse at least; I think you had best have a galloway to ride, and if not each one, at least one amongst you in case of weariness or any accident, and take guides as often as there is a chance of wanting them. I shall grudge nothing for your health and safety.

You are much obliged to your aunt for introducing you to so many respectable friends and acquaintances, in particular Lady Reay, who seems a very desirable one; but as you are in the midst of so many agreeable young ladies, take care of your heart; be least with and think least of those you like best. Excuse me if I repeat cautions on this subject unnecessarily till you have a prospect of settling.

Adieu, my dear Son!

JAMES SMITH.

In the beginning of June the student made an excursion to Kirby Lonsdale in Westmoreland, to visit his friend Mr. Batty, and into Yorkshire to meet his father. "I have many inducements," he tells him, "to take this journey; and first, the company of an agreeable and ingenious Frenchman, Dr. Broussonet, who has been in Edinburgh a week to see the place, and with whom I have been very happy. He is an eminent naturalist, and intimate with Sir Joseph Banks."

The following letter from Mr. Smith is written after they had met, and his son had returned to Edinburgh.

Mr. Smith to Mr. James Edward Smith.

My dear James, Norwich, August 5, 1782.

Your letter from Kirby Lonsdale, and that lately from Carlisle, afforded us great pleasure, to hear you were well and so far on your return to Edinburgh. I should be very happy indeed to visit Cumberland and Westmoreland with you, and nobody knows but it may fall out so at some time or another: it is easy to conceive the want of a companion must abate a great deal of the pleasure in viewing the beautiful and romantic scenes; there is even a degree of horror in the grand and majestic prospects of nature, in solitude.

I will not say what flattering hopes I form, but I am much mistaken if kind Providence has not put your fortune in your own power, and that you have little to do besides pursuing the track you are travelling with so much success, but to shun the most obvious dangers and mistakes in life. The pleasure and comfort your meeting gave me is inexpressible. On our return home we came to Lutterworth: here we ascended the pulpit in which the first English reformer, Wickliffe, used to preach, and sat in the chair, still preserved, in which that eminent man died*.

I am, your affectionate Father,

JAMES SMITH.

* "*The Lives of John Wickliffe, and of the most eminent of his Disciples, Lord Cobham, John Huss, Jerome of Prague, and Zisca.* 'After the way which they call heresy, so worship we the God of our fathers:' Acts, chap. xxiv. ver. 14. By William Gilpin, M.A. 1766, 8vo. 372 pp.—The writer of this very entertaining work has shown himself a man of abilities, a gentleman, a scholar, and a friend to truth and religious liberty; and the very great men he hath chosen for the subjects of his pen are worthy to be well considered in all ages, and afford very many useful lessons to all succeeding times, of the amazing force of truth.

Mr. James Edward Smith to his Father.

Honoured Sir, Edinburgh, August 6, 1782.

I left Carlisle on Tuesday night at eight o'clock, and arrived at Moffat next morning by five. This is a neat pleasant town, where there is a sulphureous spring much resorted to: the town is at present full of genteel company, and they have dancing almost every night. Here I found Dr. Walker, as I expected: he has a good house and noble garden here, which he will leave in November, as he will then remove to a place three miles only from Edinburgh, where he has got a living in exchange for Moffat. I spent that day and the next very happily with the Doctor: he is a very agreeable man, the life and soul of

Wickliffe seems to have been in religion, what Lord Bacon was in philosophy; that is, the first light, and of the most amazing brightness. Huss was a man of uncommon virtue and great parts; Jerome, of more refined abilities and greater learning; and Zisca, a most extraordinary military reformer, and of talents and capacity to war, equal to any man that we read of in history. So is his history one of the most uncommon, and the fullest of great events to be met with. He was the founder of the city of Tabor in Bohemia on the river Maldaw, which was his strong retreat, and from which his sect was called Taborites.

"It would be wrong to omit the mention of Lord Cobham, who is a shining example of a military man in a very high station and of eminent abilities, converted from the irregularities of such a life and the errors of popery, by the force of the truths delivered by Wickliffe, and most heroically maintaining his virtue and his truth at the expense of his life. His magnanimous and pathetic behaviour at his examination and tryal before the convocation, afford a most interesting, noble, and moving scene."—*From Mr. Smith's common-place book.*

Moffat; his loss will be equally felt by the gay, the industrious, and the unhappy. I reached Edinburgh on Friday evening, and have had great congratulations on my return. Dr. Hope paid me the very high compliment of saying he had wanted me to keep him from falling into many mistakes: I supped there on Sunday, and talked with him about the medal, among other things; he said he had not published it in his class, as he saw nobody there who was likely to try for it, except those who were members of our Natural History Society, where it had already been published. I think this was wrong; he ought to have made it as public as possible: he concluded by saying he thought I had it all in my own hands.

I am, honoured Sir,

Your obedient and affectionate Son,

J. E. Smith.

Mr. James Edward Smith to his Father.

Honoured Sir, August 29, 1782.

My late tour was shorter than had been proposed, owing to bad weather, but was very agreeable and successful. We set off on Tuesday the 20th instant, and got to Glasgow that night. Mr. Hope introduced us to some of the professors, who were very polite; and I called on Mr. Grant, who was very glad to see me. Glasgow is, I think, one of the finest towns I ever saw: the buildings in the two principal streets (which cross each other) are very noble, in the style of Queen Elizabeth's time; these streets are very broad, and crowded with people, like Cornhill: in other parts of the town are a great number of superb modern houses. The Green, which is between the town and the Clyde, is very delightful; it is about two miles long and half a mile broad, and planted with very large trees: here the women wash their linen in the open air, having fires in small iron stoves. On Wednesday evening we went in the diligence to Dumbarton, where is a castle built on a very singular rock, from which is a view down the river worthy of Italy; yet this is called "a barren land without a tree"! From hence we took a chaise to Luss, fourteen miles, through a sweet country on the shore of Loch Lomond; rode by the pillar in memory of Smollet. Loch Lomond is full of beautiful islands; but though its borders be clothed with wood and ornamented with towns and gentlemen's seats, yet there is not that picturesque variety about it which the lakes of Cumberland afford.

Mr. Stewart, minister of Luss, was the companion of Mr. Pennant in one of his tours. I became acquainted with him at Edinburgh, where he was made an honorary member of our society on my nomination: he is a first-rate naturalist, and remarkable for his modesty and simplicity of manners. On Friday he accompanied us to Ben Lomond: we took a boat, and sailed across the lake, five miles; the weather was fine, and it was a most delightful voyage; a fine eagle was soaring above our heads. After landing we began to ascend the mountain, whose top is full five miles from the shore of the lake, and whose perpendicular height is 3240 feet. Two men went out with us, who carried provisions and rum, of which I drank, during the whole of our expedition, *three wine-glasses* with great advantage, for at the top of the hill it was extremely cold, and rain came on, so that we were, nearly all the time we were at the top, wrapt in clouds: we ascended the highest point, on which we stood as on an island in a sea of clouds; from time to time, however, we had transient views of the country below us, as if by enchantment; on one side Loch Lomond chequered with islands; on the other, a sweet valley with the Forth winding through it in the most fantastic manner; on the north-east side, the mountain is absolutely perpendicular, and we looked straight down on the river at its foot. We found a great number of very rare plants, which amply rewarded us for our journey, and about six o'clock began to descend again, and got back to Luss before nine.

On our return home we saw Carron iron-works, which are really stupendous, and the ancient town and castle of Linlithgow: the country most of the way is barren and dreary, consisting of fields of starved barley, and turfy moors.

I have had a very obliging letter from Dr. Broussonet: he says he has sent me a copy of a book he has lately published on fishes. He is now at Paris.

Your affectionate,

J. E. Smith.

Mr. James Edward Smith to T. J. Woodward, Esq.

Dear Sir, Edinburgh, Sept. 28, 1782.

Although I have at present much upon my hands, I would not neglect answering your kind letter, as you wish to have an account of my journey. We were out but a week, and went no further than Loch Lomond; we ascended Ben Lomond in company with Mr. Stewart, so often mentioned in Lightfoot's book. The weather was cloudy, and we could see nothing of the country, but have great reason to be satisfied with our botanical success. The best things we found were *Sibbaldia procumbens; Azalea procumbens; Alchemilla alpina; Polygonum viviparum; Saxifraga stellaris, nivalis, oppositifolia, hypnoides* and *autumnalis; Juncus spicatus* and *triglumis; Rubus Chamæmorus*, in fruit; *Silene acaulis* in seed, except one specimen which had a flower, and by which we saw it was truly a *Silene; Gnaphalium alpinum* of Lightfoot, i. e. *supinum* of Linnæus; *Salix herbacea; Lichen crinitus, torrefactus, polyphyllus, ventosus, ericetorum, cæruleo-nigricans*, and many other plants, which would have transported me a few months ago, but I made great acquisitions in Westmoreland. From Mr. Stewart's garden I got *Juncus biglumis, Anthericum calyculatum, Salix lapponum*, and some others; *Vaccinium Vitis-Idæa* in fruit; *Cornus herbacea*, roots; *Astragalus uralensis*, seeds, &c. From this gentleman I hope to receive some more specimens soon, so that I shall have nearly all the rare Scotch plants. I have lately added to the Flora Scotica, *Polygonum pensylvanicum, Senecio saracenicus, Lichen pyxidatus,* β of Huds.; *Byssus rubra?* Huds.; *Agaricus deliciosus* and *viridis,* and *Lycoperdon epiphyllum* of Linnæus, not of Lightfoot or Hudson.

I have done very poorly in *Fuci,* &c., having repeatedly gone down to the sea-shore, when the wind has nearly blown me off my feet, without finding scarcely a morsel of vegetable matter. This is a very bad coast for sea plants, not comparable to ours. I got no fossils in my tour: the Asbestos is very rare, and perhaps, were you to see it, would disappoint you as it did me. The *Arundo* in Earsham wood is certainly the *epigejos.* I have the true *calamagrostis.* I am extremely obliged to you for your excellent drawing of *Lycoperdon coliforme,* and for your valuable remarks on that genus. The *L. pedunculatum* I have seen at Dr. Hope's; it was found in Scotland: I have formerly observed the circumstance you mention in the *L. fraxinum.*

I found the *L. stellatum* in a young state; it is now in the form of a white ball, the volva being entire, about the thickness of a leather glove, and covering the head; the perforation in the head is as distinct as when the fungus is ripe. I have a species which I take to be new, and have sent Mr. Dickson a specimen: it is very small, and grows in clusters on moss; I call it *Lycoperdon fragile,* and define it, *L. parasiticum pyriforme, fragile, nitidum, badium, farinâ fuscâ*: 'tis not a *Sphæria.*

Dr. Walker has found a number of plants in Scotland which are not in Lightfoot: among them are ***Rubus arcticus***, ***Lysimachia thyrsiflora***, and the indubitable ***Elymus arenarius***, which I have seen and examined, and of which I have a specimen brought from the Göttingen garden, and is the same with the Doctor's; that, therefore, brought from London by Mr. Crowe is a new English plant, if really found in England.

I am, dear Sir, yours, &c.

J. E. SMITH.

Mr. Smith to Mr. James Edward Smith.

My dear James, Norwich, Nov. 3, 1782.

Mr. Woodward called here last week on his way to Narford, where he is gone to spend a few days.

Mrs. —— is vastly pleased with your letter, and we are pleased with it too; you have a better knack at *la badinage* than I imagined. She is a lady very proper to correspond with, to introduce a young man into that kind of style which has its *agrémens* as well as utility. I suppose you know it is the way in France for every young gentleman to have such a female friend as will introduce him into the world in every sense of the word, and I need not tell you how far they carry it. She is not only his correspondent to form him to an easy, familiar, polite and gay style in letter-writing, to teach him the graces in company and conversation, but she is his tutor in gallantry and the knowledge of the character, the tastes, the foibles of the fair-sex;—and it is a scandalous corruption, for she is too often their betrayer, at least by giving him lessons to employ to their injury. The famous Ninon de l'Enclos' correspondence with the Marquis de Sevigné is a very curious and well-known instance of it. He was initiated by her into all the mysteries of intrigue and gallantry; whilst his mother, who was a woman of uncommon fine understanding and virtue, was endeavouring by her letters and instructions to form him to good morals, virtue, and piety. I believe he did not make the choice of Hercules, but chose Pleasure for his deity, and consequently made no *figure in the world,* and if I am not mistaken was *unhappy,* though born with every quality and a good fortune to make him otherwise.

You cannot think I mean to carry the comparison any further than that Mrs. —— will make you a cheerful, sensible, pleasing correspondent, and give scope to the *style enjouée* in your epistolary correspondence. I trust you will acquit yourself well, and want no advice from me to urge you to perseverance in the difficult roads of knowledge and honour, nor, when attained, what conduct is necessary to procure and preserve the *esteem and love of mankind,* which are among *the most solid advantages of life.* I cannot describe the pleasure it gives me to reflect that you have only to avoid stepping aside from your natural disposition, and not to torture your character, and you are formed to be respected, and, what is of more value, to be loved by mankind as well as by your most partial friends.

What you relate concerning —— does high credit to you, and some to him: 'tis happy indeed when the imprudences of young men serve so good a purpose as to bring them to the love of virtue and of truth (and he seems in earnest, pray Heaven he may!). It was not much to be expected that his education and company and way of life in this age should permit him to escape the follies he speaks of; but if they serve to show themselves to him in their true colours, and he loves virtue the more for it, I could almost say they will do him honour. 'Tis noble to reform, though not so great, so estimable as to be innocent, nor can ever be so happy.

JAMES SMITH.

The Rev. Henry Bryant to Mr. James Edward Smith.

Dear Sir, Heydon, Nov. 11, 1782.

I congratulate you on your safe return from your little northern tour, and am sorry the weather proved so unfavourable as to spoil much of the pleasure of it. I have sent you a specimen of my *Lichen parellus*. I think it differs from yours, though perhaps yours may be right and mine wrong.

We have had a very wet, uncomfortable, sickly summer, and I suffered much from the epidemic influenza, consequently have done but little in the botanical way: the chief things I have found are these, viz. *Scirpus pauciflorus*, Lightfoot; *Galium erectum*, Hudson; *Scutellaria minor; Peucedanum Silaus; Leonurus Cardiaca; Riccia fluitans*, and, I believe, *Targionia*.

Our turnips in Norfolk this season have suffered greatly from a species of black caterpillar; thousands of acres have been destroyed by them, and no method could be found so effectually to stop their ravages as the employing women and children to pick them off, either by the day, or at three halfpence the pint. I have taken great pains to find out the fly which produceth them, but to little purpose.

Albin has figured the caterpillar well, and says, "These black caterpillars, of the Ichneumon kind, were found feeding on the turnip-leaves in the beginning of September 1719, they being so numerous at that time about London, that they destroyed whole fields of them: about the latter end of September they went into the earth, and spun themselves up in a transparent case, and changed into a chrysalis, and in May following came forth a small Ichneumon fly, as in the figure annexed*."

Now the figure of this fly is not of the Ichneumon kind; it has only two wings, and belongs to the genus *Musca;* but none of that genus spring from caterpillars, but from maggots. I have offered and given premiums to many persons to pick up and bring me all the different sorts of flies which they find upon the turnips, but can gain no satisfactory knowledge about them; they have all, or most of them, brought me a species of fly with four wings, with black and yellow intermixed upon their bodies and legs, which have been prodigiously numerous; but I know not what it is, unless it be the *Tenthredo Rosæ.* I have inclosed you two specimens, and beg you would get me the best information you can about it, and at the same time to find out,

* Albin's History of English Insects, tab. 62.

if you can, what fly Albin's black caterpillar really produces.

HENRY BRYANT.

Mr. James Edward Smith to his Father.

Honoured Sir, Edinburgh, Dec. 31, 1782.

Our Natural History Society goes on increasing. I believe we shall have four annual presidents chosen; if so, I hope to be one of them. I have just given in a paper on the Phænomena of Vegetable Odours, which was well received. I have spoken often in the Medical Society. I am to have Dr. Hope's medal, but 'tis not yet come from London. There were no other collections given in besides mine; it has been examined, and thought worthy; indeed Dr. Hope paid me very high compliments upon it.

January 3, 1783.

I have just been at the funeral of an acquaintance, whose death gives me great concern: he was the son of Dr. Reid of Glasgow, author of a celebrated work on the human mind; he was a young man of the first abilities and accomplishments, but of the greatest modesty and diffidence. I had flattered myself with the hopes of being intimate with him; this was his first winter here, and we have been much together. He is the last of a numerous family, who have all died about the same age, just entering into life. His father bears it like a philosopher. I cannot help comparing him to a venerable oak that has been bowed before many a blast, and stripped by degrees of its leafy honours, but that has now nothing to lose, and braves the fury of the storm inflexible. The mother is not so tranquil. They are both in Edinburgh.

Many young men have had fevers, but have all recovered except Mr. Reid; he died of a very peculiar disorder, which came on at the crisis of the fever; it is called tympanites, and is a collection of air on the outside of the intestines in the cavity of the abdomen; none of the Professors ever saw it before; the hole in the intestine through which it passed was so small as to be found with difficulty on dissection.

J. E. Smith.

Mr. Smith to his Son.

Dear James, Norwich, Jan. 14, 1783.

I cannot help congratulating you on your obtaining the botanic medal, and at the same time lamenting the loss of your friend Reid, and we daily return God thanks you are well: *not to fear* infectious disorders Mr. Martineau says is the best preservative. We are every where and every moment surrounded by dangers; and you and we are taught to trust in a good Providence for our protection, always meaning that whatever we would obtain, whatever avoid, our own endeavours and prudence must be exerted for the purpose; those very powers we have capable of contributing thereto are part of the means, we may rationally suppose, that our Creator and Preserver designs and uses for those ends, the ends of his government in the world; and not a sparrow falls to the ground without his permission.

Yours,

JAMES SMITH.

Mr. Smith to Mr. James Edward Smith.

My dear Son, Norwich, Feb. 28, 1783.

We have had the pleasure of hearing you are well, and are perfectly happy on the score of your health. I need not add that we are so also upon the further addition to your academical honours, in being chosen first President of your Natural History Society in so distinguished a manner as will leave an undoubted testimony of your being its founder and supporter. I cannot account for your having such a preference. These distinctions I flatter myself are prognostics of the eminent rank you will by and by stand in, and the use you will be of to yourself, your friends, and to mankind,—pleasing reflections indeed to parents who have your happiness so much at heart.

We begin to think how near the month of May is; and although we cannot expect you have formed your plan for leaving Edinburgh and coming to Norwich, I cannot help mentioning it, and that when you have thought ever so little upon it, you will give us the hint.

Our compliments wait on your friends, and our tender love on you.

I am, dear James,

Your affectionate Father,

JAMES SMITH.

Mr. James Edward Smith to his Father.

Honoured Sir, March 6, 1783.

I cannot help expressing to you the dissatisfaction which I have experienced in my inquiries into the theory and practice of physic. I really believe medicine, if it deserves the name of science at all in its present state, is in the most barbarous condition of any science, and only now emerging from the greatest darkness and absurdity. It is commonly declared by all practitioners, that theory is nonsense, and that experience, that is empiricism, is everything. Cullen's theory is visibly going into the same state of contempt as Boerhaave's has been reduced to, and his lectures are by no means consistent with it, though admirable as mere practical lectures. These considerations and some other have induced me to attend Browne this winter; and I am happy in having done it, for his system and view of the human œconomy are certainly the most philosophic of any, and are gaining ground in a wonderful manner: perhaps, however, he may have only his day. He has many of the most respectable pupils, and behaves very well to us. I am happy to have procured the admission of my friend Dr. Broussonet of Montpellier as an honorary member of the Medical Society: he was admitted unanimously on my recommendation, of which I am not a little vain. The very day of his admission I received a letter from him expressing his desire of that honour, and offering to procure in return my admission into some of the French academies, of which I gladly accepted. I also procured him an honorary seat in the Natural History Society; and I believe he will be admitted to the Philosophical Society, which is composed of the first *literati* in Scotland. Lord Kaimes (of whose death I suppose you have heard) was president.

I cannot help sending you a copy of a letter which I received, as president of the Natural History Society, from Lord Buchan, on his admission to it.

Your affectionate and dutiful Son,

J. E. Smith.

To J. E. Smith, Esq., President of the Society for the investigation of Natural History.

Sir,

The notification you have done me the honour to transmit to me of my election as an honorary member of your literary association for the investigation of natural history demands an early and respectful acknowledgement. I give it with alacrity and gratitude.

From my earliest infancy all my thoughts have

been set on public good, and I have thought no sacrifice too great for its promotion. I consider your association as connected with the darling occupation of my life, and your society will find in me a sincere and active friend. I know the merits of your professor; he has been an object of my esteem and literary regard for eighteen years past; and I am persuaded that you will find in him an assiduous and successful commentator on the subjects which your youthful ardour may engage you to explore.

I entreat of you, Sir, to convey to your brethren the thanks of a member of the great republic of letters, who, at no advanced age, begins to grow old in the service of that community which seems to have adopted him more heartily than any other*.

I am, Sir, with regard,

Your obliged and obedient humble servant,

(Signed) BUCHAN.

Edinburgh, 1783.

Mr. Smith to Mr. James Edward Smith,
Mr. Bickersteth's, Kirby Lonsdale.

My dear Son, May 18, 1783.

I am very certain you are able to form some judgement of what we felt when we heard you had been so very ill; indeed, my dear James, a very, very great

* This alludes to his lordship's disappointment in not being elected one of the sixteen peers, on the death of Lord Breadalbane.

share of the happiness that may remain to us to be enjoyed in the rest of the years we have to continue in this world depends upon your life, and the health and happiness you will enjoy in it. No child certainly was ever dearer to parents than you are to us, for every reason that can affect the human soul: you are seldom absent from our thoughts; you are the dear object of our fondest wishes, the never-forgotten subject of our prayers to Heaven. 'Tis not for our own happiness in the expected enjoyment of your much valued company, 'tis for your own sake, we so earnestly pray for your life and health: we forebode there is a noble and pleasing career for you to run, in which your happiness and your honour will be great; and we flatter ourselves you will be useful to your fellow-creatures. With these hopes, these prospects, and these tender sentiments, we must be, we are, inexpressibly sensible to everything that relates to your health and safety, and must urge with all the force in our power that you will run no sort of risk of either on your journey home, nor suffer any temptation to botanize, to see the lakes, mountains, or any other natural curiosity, lead you into the least possibility of taking cold; but come home as soon as you can safely, and bless your dear affectionate mother and me with a sight of you.

Mr. Windham is appointed secretary to Lord Northington, lord lieutenant of Ireland; and he has appointed Mr. Repton, your friend, his secretary, which you will be glad to hear. Adieu, my dear! May it please the gracious God to confirm your health, and bring you safe and well to the embraces of your most affectionate parents, and the bosom of your family!

I am, your ever affectionate

JAMES SMITH.

Mr. James Edward Smith to his Father.

Honoured Sir, Kirby Lonsdale, May 23, 1783.

Your most valuable letter reached me yesterday. I cannot sufficiently express my gratitude for the sentiments it contains, nor the feelings I experienced on reading it. How much soever I may fall short of the flattering height of honour and happiness which your paternal partiality has set before me, I hope, if I can trust my own heart, I shall never be deficient in a return of affection and duty to those parents in whom I am so peculiarly happy, and to whom I am indebted for the foundation of every good inclination which it has pleased God to help me to cherish. You will by this time have received letters from me, and will see I continued to mend as fast as possible. Nothing can exceed the attention and genuine hospitality which I experience here. I shall make no visits on my return, for by that time every mile will seem ten till I see you at Norwich.

I am, your dutiful Son,

J. E. SMITH.

The whole foregoing narrative, in familiar letters to his parents, discovers the humility and tenderness of a child towards them; while by the sole influence of personal character and conduct he at once made his way into the best society, and planted himself in a niche in the temple of science, when others, with more apparent advantages, were consuming their time in "chinking useless keys, and aiming feeble pushes against the inexorable doors."

He spent about two years in that accomplished community, in a well regulated course of useful discipline and studies, and in the agreeable and improving commerce of gentlemen and scholars; in a society, where emulation without envy, ambition without jealousy, contention without animosity, incited industry and awakened genius; where a liberal pursuit of knowledge, and a genuine freedom of thought, was raised, encouraged and pushed forward by example, by commendation, and by authority.

Of the warmth and goodness of his heart, his early letters bear unequivocal testimony, as well as of that peculiar tendency in his nature to form attachments, which he carried with him through life: and wherever these were placed, nothing on his part ever changed their force, he thought no sacrifices too great, and no expressions too strong, to attest his regard.

In these partialities he was influenced by his love of genuine nature, and the appearance of confidence and dependence upon him. "My heart," he says in a letter written in 1783, "is formed for social enjoyments; but how often have its warmest affections been torn asunder when just most fully developed! I shall hardly ever dare to fix, for fear of a disappointment."

This is the language, and such are the feelings of ingenuous youth: but such a heart, whatever resolves the head might make, must seek its happiness in new affections, and his were not long condemned to solitude, or withered by despair.

The following letter, written the next year to his earliest friend and near relation, (of whose connection with him he was always justly proud, and whose friendship lasted unimpaired through all the vicissitudes of absence, habits, and a different clime,) may give a lively idea of that ardent temperament which has just been described.

*Mr. J. E. Smith to N. E. Kindersley, Esq.**

Tinnevelley, 1784.

Why, dearest friend, do you think I am changed? How can you blaspheme the name of reason, so as

* Author of the following work:

" *Specimens of Hindoo Literature, consisting of Translations from the Tamoul Language of some Hindoo Works of Morality and Imagination; with explanatory Notes, to which are prefixed introductory Remarks on the Mythology, Literature, &c. of the Hindoos.* By N. E. Kindersley, Esq., of the Hon. East India Company's civil service on the Madras establishment: 1794. 8vo. 335 pages.—This work, by a most worthy and esteemed relation, is very curious, and appears to be extremely well executed. It will gratify the inquirers into the Hindoo religion, and the manners, ideas and literature of that ancient extraordinary people. It is a beautiful edition, and has five curious plates, which are not mentioned in the title-page."—*Mr. Smith's common-place book.*

to suppose she can have taken away my honesty and openness of heart? Thank God, this is not the case.—How often do I think of the days of our childhood, when we have so often by sympathy retired from the social scene, to relieve ourselves from that gentle restraint which the presence of our partners or friends (though the most indulgent in the world) laid us under, to pour out all our thoughts on each other's bosom, to communicate our little discontents or our joys, our childish observations, and our innocent merriment; but chiefly to indulge that sympathy of soul which appeared so early, and which may Heaven still cherish! How often have we been forced to have recourse to the mute expression of looks or embraces, when our young bosoms swelled with feelings which our artless tongues could not utter! My heart exults with conscious dignity at the idea. I recollect with no less pleasure the few but most happy days we have passed together in the course of our riper youth.

A thousand Norwich and Yarmouth scenes arise to my mind;—our evening walk by the sea-shore; our more cheerful excursion in the Yarmouth cart; our dancing parties, and the conversations which passed after we had left them; our last sad parting; —can we forget these? If we do not forget them, can we ever think otherwise of each other than we do? Let us then not suspect each other, or if we do, let us communicate our suspicions.—Friendship is an intellectual marriage, and the same turn of mind and character which makes us lastingly happy in the latter condition, can alone make us so in the former.

I lately received yours of May the 29th, and most heartily congratulate you on your advancement. May you ever be happy, and escape the snares which surround you! I do not fear it. Surely I knew your heart once; and surely such a heart cannot be corrupted while there is a Providence watching over the well-disposed; and that there *is*, I would not give up to attain the literary reputation of all the ingenious perverters of reason that ever lived. Still you express a fear of my becoming a literary coxcomb or a fastidious man of the world! Read the first part of this sheet; it was written many months ago: I have looked it over in many different moods: I now deliberately send it you. You may perhaps smile at it, think it boyish, too warm to be sincere: but I will not suppose such things; if it gives you half the pleasure in reading it that it did me in writing, I shall be happy indeed.

I am your,

J. E. S.

In another letter he tells his cousin, "What a pleasing picture does your letter give me of your mind! I am fully sensible of the value of a true friend, and will always be quite open with you. I have been happier in my friendships than most people: but with you, and in one instance besides, I've enjoyed that true union of hearts and mind which is the essence of friendship; I mean in an intimacy I have formed with a young man of my own profession. We became acquainted at Edinburgh. We must soon be separated, and I shall be afraid to form another so close connexion."

The subjoined letter is addressed to the friend here spoken of.

Dear Batty,

You are, perhaps, like me, too apt to regret past pleasures, and neglect present ones. This disposition should not be too much indulged; for when the object of our regret is really important, our distress might be increased to an intolerable degree. You, my dear friend, have in your own power an inestimable source of happiness, in the amiable sensibility which you possess in so eminent a degree; yet this choice gift of Heaven may occasion its possessor as much misery without the direction of reason, as it would happiness with it.

It is a most discouraging thing to a young man entering into life,—his heart, without reserve or suspicion, overflowing with the "milk of human kindness,"—to be told by those who have gone before him, that his ideas of friendship, love, honour, are merely romantic, and not to be realized in a commerce with the world; that there, self-interest, ambition, avarice, and lust, reign with absolute sway; that those feelings which (if he be not a villain) have chiefly contributed to his happiness hitherto, must now be restrained by prudence, and be perfectly obedient to the dictates of interest and worldly advantage. They tell him, that now

"The wild romance of life is done;
Its real history is begun."

I would fain hope this is exaggerated: not that I would by any means reject the use of due caution and prudence in *forming friendships*. I am perfectly convinced that on this depends the existence of those very feelings; and perhaps the persons who compose most of the worthlessness of the world, are those who for want of this *proper* care have had their dearest hopes and expectations deceive them. Let us therefore, when we hear these complaints, carefully consider from whom they come; whether from such an one as I have just mentioned, or from a person, who, having sacrificed his own feelings to interest, wishes to reduce all mankind to the same level; or from one of a fretful, peevish temper, who expects too much from others, far more than he will grant them in his turn; or, lastly, from one who has naturally no feeling at all.

I trust there is more virtue in the world than we are generally told of. Those lovely dispositions that glow in the youthful heart, may perhaps be generally in some degree concealed by various means amid the busy pursuits of active life, and sometimes may be clouded by a degree of ambition or self-interest. But in the decline of life we see the social feelings revive: then old friendships are renewed; children are doated on; a thousand little offices of love are mutually performed; and I confess I do not know an object of more respect and admiration, instead of contempt and ridicule, than an old person taking pleasure in recollecting and relating the scenes of his past pleasures, and cherishing every idea of his former friends. I have indulged myself in a little prolixity on this subject; but I hope you will excuse it, as I trust it is an interesting one to both of us. Nor am I afraid you should be severe in your criticisms on what I have said.

I attend the infirmary with much pleasure, there is so much room for observation and reasoning, and I have got over the disgust.

Dr. Hope honours me with his notice, attention, and assistance. The more I see of that excellent man, the more I adore him. I admire his botanical lectures; his delivery is agreeable, with as many "behoves" as Dr. Walker; his politeness and condescension unparalleled. How happy should I be to call you out of bed in a morning to go to him! Tho' I cannot do this, something unavoidably takes me down Robinson's Close; and I cannot help looking at that gloomy dwelling which I have so often visited with a most cheerful step when it contained my friend. Your fine ears would be dreadfully shocked by the instrument used by Dr. Hope to call us together.

I am your faithful and affectionate Friend,

J. E. Smith.

To the same.

God bless you, my dear Batty, for writing me so early so kind a letter! You've amply repaid me for *my trouble* as you call it. You must come the very first of October and be examined a little. But why, my delicate, scrupulous friend, do you say so much about trouble and obligation to me, who am far more obliged to you?

Nothing would, I think, be too much to serve the friend I really loved: yet I would avoid the extreme which some very good-natured people run into, who are *fetchers and carriers* to all the world. Your partial opinion of me shows your own good heart. Indeed, Batty, I must confess I am capable of very strong attachments, and sometimes perhaps of too strong dislikes; one often takes prejudices which time either confirms or removes. When I first was with you, I was prejudiced in your favour: I soon thought I saw you had feeling (the foundation of all that's good); and soon, that I saw real merit through that amiable modesty and diffidence which are so great an ornament to the brightest abilities. I determined to be more acquainted with you. I saw your friendship was better worth my cultivating than that of the more forward or splendid; and as I have known you better, I have bound you to my heart as an inestimable jewel. Think not this is all pure disinterestedness: the first prejudices we form concerning persons and things border on weakness; we are therefore peculiarly happy to have them confirmed; for the human mind is confessedly more tenacious of its weaknesses than of anything else that belongs to it. I offer you, my Batty, a warm, an honest heart: if I had more ability to be of use to you, I should be more happy. I shall *trouble you* as often as you can be of use to me next winter. Come then, my friend, let friendship give a relish to our studies and heighten the pleasure of our relations: let us assist each other in avoiding the snares of the profligate, and the still more dangerous solicitations of the good-natured and inconsiderate; both of whom, though for opposite reasons, wish to make all mankind like themselves. One method of obtaining this end is to represent all mankind as being already so, to question the existence of any virtue, and to talk of the appearance of it as grimace. This is done, more or less openly or artfully, by all who leave its paths. I have seen many such characters; particularly one, since I came here, whose machinations I have carefully observed. The Italians say *Volto sciolto e pensieri stretti:* one may be as free and compliable as possible in unimportant matters, and yet inflexible in things of consequence, without incurring the charge of formality; whereas hypocrites are always peculiarly nice about the smallest matters.

But where am I going? How do I pester you with my scrawling! I wrote the above, in the fullness of my heart, immediately on the receipt of your letter, which I have read over and over with the greatest pleasure.

Believe me yours,

J. E. Smith.

These are specimens of his pure affections and high moral worth, at a period of life when they may be expected to show a vigorous growth; but that unsuspicious simplicity and depth of feeling which marked his early years was never obliterated; and if it was a failing, undoubtedly it was one "which lean'd to virtue's side." Inclined by nature or a happy mental temperament to duty and integrity, it must be acknowledged, nevertheless, that Sir James was indebted greatly to an education among those who placed the standard of virtue upon higher ground than usual; who considered errors which too frequently, through false indulgence, through indolence, through extreme folly, are overlooked as pardonable or inevitable because they are common, as evils to be carefully avoided, because they surely lead to misery.

The high principle of rectitude which appears in the next letter, from his relative of the same age, is worthy of record and imitation, and confers honour on both parties.

N. E. Kindersley, Esq. to Mr. J. E. Smith.

My dearest Friend, Tinnevelley, 19th June.

I had the pleasure to write to you a few days ago, but now resume my pen to attempt to express the uncommon satisfaction I received two days ago in reading a letter from you.

It gave me a more than usual pleasure, not only to see the very great confidence you repose in me, but more especially as I am now assured you are not one of those innumerable multitudes of young men who are insensible to religion and virtue; so much greater in number (beyond all comparison) than the good, that you will not, I hope, condemn me as uncharitable, when I thought it *highly probable* even *you* were one of them. I congratulate you on your happiness with the greatest pleasure, and thank Providence that the person who had the greatest share in my friendship of any man in the world, is at the same time of my way of thinking in matters of religion.

In my last I informed you of my present situation, and that I was much pleased with it. I am assistant to Mr. Eyles Irwin, whose Travels overland, and his Eastern Eclogues, you may have read*. The Nabob's countries being assigned over to the Company in the year 1781, for five years, has given them the absolute sovereignty of it for that time. This district is the most southern of the Nabob's country, and is one of the largest and finest. Mr. Irwin is intrusted with the entire government of the revenue, and has with that all the judicial authority in his hands of a country —— miles in circumference. He is exceedingly worthy of his trust, and is perhaps the brightest example of integrity and zeal in office that ever graced India. He might with the utmost ease, and secrecy too, make one or two hundred thousand pounds, but he will not make a sixpence beyond his pay;—a married man too, with a family. His whole time is taken up in making the country produce such a revenue, and establishing such a government, as may do credit to the Company's management. You see there are *some* honest men even on the corrupt shores of Hindostan;—would to God there were enough to make me charge you with uncharitableness in your opinion of our general depravity! Under such a

*Also author of Adventures in a Voyage up the Red Sea, and in a Journey through the Desarts of Thebais.—See *Annual Register for* 1780, *p.* 40-54.

man you will not think I have much reason to arrogate merit to myself, if I am uncorrupt. Indeed, James, I never doated on money; and were I not influenced by any motive of conscience, I believe the principle of honour would prevent my being very rapacious; though perhaps it would not of itself make me so rigid as I now feel myself obliged to be.

I have caught the honourable infection from my superior, and enter into the interests of the Company in this quarter, and into the prosperity of the provinces, with a force very uncommon. I have a great deal to do, but I do it with pleasure, as it is business of consequence; and as I find that with ordinary abilities, a degree of activity and diligence, with *good dispositions*, and great power to perform your wishes, a country like this may be made happy in itself, and productive to its owners.

It has been the want of strict integrity that has hurt the English name more than any one thing. The black people, though *corrupt to a man*, have the highest opinion of integrity. They consider Irwin here as a wonder; and did they not know his abilities and labours in this country, they would believe he was crazy for not accepting the presents brought him. The rents are now as regularly collected as in Europe; a thing before unknown. It was always the custom for the Nabob to let his country at a high rent; but what with the constant elopement of renters, and the many hands the revenue passed through, the Nabob was happy with one half, and often I believe a quarter, of the nominal rent. You will easily apprehend that I am now in a way of making a great deal of money; for Mr. Irwin has an entire confidence in me, and leaves much of the collection of the rents, as well as decision respecting disputed lands, &c. in my hands: but I hope you will as easily believe that I cannot descend to such dirty ignoble means of enriching myself. I do assure you, upon my honour, that I never yet received a bribe in *any shape whatever;* though in the six months I have been here, I may safely say I could have made at least 1000*l.* sterling, were I bent upon it. With integrity and diligence we are almost sure of succeeding. I can lay my hand upon my heart, and solemnly assert that what I have, has been acquired as honestly and humanely as that of your worthy father. The people here are surprised, that have made me offers; however, I find the good effects of it. Mr. Irwin knows that the great powers he has intrusted me with are exerted for the public good: the people respect those orders, and admire those decisions, which they know are not dictated for private views. And the poorest (a thing very uncommon in India) have the confidence to complain of injuries from the rich, because they know they must have justice, and need not apprehend a wrong decision from any thing but inability.

Your mind is too noble not to conceive with ease, that being looked upon in this light is much more satisfactory to my pride (if you please) than the possession of money, bought at the expense of a breach of my solemn engagements with my employers, and the reward of neglect in doing justice, if not of absolute injustice. My pay is very handsome for a young man; I keep a palanquin, and I indulge myself with a handsome horse to ride,—an exercise I have found of great benefit.

The Tinnevelly country is a very delightful one; the heat for some time, till the rains set in, is great beyond imagination. I live at a beautiful garden on the banks of the river, which adds to a most charming prospect. At forty miles distance are inaccessible hills, but very apparent. The rain first falls on these hills, which are hid under clouds for some days, which is a sign that the river will soon be full; for the water runs down with such astonishing rapidity, that I can assure you, from my own frequent knowledge, that the river which in the morning was not a foot deep, will before noon be impassable by men, and so rapid that bullocks, &c., which in crossing, at first were only half below water, have before they got over been carried away and often drowned in the stream. This river is called the Tummer bunny. The water is very sacred among the Bramins, and is the great source of the fertility of this province. You can hardly think how interested I am in the cultivation and prosperity of this country. I take the greatest pleasure in seeing the river fill and spread itself into tanks, ponds, &c. which it does in two or three hours.

Before the rains fall on these immense mountains, the weather is so hot, and the earth so dry, as to afford a wonderful phænomenon that I have often been witness to. In this season I have seen immense fires on the inaccessible mountains, which are occasioned by the friction of the bambo trees against one another, which always happens in high winds;—so dry is this country before the rain.

The knowledge I have of the people makes one often smile at the ideas of Europeans concerning Indians. I assure you that the "innocent Indians" are the most depraved people in the world that I know of; and there are not more impudent, debauched, arch-villains than the "holy harmless sons of Brama:" they are insolent and tyrannical to a degree. I had a little adventure with them the other day. While we have the Nabob's country, we pay all the church expenses, which the superstition of these people has made very great. A great feast is now coming on, for which they got leave to cut a certain kind of wood. A single tree of this was in the garden of a gentleman, which I would not suffer them to cut. They then all shut themselves up in the pagoda, and from the top of it held out a flag of defiance and mutiny. They charged the people to rebel, and all the shops to be shut up. They were obeyed in the last instance. Mr. Irwin was up the country about fifty miles distant. I rode over to a place called the Cutchery (the court of business and justice), of which I am, in Mr. Irwin's absence, sole president and governor. In my way I was saluted with more exclamations of mutiny. I sent for the Bramins, who said they would come down and submit if I would give them this tree. I would make no composition; but told them to go back, and I would follow them in five minutes;

when if the flag of mutiny was not taken down, and the pagoda door opened, I would put on the Company's seal and starve them. They saw I was resolved, and instantly obeyed. Under one of their own princes they would have succeeded.

An extraordinary event requires my presence at the fort of a polygar, where the commander-in-chief of the army to the southward is arrived. I am to endeavour to prevent the matter coming to any extremities, the man being under the Company's protection. This may be the subject of another letter. Till then believe me, with increased fervency,

My dear Cousin, yours most affectionately,

NATHANIEL EDWARD KINDERLEY*.

* The Kinderley family having been mentioned in a former page, it may not be uninteresting in this place to relate the following anecdote, which an old servant who had lived fifty-two years with Mrs. Kinderley and her daughter Mrs. Smith, frequently repeated as a fact with which she was well acquainted, and in part a witness of.——The Rev. John Kinderley's connexion with Scotland had procured him the acquaintance of several families in the North, among whom Lord D—— was one of his most intimate friends. This nobleman had met with a lady at Bath, both young and attractive, and who passed for the widow of an officer. His lordship becoming attached to this lady, he married her, and they soon after left England to reside on the Continent. Here, after a few years, she was seized with an alarming illness, and earnestly desired her lord, in case of her death, that she might be conveyed to England and interred in a particular church, which she named. Upon this event taking place, Lord D—— accompanied the body in the same ship, and, upon landing at Harwich, the chest in which the remains of his lady were inclosed excited the suspicions of the Custom-house officers, who insisted upon ascertaining its contents. Being a good deal shocked with such a threat, Lord D—— proposed that it

should be removed to the church, and opened in the presence of the clergyman of the parish, who could vouch for its containing what he assured them was within :—accordingly the proposal was yielded to, and the body conveyed to the appointed place, when, upon opening the chest, the attending minister recognised, in the features of the deceased, *his own wife!* and communicated the unwelcome discovery to his lordship on the spot. It appeared, upon further conversation, that Lady D—— had been married against her inclination to this person, and, determining to separate entirely from him, had gone he knew not whither, and under an assumed name and character had become the wife of Lord D——. The two husbands followed her remains to the grave the next day; and on the same evening Lord D—— in great distress of mind, attended by one servant, came to his friend's house in Norwich for consolation. It was winter, and about six o'clock when he arrived. Mr. Kinderley was called out to speak to a stranger, and returning to his wife, desired her to leave them together, pretending that a stranger from Scotland was arrived on particular business. Lord D—— sat up with Mr. Kinderley the whole night, to unbosom his affliction and extraordinary fate to his friend, and at day-break, in order to avoid any interview with his host's family, for which his spirits were unequal, he departed.

The following affecting letter, given to Mr. Kinderley by the Rev. Thomas Pyle of Lynn, although it contains nothing relative to the family, yet being written by so celebrated a man as Dean Swift, may be found not void of interest, and has not before appeared in print.

The Rev. Dr. Jonathan Swift to the Rev. Thomas Pyle, Lynn, Norfolk.

"Sir, London, Dec. 26, 1711.

"That you may not be surprised with a letter from a person utterly unknown to you, I will immediately tell you the occasion of it. The lady who lived near two years in your neighbourhood, and whom you were so kind sometimes to visit, under the name of Mrs. Smyth, was Mrs. Ann Long, sister to Sir James Long, and niece of Colonel Strangways. She was of as good a private family as most in England, and had every valuable quality of body and mind, that could make a lady loved and esteemed. Accordingly she was always valued here above most of her sex, and by the most distinguish'd persons. But by the unkindness of her friends, and the generosity of her own nature, and depending upon the death of a very old grandmother, which did not happen till it was too late, she contracted some debts that made her uneasy here, and in order to clear them, was content to retire unknown to your town, where I fear her death has been hastened by melancholy, and perhaps the want of such assistance as she might have found here. I have thought fit to signifie this to you, partly to let you know how valuable a person you have lost, but chiefly to desire, that you will please to bury her in some part of your church, near a wall, where a plain marble stone may be fixed, as a poor monument for one who deserved so well; and which, if God sends me life, I hope one day to place there, if no other of her friends will think fit to do it. I had the honor of an intimate acquaintance with her; and was never so sensibly touched with any one's death as her's. Neither did I ever know a person of either sex, with more virtues or fewer infirmities: the only one she had, which was the neglect of her own affairs, arising wholly from the goodness of her temper. I write not this to you at all as a secret, but am content your town should know what an excellent person they have had among them. If you visited her any short time before her death, or knew any particulars about it, or of the state of her mind, or the nature of her disease; I beg you will be so obliging to inform me. For the letter we have seen from her poor maid is so imperfect, by her grief for the death of so good lady, that it only tells me the time of her death: and your letter may, if you please, be directed to Dr. Swift, and put under a cover, which cover may be directed to Erasmus Lewis, Esq. at the Earl of Dartmouth's Office at Whitehall.

"I hope you will forgive this trouble for the occasion of it, and give some allowances to so great a loss, not only to me, but to all who have any regard for every perfection that human nature can possess; and if in any way I can serve or oblige you, I shall be glad of the opportunity of obeying your commands.

"I am, Sir, Your most humble servant,

"J. Swift."

CHAPTER II.

Sir J. E. Smith leaves Edinburgh.—Lodges in Great Windmill-street, London.—Dr. John Hunter.—Sir J. Banks informs him that the library and collections of Linnæus are upon sale.—Writes to Dr. Acrel.—Correspondence with his Father.—Letters of Dr. Acrel, and catalogues of the collection.—Agrees to become the purchaser.—Letters from Mr. Pitchford, Dr. Withering, Dr. Stokes, Professor J. Sibthorp.—Rev. H. Bryant.—Chosen Fellow of the Royal Society.—Letter from Dr. J. Hope.—Preparation for going abroad.

UPON leaving Edinburgh, Sir James's next object was to fix himself in London for the purpose of attending the great school of anatomy of which Dr. John Hunter was the head, and to avail himself of the medical instruction of Dr. Pitcairn; and on the 25th September, 1783, he took lodgings with his fellow-student, Mr. Batty, in Great Windmill-street, at the top of the Haymarket. "Mr. Baillie," he tells his father, "Dr. Hunter's nephew, is very civil to us; but we are charmed with John Hunter, he alone is worth coming to live in London for. I shall devote myself," he continues, "chiefly to dissection as long as I find it necessary, and afterwards, I believe, St. Bartholomew's hospital will be worth my notice. I will confess that at my first seeing the dissecting-room, which is abominable and horrid beyond conception, I found it very easy to persuade myself that I could do without being any thing more than a spectator there*." Nevertheless, he appears to have settled with great comfort to himself, and to the satisfaction of his friends.

Of his conduct while he resided here, no better testimony can be given than that which immediately follows from a man of strong sense and principle, still more anxious about the reputation of his son than for his worldly advancement.

* These were the sensations which revolted him at first, although they abated as his curiosity was awakened to the study; and thus has he not unfrequently contrasted them, in the vivid language of the Seventh Promenade of Rousseau :—

"Quel appareil affreux qu'un amphithéatre anatomique, des cadavres puants, de baveuses et livides chairs, du sang, des intestines dégoûtans, des squelettes affreux, des vapeurs pestilentielles ! Ce n'est pas là, sur ma parole, que j'ira chercher mes amusemens.

"Brillantes fleurs, émail des près, ombrages frais, ruisseaux, bosquets, verdure, venez purifier mon imagination salie par tous ces hideux objets."

In a letter written at this time to his father, he tells him, "We have dined with Dr. Osborn, the colleague of Dr. Denman ; he is a man of rank in the literary world. At his house we met the famous, or rather infamous Dr. Shebbeare, a most entertaining and lively companion ; the best teller of a story I ever heard, beyond all comparison ; but a most malicious violent-tempered man : being an Irishman, his most predominant hatred is against the Scotch. He can counterfeit any dialect whatever ; his Scotch is the most accurate I ever heard out of that country. It is curious to observe (as I remarked to my companion at the time) in these gross strong-featured minds, as in a microscope, the workings of those passions and dispositions, which in common characters are so faintly and confusedly marked, that we can seldom trace them to their sources, or observe their various connections and dependencies upon one another."

My dear Son,

Norwich, Nov. 24, 1783.

In regard to your mode of living, it is as moderate as can be, and you seem to be very good managers; you may be allowed more of course for diversions and amusements, which I persuade myself will not go beyond the bounds of prudence and moderation, still less of virtue. Nothing can be a greater cause of joy to me than that a son, in whom I have from his infancy had so much pleasure and comfort, and on whom my firmest hopes of happiness in my declining years are founded, gives me no anxiety and uneasiness, though without control, at a critical time of life; situated in the midst of all the fascinating pleasures, and the most alluring temptations, that the most sumptuous, most luxurious, and most vicious and corrupt capital in Europe, or perhaps in the whole world, can produce.

May the Almighty God protect him in every danger, and deliver him from the evil of every temptation!

I am, your affectionate Father,

JAMES SMITH.

Dr. Pitcairn entertained a warm friendship for his pupil, and considered him remarkably acute in the detection and nature of internal diseases; and there seems little doubt that the practice of physic would have become the pursuit of his after life, but for the unexpected events which put him in possession of the collections of Linnæus.

His passion for natural history continuing paramount in his mind, the house of Sir Joseph Banks was at all times the place of resort most attractive to him; and here he first heard that the museum and library of the celebrated Swede were upon sale.

Upon the demise of young Linnæus, Dr. Acrel, Professor of Medicine at Upsal, had written to D . Engelhart, who was then in London, offering the whole collection of books, manuscripts, and natural history, to Sir Joseph Banks, for the sum of 1000 guineas. "It happened," adds Sir James, "that I breakfasted with Sir Joseph upon the day the letter arrived, which was the 23rd of December, 1783; and he told me of the offer he had, saying he should decline it; and, handing me the letter to read, advised me strongly to make the purchase, as a thing suitable to my taste, and which would do me honour." Being thus encouraged by Sir Joseph, he went immediately to Dr. Engelhart, with whom he had been intimately acquainted at Edinburgh, and made his desire known to him; and they both wrote the same day to Professor Acrel, Dr. Engelhart to recommend his friend, and the other desiring a catalogue of the whole collection, and telling him if it answered his expectations he would be the purchaser at the price fixed.

On the following day the young student of physic made these occurrences known to his father, and thus entreated his assistance:—

Honoured Sir, Dec. 24, 1783.

You may have heard that the young Linnæus is lately dead: his father's collections and library and his own are now to be sold; the whole consists of an immense hortus siccus, with duplicates, insects, shells, corals, materia medica, fossils, a very fine library, all the unpublished manuscripts; in short, of every thing they were possessed of relating to natural history and physic: the whole has just been offered to Sir Joseph Banks for 1000 guineas, and he has declined buying it. The offer was made to him by my friend Dr. Engelhart, at the desire of a Dr. Acrel of Upsal, who has the charge of the collection.—Now, I am so ambitious as to wish to possess this treasure, with a view to settle as a physician in London, and read lectures on natural history. Sir Joseph Banks and all my friends to whom I have entrusted my intention approve of it highly. I have written to Dr. Acrel, to whom Dr. Engelhart has recommended me, for particulars and the *refusal*, telling him if it was what I expected, I would give him a very good price for it. I hope, my dear sir, you and my good mother will look on this scheme in as favourable a light as my friends here do. There is no time to be lost, for the affair is now talked of in all companies, and a number of people wish to be purchasers. The Empress of Russia is said to have thoughts of it.

The manuscripts, letters, &c., must be invaluable; and there is, no doubt, a complete collection of all the inaugural dissertations which have been published at Upsal, a small part of which has been republished under the title of *Amœnitates Academicæ*; a very celebrated and scarce work. All these dissertations were written by Linnæus, and must be of prodigious value. In short, the more I think of this affair, the more sanguine I am, and earnestly hope for your concurrence. I wish I could have one half hour's conversation with you; but that is impossible."

To this appeal there came an admirable reply, regarding the expediency of the purchase, in every point of view; but so much approaching to a refusal as to produce great uneasiness and alarm: and another letter, reiterating the persuasions in the first, had but little more effect. Repeated efforts, however, produced some change of opinion; and in a subsequent letter, Sir James tells his father, "I have learnt from Mr. Dryander what the collection consists of; he has often seen it: it was kept in a room built on purpose by itself, for fear of fire. One side of this room was quite occupied by the cabinets of fossils, which are very fine; in another part was a large cabinet of corals, and some animals, as he thinks: there was also a very large collection of insects and shells. The dried plants of the elder Linnæus were about 8000; and his son's collection in his travels, from Sir Joseph, and in France, about as many more. There were many cabinets round the room, and also a few books for common use; but his principal library was kept in another place, and this Dryander never saw: he tells me it was considered to be a good one."

Mr. Smith to his Son.

My dear Son, Norwich, 12th January, 1784.

The dutiful and affectionate light in which you now see what has passed between us upon the subject under consideration, you may be assured makes a very deep and feeling impression upon my heart. I am almost sorry for what my affection and duty to you and my family seemed to me to force me to write, as I knew it would give you pain; but now that I perceive a cooler expostulation would have wrought the same effect, I make some reproaches to myself for having given you more uneasiness than was necessary; and that must and does hurt a father who loves you to excess, who wishes and strives and prays most fervently that all your days may be tranquil, and all your undertakings successful. My soul is full of parental tenderness at this moment, and would fain expand itself upon this subject; but I have not time. Suffice it now to say, that we both think and feel as we ought to do upon the occasion, and that you have satisfied every sensation of my mind that regarded the relation you stand in to me: it shall be my care to strengthen your confidence in my solicitude and unalterable regard for your welfare. But the thing that strikes me very forcibly in your last, seems to confirm an opinion I took up at first; that is, *the bulk* of the collection. Here is a room (no doubt of large dimensions) built on purpose to contain a great many cabinets, and a few books; the principal library was in another place,—no doubt a very large one too:

we both know a large library takes up a deal of room. All this, and a great deal more that must be supposed, convinces me that it will require no small nor inelegant house to place so capital a collection and library in a commodious manner, such as will answer your design in the possession. Indeed I perceive that, however probable the possession of this and your plan is to prove advantageous, I am afraid it is out of the reach of our abilities to attain.

Had I but *you*, I had not hesitated one moment; every shilling of mine should be at your devotion to serve any good purpose; and your dear mother would be as contented as I should be, to retire upon the moderate income of our real estate, till Providence, withdrawing us from the world, should leave you in possession of that also. That you have consulted Dr. Pitcairn and other judicious friends, I much approve; they and your own prudence will advise you about the intrinsic merit and value of the collection, how to have it examined, and every thing relating to that part of the business: but none of them can know how far the purchase would be expedient in our circumstances and situation;—our own wisdom must guide us there.

I cannot but suppose that the library of such a man as Linnæus, and which is called a fine library too, must be worth a great deal, perhaps all the money advanced; but upon the subject of its great supposed value, the character and esteem of its collector in his native country, who must be proud of him and everything tha belongs to him,—I can hardly conceive they will suffer it to depart from

Sweden for so paltry a sum, considered in a national light, and from an university whose reputation he has contributed so much to raise.

The kingdom of Denmark, and all Germany, and Holland, France, and Switzerland, from taste and learning; as well as Russia, from ostentation or improvement,—will be competitors for it, as well as England. And we cannot but suppose, if it is to be peremptorily sold, the object being the money it will fetch, they will make use of their whole endeavours to get as much for it as it is worth, if they are so honest as not to desire more.

Without calling Dr. Acrel's or any other person's honour, known or unknown, in question, ask yourself how seldom it is the case, without great ignorance about the true value of the thing disposed of, where there are many desirous of purchasing, that it does not fetch at least its full value. So that I am inclined to think, that after all your anxiety about it you may not be able to obtain it; and I hope you will not fix your mind so strongly upon it as to create you uneasiness if you miss it, from any cause whatever. But wait calmly the answer to your letter to Dr. Acrel, till you see and examine the catalogue with care, and then determine as circumstances require; and I hope it will please Heaven to direct you for the best in a matter of so very great importance.

I would caution you against the enthusiasm of a lover, or the heat of an ambitious man.

I need not surely now tell you how dear you are to me, how much I esteem you, nor what I hope from you. If you are a stranger to these things, Nature does not write a legible hand, or you have not learned to read her writing; but I know you have, and that you do my great love of you justice. Adieu, my dear son! May Heaven direct all your steps, and shower its choicest blessings on your head!

For ever your affectionate Father,

JAMES SMITH.

Dr. Acrel to Mr. James Edward Smith.

Vir nobilissime, Upsaliæ, d. 9 Februar. 1784.

Humanissimas tuas litteras, Londini d. 23 Decemb. a. p. datas, accepi; et licet petitioni tuæ, in mittendo catalogo rerum natur. p. def. Linnæi, ob temporis angustiam hodie satisfacere non possim, responsionem tamen differre non debui: me vero quam proxime missurum polliceor. Permittas etiam ut in scribendo me lingua Latina inserviam, dum linguam tuam maternam, quam amo et intelligo, tantummodo ad legendos libros calleam. Bibliothecam Linnæanam perlustravi, titulosque librorum adnotavi: superet voluminum numerus 1500: exceptis manuscriptis et litteris, quas non adhuc in ordinem redegi. Permulti eorum sunt ligatura Gallica; reliqui plurimi et quidem majoris momenti libri bene custoditi. Defecti, uti vocant, fere nulli. Omnes vero ad completam bibliothecam botanicam, ab antiquissimis ac nunc rarissimis operibus, usque ad recentissima eadem, pertinent: figuris nitidissimis ornati. Libri medici fere nulli. Cetera catalogus monstrabit.

In eo jam occupatus sum, ut quantum fieri potest, numerum ineam herbarum, in ditissimis herbariis Linnæanis, scilic. patris et filii, occurrentium. Superant vero fere numerum. Observandum, hæ collectiones dividi deberent, 1. in herbarium Linnæi patris, et, 2. filii: quarum—

1. Collectio patris consistere ex circiter 14000 plantis, in tribus thecis sub numeris *A*. *B*. & *C*. —Ut vero vir illustrissimus, dum vixit, nihil ad ostentationem habuit, omnia vero sua in usum accommodata; ita etiam in hoc herbario, quod per xl. annos sedulo collegit, frustra quæsiveris papyri insignia ornamenta, margines inauratas, et cet. quæ ostentationis gratia in omnibus fere herbariis nunc vulgaria sunt. Et dum aliorum herbaria, recenter collecta, similitudinem habent libri cujusdam novi e manu bibliopegæ illico assumti, sic herbarium Principis Botanicorum usu magis tritum, annotationibus ubicunque refertum; minime vero blattis tineisque corrosum, vel alio quodam modo inutile factum.

2. Herbarium Linné filii, splendidum magis ac nitidum prout recentius collectum. Numerus specierum adhuc incertus. Insignis etiam earum numerus, quas in Anglia, sub itinere suo, nec non in Gallia, e collectionibus Smeathmanni, Massoni, Aubleti, Sonnerati, Dombeyi, aliorumque, deprompsit; plura etiam de hac collectione catalogus monstrabit.

Eminet inter cetera collectio illa pulcherrima,

ditissima *Conchyliorum*, adhuc patre vivente, inter pretiosissimas numerata, postea a Linné filio insigniter aucta et ditata.

Insectorum collectio, si separatim divenderetur, licitator quidam Suecus 170 ducatos aureos impendere non hæsitabit.

Corallia et cetera producta marium insignia sunt.

Aves, circiter 50, in thecis vitro clausis servantur.

Pisces, tam in sp. vin. quam siccati supra chartam glutine affixi, permulti sunt.

Cetera, ut fructus, nidus avium et insectorum, vestimenta Indorum et incolarum Maris Pacifici, ut taceam: *Materia Medica* nulla. Litteræ clarorum virorum ad Linnæos scriptæ superant numerum 3000, quæ, una cum manuscriptis, et, inter ea, *Nova Zoologiæ Fundamenta Linnæi Filii*, materiam dabunt operum posthum. Linnæorum uberrimam, et a bibliopolis et typographis avide expectantur.

Ea tamen est hæredum in defunctos Linnæos pia memoria, ut manuscripta et litteræ non nisi ad doctum quendam et honestum virum venumdarentur; ne typis mandaretur quidquam, quod publico scire non interest. Eam etjam fiduciam in te habent, vir nobilissime, si collectionum emptor fueris. Collectio illa sub titulo *Epistolæ claror. Virorum ad Hallerum* pecuniam certissime auctori convexit, parum vero honoris.

In ceterum tibi sanctissime polliceor, ex omnibus collectionibus, in quocunque genere, ne minima quidem particula, me sciente vel volente, distrahi vel abalienari; et quantum in me erit, omnia sarta tectaque ad te venire curabo. Catalogum quam proxime exspectas. Vale interea, atque fave

Nominis tui nobilissimi

Cultorem fidelem

JOH. GUSTAV. ACREL.

Dominum D^{m}. Engelhardt, amicum meum sincerrimum, permultas dicas salutes.

Mr. Smith to Mr. James Edward Smith.

My dear Son, Norwich, March 1, 1784.

The frost broke up, the snow gone, and waters run off, I mean to set out for London next week with your brother John: we shall come on horseback, or I shall put the horses in a chaise, I don't fully resolve which. It will be proper to know whether you can meet with a lodging for us, the town is now so full of people attending parliament. Again I shall want a livery stable for two horses.

I have daily expected to hear you have had an answer from Upsal: if all be right, 'tis as well it stays till we can travel more comfortably. I am glad to hear you are prepared to meet whatever may be the event of the Linnæan collection. Things are so uncertain in this life, we cannot tell what will be the best; but a due exertion to improve the circumstances we find around us will most generally be attended with success, and I entertain very flattering hopes of yours.

P.S. I am told the post has come constantly from Sweden notwithstanding the frost.

Yours,

JAMES SMITH.

Mr. James Edward Smith to his Father.

Honoured Sir, London, April 10, 1784.

I was a little impatient to hear how you got home. On Tuesday I received the wished-for catalogue from Sweden: it is very full and exact, much better than I expected. There are many valuable books: such as the King of Denmark's book of shells, like that at Cambridge; Sloane's Jamaica, worth ten or twelve guineas; and many others worth from five to ten pounds; besides a complete collection of the most useful books in natural history, and many medical books. The greater part are Latin, many French, English, Italian, some Swedish, &c. There are also a few books which from their extreme scarcity sell for an exorbitant price; one little book on insects, coloured, for a copy of which Sir Joseph Banks gave books to the value of thirty pounds, and which has long been sought for in vain for the Royal library. The whole number of works is about 1600, of volumes above 2000. The manuscripts also are very valuable, full as much as could be imagined. Plants 19,000: insects, shells, &c. are said to be very valuable and numerous, but of these I am soon to have a further account.

There is a collection of plants called the small herbarium of young Linnæus, which was collected before his father's death, and contains nothing but what is in the great herbarium: this he desired before his death might go to Baron Alströmer to satisfy a debt of 200 rix dollars (fifty-five pounds) which he owed him. The executors, unwilling to separate anything from the collection, offered the whole to Baron Alströmer for 1000 guineas; they have not had his answer, nor do they expect he will buy it, as he is quite paralytic, and can neither read nor write at present: they therefore do not doubt his taking the small herbarium only, and in that case they offer me all the rest for 900 guineas, as the very lowest price: they have had offers of an unlimited sum from a Russian nobleman, but have declined treating with him till they had my final answer.

I wrote last night, remonstrating against their having made such an offer to Baron Alströmer, as nothing was said of it in Dr. Acrel's first letter to me, and I agreed to take the whole without the small herbarium for 900 guineas, or with that, and pay Baron Alströmer's debt. I hope this will receive your approbation.

The executors demand 500 guineas to be paid as soon as the bargain is concluded, the rest six months after.

JAMES EDWARD SMITH.

Dr. Acrel to Mr. James Edward Smith.

Vir nobilissime, Upsaliæ, d. 6 Martii, 1784.

Catalogum bibliothecæ Linnæanæ jam tibi sisto, vir nobilissime, eum in finem, ut judices quid in re herbaria valeat selecta hæc librorum collectio: continet enim libros permultos quos ob raritatem alibi frustra quæsiveris.

Herbarium illud magnum Linnæi patris, in thecis v. asservatum, consistit plantis circa 19,000; nam his numeranda sunt etiam ea specimina permulta, quæ sub itinere suo in Anglia et Gallia, e herbariis Commersoni, Dombeyi, Sonnerati, aliorumque, collegit illustr. Von Linné filius, et quæ chartis glutine adfigere viri illustris cita nimis mors prohibuit.

Alia quædam plantarum collectio, sub nomine *herbarii parvi Linné filii*, adest, continens tam indigenas quam exoticas, quas in juventute collegit Linné filius, ideoque nihil aliud est quam tirocinium suum, sub auspiciis patris. Ita autem disposuit Linné junior, ante obitum, ut hæc collectio, quæ minoris est momenti, nec alias plantas habens quam quæ in magno herbario præsto sunt, traderetur viro nobilissimo Baroni Alströmer, ob contractum æs alienum, inque ejus solutionem. Dispositionem hanc pie defuncti abnuere non voluerunt hæredes, nec facile concedere, ut, quidquid sit, etiam minimi momenti, e collectionibus distraheretur, ratum duxerunt, omnes historiæ naturalis collectiones cum bibliotheca et manuscriptis Baroni Alströmer ad emendum offerre, eodem scilicet pretio, 1000 guinearum. Nescimus adhuc, si pecuniam solveret, nec ne: ego vero valde dubito Baronem Alströmer, ob gravem morbum articulorum, qui eum nec scribere vel legere, multo minus naturalia tractare, concedit, emtorem fieri. Nobilis Moscovita, nuper etiam ad solvendam non determinatam, pecuniæ summam sese obtulit, si possessor fieret collectionum; abnui ego, antequam ultimam tuam responsionem acceperim.

Catalogum insectorum, conchyliorum, et reliqu. tibi quam proxime mittam; etiamsi persuasus sis, has collectiones et optimas et numerosissimas esse.

Si *herbarium parvum* Baroni Alströmer traditur, ad solvendum debitum 200 thaleror. imperialium monetæ nostræ, reliqua quotquot sunt, a maximis ad minima, non infra 900 guineas divenderentur: sed emtori id certum sit, ut, ita me Deus! nihil eorum ab aliquo distraheretur.

Si igitur *Linnæana Opera et Collectiones* desideres, vir nobilissime, mihi id notum facies proximo tabellarum die, et simul, legato Magnæ Brittaniæ, illustrissimo D°. Wrougthon, Stockholmiæ, vel secretario missionis, litteras mittas, ut si forsan quis scrupulus te urgeat, (qui inter honestos viros non facile contigit,) rem tuam apud nos peragent, bonaque tua observent.

Emtionem festinat, præter alias rationes, etiam ea, quod domus, ubi magnæ hæ collectiones asservantur, professori successori traderetur, et vereor, ne ex demigratione e domo in aliam, e collectione subtilioribus speciminibus damnum inferreretur.

Si nos inter convenietur, pecuniæ dimidiam summam solvere grave tibi non duxeris spero, et alteram dein post vi. menses.

Manuscripta adhuc perplurima deterrui, quam quæ in catalogo adnotata sunt; ita etiam dissertationes, aliaque scripta, magno numero.

Vale, vir nobilissime, et amicorum amico D°. D°. Engelhardt salutes.

Nobilissimi tui nominis

Cultor observantissimus,

J. G. ACREL.

N.B. Ab Upsalia usque ad pontem navalem Londinensem, collectiones per mare transferri possunt, adeo ut facilis et quassationi expers sit earum transportatio.

Mr. James Edward Smith to his Father.

Honoured Sir, London, April 23, 1784.

Yours of the 12th made me very happy. I hope we shall have no reason to alter our sentiments about the step I have set. As it is now pretty generally known, I hear more of the opinion of people on the subject than I did before, and am very much encouraged by them. I thought it a piece of respect due to my old botanical friends to inform them of my purchase before they heard of it by any other means; I therefore wrote lately to Messrs. Woodward, Bryant, and Pitchford, to tell them, and desired the latter to inform Mr. Crowe and Rose. I mentioned particularly that my medical studies were to go on as before.

The collection comes every inch of the way from Upsal by water.

Your dutiful Son,

J. E. SMITH.

Mr. Pitchford to Mr. James Edward Smith.

Dear Sir, Norwich, May 2, 1784.

I am favoured with your letter, for which I think myself much obliged to you, as I began to fear I was struck out of the list of your correspondents; but this letter is a letter indeed! and makes ample amends for past deficiencies. I sincerely give you joy of your purchase as a matter which will afford you a fund of instruction and amusement in natural history, and I should imagine will be a means of making you much known. Mr. Woodward was here this week, and acquainted me with some essential particulars not mentioned in your letter. You may imagine the surprise we were all in. We dined at Mr. Crowe's, who I imagine will talk with you a great deal about it. He was for desiring me to write to you immediately, to beg you would by no means make any agreements as to the disposal of your purchase; but as he sets off for London on Monday, May the 3rd, he can better make you acquainted with his intentions himself. Poor Mr. Rose (who has lost Mrs. Rose) commissioned me in particular to return you his thanks for your remembrance of him.

You certainly make a very proper remark in saying that this purchase is not to interfere with your medical pursuits, as the cultivation of natural history cannot be pursued with vigour but by persons of independent fortunes; for others it must only be as an amusement, or relaxation from other studies.

I shall now look up to you as to a second Linnæus, and without any compliment I think you highly deserving of being the possessor of such remains: at the same time I am afraid your other more serious pursuits will not suffer you to make the use of them your abilities would otherwise entitle you to. The English botanist will now have an opportunity of knowing what natives of his own country are in the *Sp. Pl.*

In hopes of hearing from you a little oftener,

I remain, dear Sir,

Yours sincerely,

J. PITCHFORD.

Dr. Withering to Mr. James Edward Smith.

Dear Sir, Birmingham, May 27, 1784.

I thank you most sincerely for the very judicious and liberal criticisms you have made upon the sheet*, and am happy to inform you that the work in its new form meets with the approbation of all who have seen it. Your remarks upon the difficulties of making the references are unfortunately too true; but as nothing is taken upon authority, I hope the actual examination made upon this occasion will do away more error than it will introduce.

Our friend Dr. Stokes has undertaken this part of the business, and I think his accuracy and industry, as well as his experience, will not be readily

* Of the 2nd edition of Dr. W.'s *Botanical Arrangement*, the 1st edition of which was published in 1776 in two volumes.

outdone. The names, both trivial and generic, are accented through the body of the work, as most people seemed to concur with your opinion. The times of flowering are marked as accurately as I knew how to mark them; those of ripening the seeds would certainly be useful, but I know not any source from whence they could be derived. The same difficulty occurs too as to the opening and shutting of the flowers. The budding of the leaves differs so much in the southern, the midland, and the northern parts of this island, and all these differ again so widely in particular years, that the task of marking them would be endless. I have seen the gooseberry trees in Scotland naked, and in six days' ride to the south found them in full leaf. Trees which in this place were in full leaf on the 18th of April last year had not a bud unfolded on the 10th of the present May.

I apprehend that many of our plants supposed to be Linnæan may be in the same predicament with the *Solidago Virgaurea*, but certainly can no otherways be ascertained than by an actual comparison of our dubious plants with his, and this we were in hopes of doing had young Linné lived; but we hear the whole collection is coming to England, though ignorant into whose hands it has fallen. You can probably inform me, and likewise of the probability of procuring aid from that quarter.

I remain, Sir, with great respect,

Your obliged

W. Withering.

Mr. James Edward Smith to his Father.

Honoured Sir, London, June 18th, 1784.

This day I received the long wished-for letter from Sweden. It contains an accurate inventory of the insects and shells, with the number of species in every genus, by which it appears that these collections are truly noble, even beyond what I could expect. The species of insects are in all 3198; of shells 1564, and 200 more not arranged: there is also a fine collection of minerals; of these there are 2424 specimens; among them are 108 silver, and 31 gold ones, &c. &c. There are 45 birds in glass cases.

The bargain is concluded with me on these terms, —Baron Alströmer is to have the small herbarium, and I am to give 900 guineas for the rest.

J. E. Smith.

Dr. Acrel to Mr. James Edward Smith.

Vir nobilissime, Upsalia, Maij 1784.

Binas tuas accepi litteras, die 9 Aprilis et 20 ejusdem mensis datas: debui jam antea responsionem dare, etiam dedi, nisi intensissimum frigus variaque alia negotia me retardarunt catalogum rerr. naturall. Linnæorum perducere ad finem. Fateor ingenue, me nondum omnia in ordinem et numerum reduxisse, negotiis præcipue academicis ut et aliis obrutus: quæ vero hac vice tibi ob oculos pono, sufficere ad judicandum pretium puto. En igitur ea, quæ post bibliothecam et herbaria sequuntur, de quibus antea scripsi:—

INSECTA

in duobus thecis conservata, pulchra et omnino splendida: ex his continet sequentia:

Coleoptera.

Genera.	Numerus Specierum.
Scarabæus	139
Lucanus	7
Dermestes	44
Hister	7
Byrrhus	11
Ptinus	14
Bruchus	9
Cassida	24
Silpha	47
Coccinella	58
Chrysomela	148
Hispa	4
Gyrinus	3
Atelabus	15
Curculio	138
Cerambyx	91
Leptura	43
Necydalis	10
Lampyris	6
Cantharis	36
Elater	38
Buprestis	37
Cicindela	19
Carabus	51
Dytiscus	31
Tenebrio	63
Meloë	26
Mordella	8
Staphylinus	24
Forficula	2
	1153

Cl. II.

Hemiptera.

Genus.	Species.
Blatta	7
Mantis	16
Gryllus	55
Fulgora	7
Cicada	48
Nepa	7
Notonecta	4
Cimex	164
Coccus	7
	315

Cl. III.

Neuroptera.

Genus.	Species.
Libellula	17
Hemerobius	9
Myrmeleon	7
Ephemera	8
Phryganea	19
Raphidea	3
Panorpa	3
	66

Cl. IV.

Lepidoptera.

Genus.	Species.
Papilio	239
Sphinx	59
Phalæna	625
	923

Cl. V.

Hymenoptera.

Gen.	Spec.
Tenthredo	78
Sirex	7
Ichneumon	87
Sphex	50
Chrysis	11
Cynips	5
Vespa	51
Apis	53
Mutila	12
Formica	8
	362

Cl. VI.

Diptera.

Gen.	Spec.
Asilus	19
Bombylus	5
Conops	14
Hippobosca	3
Œstrus	6
Empis	6
Tabanus	17
Musca	156
Tipula	36
Culex	4
	266

Cl. VII.

Aptera.

Gen.	Spec.
Aranea	18
Phalangium	6
Scorpio	4
Oniscus	20
Scolopendra	15
Iulus	6
Monoculus	4
Acarus	4
Pediculus	6
Cancer	30
	113

Summa Specierum, 3198

CONCHYLIA

consistunt sequentibus Gener. et Speciebus cum variationibus:

Gen.	Spec.
Chiton	3
Lepas	6
Pholas	14
Mya	13
Solen	15
Tellina	54
Cardium	44
Donax	13
Venus	89
Spondylus	6
Chama	14
Arca	24

Ostrea	77	Strombus	62
Anomia	32	Murex	145
Mytilus	26	Trochus	88
Pinna	8	Turbo	57
Argonauta	2	Helix	84
Nautilus	20	Nerita	61
Conus	74	Patella	94
Cyprea	85	Haliotis	14
Bulla	20	Dentalium	10
Voluta	49	Serpula	20
Buccinum	154		
Do.	89	Summa Testaceorum,	1564

N.B. Præter hæc Testacea in theca conservata, etiam alia eaque numerosa collectio Conchyliorum adest; cujus tamen numerus mihi incertus: quantum vero ex sola inspectione judicare licet, eam numer. 200 circiter diversarum specierum continere, facile credo.

MINERALIA,

in 2 thecis conservata; quæ collectio sequentia habet

Genera,	et Num. Specierum.	Genera.	Species.
Schistus	76	Nitrum	103
Marmor	150	Crystallorum figuræ	8
Gypsum	26	Natrum	44
Stirium	16	Borax	22
Spatum	25	Muria	8
Talcum	59	Alumen	9
Amianthus	45	Vitriolum	10
Mica	35	Succinum	8
Cos	50	Bitumen	40
Quartzum	28	Pyrites	125
Silex	39	Arsenicum	28
Saxum	139	Hydrargyrum	9

Molybdenum	8	Entomolithus	23
Stibium	23	Helmintolithus	162
Zincum	35	Phytolithus	27
Vismuthum	6	Graptolithus	21
Cobaltum	14	Calculus	10
Stannum	20	Tartarus	2
Granatum	10	Actites	30
Plumbum	232	Pumex	11
Ferrum	184	Stalactites	35
Cuprum	206	Tophus	76
Nickelum	4	Ochra	61
Argentum	108	Arena	4
Aurum	31	Argilla	86
Platinum	1	Calx	20
Zoolithus	2		
Amphibolithus		Specier. mineral. summa,	2424
Ichthyolithus	10		

ANIMALIA.

Aves in thecis operculis vitreis ....No. 45
Rostra avium.

Pisces, siccati et chartis glutinati........No. 158
Quædam in spir. vin.

CORALLIA,

magno numero, sed non adnotata.

Quæ restant, tam naturalia, quam eorum producta, ut flores, spadices Palmarum, radices, fructus, necnon curiosa varia vestimenta et utensilia incolarum Maris Pacifici, et Indorum etc., adducere et nominare tempus non permittit.

MAMMALIA nulla.

Quæ ad materiam medicam spectant, scilicet, gummata, resinæ, radices, etc., nulla adsunt, nec adfuerunt. Plantæ vero, unde hæc omnia desumuntur, in maximo hoc herbario sine dubio adsunt.

Post illum quem antea dedi calculum, insignem herbarum siccatarum copiam inveni: sed numerum adnotare vetuit temporis angustia.

Has collectiones, omnes et singulas, una cum reliquis omnibus, a maximis ad minima, quæ in musæo Linnæano post mortem inventa sunt, tibi jam, nomine hæredum offero emenda, sequenti sub conditione, quod,

1°. Concedas Baroni Alströmer *herbarium parvum*, plantas continens quas in juventute et sub auspiciis patris collexit Linné filius; non tam ad solvendum debitum illud 200 thal. imper., quam ob pretium illud amicitiæ et adfectionis, quæ L. Bar. Alströmer et Linnæum a teneris inde interfuit.

2°. Quod summæ (900 guineas) dimidiam partem, sine mora, mihi mittas sub forma tesseræ nummariæ (bill of exchange), a mercatore quodam Stockholmiense, more eorum, solvendam.

3°. Reliquam summæ partem, (450 guineas,) post tres menses ad minimum mihi eodem modo mittas; et interea chirographum tuum debitum ostendentem, una cum litteris tuis responsoriis nobis relinquas. Si inter te et me res ageretur, ejusmodi testimonia, ubi honor et fides suprema lex est, omnino supervacanea essent; jam vero, cum jus tertii in negotio versetur, et hæc conditio ab hæredibus, matre scilicet defuncti et sororibus, postulata, non possum non, quin votis eorum respondeam, et eam tibi proponam.

Ad me quod attinet, fidem summam tibi servabo, ne hilum quidem ab eorum omnium abalienetur vel furtivis manibus arripietur. Si tibi placeant postulata hæredum, litteris quam primum mihi certiorem facies sequentia.

1°. Si in Anglia tantum vectigalis solvatur pro libris filo ligatis (sewed books) quam pro iisdem in corio ligatis (bound ones)? Si ita est, promiscue in thecis consarcinari possint: si pro corio ligatis plus solvendum, separentur, et in suis thecis diversis servari necesse est.

2°. Ab Upsalia usque ad Stockholmiam, maritimo itinere, transportentur collectiones, impensis hæredum. In the warehouse at Stockholm you must pay the duty, but I don't know how much: credidero tamen, quod ejusmodi mercedes non magni ibidem æstimantur. Interest ne quid a publicanis diffrangeretur, quod ut prohibeatur curabo.

3°. Notus mihi est nauta mercatorius, nom. *Captain Browel*, natione Anglus, civis et Londinensis et Stockholmensis, vir probæ fidei: hic quotannis semel vel bis inter Londinum et Stockholmiam iter facit propria navi; fideliori viro res tuas committere vix habeo, quamvis etiam multi nautæ mercatoriæ quotannis a portu Stockh. Londinum petunt. Will you take insurance for the sea transport?

4°. Opuscula Bergmanni tibi mittam: quæritur vero, si mihi liceat ex bibliotheca Linnæi sequentes libros a te emere? In quarto: No. 5, Banks's Voyage; No. 7, Russel's Natural History of Aleppo; No. 7, Pennant's Voyage to the Hebrides; No. 126, Cursory Remarks on Scurvy; No. 223, Dalrymple's Memoirs of the Coast of China. In octavo:

No. 10, Miller's Illustration of the Sexual System; No. 123, Letsom's Medical Memoirs; No. 303, Letsom's Traveller's Companion. Quæritur etiam quanti a te æstimantur? Vix proposueram, nisi perpassus, te aut eos habere in bibliotheca tua, aut facile obtinere. Mihi vero perquam difficiles comparatu. Dolendum sane, quod nos inter tanta difficultas est comparandi libros; nam libri Anglici in Suecia perrari, sicut Suecici in Anglia.

5°. Bibliopola quidam Upsaliensis, nom. Suederus, Angliam petere se dixit; et officium suum mihi obtulit in custodiendis omnibus, modo illi a possessore, a te scilicet, promittitur itineris impensa: est vir bonæ fidei, et tibi sub itinere necessarius, providendo, ne quid detrimenti capiant res naturales: affirmare vel negare non potui: dic quid de ejus proposito tibi videtur.

Hæc sunt, quæ jam scribenda habeo. Tu quamprimum litteras mittas. Omittendum non est, me a Domino Sibthorp binas accepisse litteras: in primis tantum herbaria emere voluit, in ultimis vero se emtorem omnium fieri scripsit, modo si illi tantum concedatur temporis spatium, ut Upsaliam perveniet. Ita ejus verba:—"If I am not too late to become a purchaser, I will immediately on the receipt of your answer set off for Upsal. If you have absolutely disposed of it, I must attribute it to the miscarriage of my former, which I think at least you would have done me the honour of attending to." Vereor ne iter suum incepit, nam duplices litteras responsorias postulavit, Oxfordiæ et Bruxellæ. Si vero collectiones emeris, te rogo et obtestor, ut illico eum, D^m Sibthorpium, id notum facies per litteras, ne honestus vir frustra et quasi dolo circumducetur. Scripsi ei, quod negotium mihi est cum alio viro, et si non placuerint proposita mea, Sibthorp ei erit proximus in negotiatione.

Insecta sarta tectaque per itineris cursum servare difficillima res est. Oportuit thecarum fundos cera oblinire, ne acus decidui fient: sed huic operi requiritur multum temporis, necnon impensorum, nam ceræ necessaria quantitas non est exigua, nec vilioris est pretii. Dic mihi, si sufficere tibi videatur, insectum quodcumque probe inspicere et ejus acus fortiter in assere ligneo impingere? Quantum in me est, curabo ne blattis tineisque corrodentur. Consarcinatio (the packing) fiet impensis hæredum: sed si tibi curæ est tuarum rerum, suadeo, ne aliquot schill. parces viro in hoc negotio occupato solvendos: plus enim diligentiæ et curæ adhibeat; sed hæc sunt minima.

I am, Sir,

with the most faithful esteem and respect,

Your most humble obedient servant,

J. G. ACREL.

Dr. Jonathan Stokes to Mr. James Edward Smith.

Dear Sir, Stourbridge, June 21, 1784.

Give me leave to congratulate you on being become the possessor of the cabinet and MSS. of the great Linnæus, and of his excellent and amiable son, whose loss I shall ever most sincerely regret as a friend, as well as a lover of natural history. I congratulate you on this acquisition, not merely as persuaded that those excellent men will find in you an able as well as faithful editor and commentator, but as an Englishman, who feels a degree of honour given to himself, in finding that one of his countrymen, animated by a love of science, has had the spirit to make the purchase, and to import into his own country so valuable a collection. But the same information adds that very great offers have been made to you from France. L'Heritier and Broussonet I much esteem, and I remember with pleasure the civilities I received from several other naturalists of that nation; but I must here confess myself so far national, as to wish that the valuable treasure which you are become the possessor of should not go out of this kingdom. If Parliament should not think proper to add it, with your permission, to our national museum, I flatter myself private individuals will be found to form a subscription, which may enable you to resist the influence which any liberal offers from France may have upon you.

I remain, with much esteem,

Your old fellow-student, &c.

JONATHAN STOKES.

P.S. If you meet with a letter of mine to young Linné may I ask you to return it.

If your private fortune should not permit you to keep the whole, and you should be induced to dispose of any of it, I should be happy to become a purchaser of what duplicates you may find of British plants, at what should be considered as a fair price. I should be equally happy to become a purchaser in like manner of the duplicates of *European* plants.

Mr. James Edward Smith to Dr. Stokes.

Dear Sir, London, Sept. 23.

I ought long ago to have thanked you for your letter, and friendly congratulations on my late acquisition; you may be assured they were extremely acceptable to me; and the principal reason for my not replying sooner to your favour was, that I expected every day to receive my treasures from Sweden, and I wished, not only to send you some more information, but to answer that part of your letter in which you speak of purchasing some of the duplicates. As yet they are not arrived; I would not, however, be any longer silent, lest you should accuse me of inattention to your friendship. At present I have no intention to dispose of any part of the collection. I have had several overtures from different quarters on this head, but have declined entering into any treaty, as I did not purchase the collection with that view. On my return from Edinburgh next year, I shall have leisure to examine the whole, and hope then to be able to make it of use to the scientific world, by removing as much as possible some of the doubts and difficulties which have always attended inquiries in natural history, and which I need not particularize to so experienced a naturalist as yourself. I find by the catalogue, there is a large number of letters from various learned men to the two Linnæi, and you say I shall find some of yours among them, which you wish to have returned; but as I shall be very solicitous not to mutilate any part of the collection, I hope you'll permit me to keep them; especially as returning them might expose me to endless solicitations of the same kind. I hope you will rest assured that no improper use shall be made of any thing of that kind which may fall into my hands.

Permit me in my turn to congratulate you on *your* late acquisition, I mean of an agreeable and accomplished partner for life. It will always give me real pleasure to hear of your happiness, or by any means in my power, to express the esteem and respect with which I am

Your most obedient,

J. E. Smith.

Dr. Stokes to Mr. James Edward Smith.

Dear Smith, Stourbridge, October 1, 1784.

It was with the truest pleasure I learnt from your own hand, that you have no intention to dispose of any part of the collection. My application had only for its object the possibility of the contrary, and the fear lest so invaluable a treasure should pass into some foreign country, where the memories of Linnæus and his truly excellent son would find a less enlightened or less faithful guardian than yourself. Happy that this invaluable collection is in the possession of one who has candour, knowledge, and enthusiasm,

I remain, with much esteem,

Yours, &c.

JONATHAN STOKES.

Dr. Acrel to Mr. James Edward Smith.

Vir nobilissime, Upsaliæ, d. 13 Julij, 1784.

Humanissimæ tuæ litteræ, d. 25 Junij scriptæ hodie mihi ad manus pervenerunt. Ex iisdem percepi, te pecuniæ dimidiam summam apud mercatorem Wilkieson et Co. jam deposuisse, ideoque primum postulatum non procul a nobis distare, sed forsan proximo tabellarum die hæredibus pervenire.

In eo jam occupatus sum, ut omnia sarta tectaque ad te perveniant; ideoque herbaria in thecis ligneis includere; insecta, acubus fortiter ad thecarum fundos adfixa, in aliis ejusmodi thecis ligneis conservare curavi; nec in reliquis, quantum fieri potest, defuturus sum.

Ob libros mihi dono datos millenas accipies gratias. Pretium eorum lubenter persolvere et volui et debui, sed nunc in tui memoriam servabo. Ut aliquantulum par pari referam, ut pignus amicitiæ recipies rogo, Opusculorum Bergmanni 2 Exemplaria, et Dissertationum Chemico-Physicarum Wallerii voll. 2. Si in aliis libris tibi comparandis usui esse possum, habebis me obstrictissimum.

A Sibthorpio iterum litteras accepi, in quibus se

nunc emtorem omnium fieri optat, et pecuniæ summam solvere velle, scripsit. Me vero imprimis movet, quod, dum in priori ejus epistola nos inter sermo erat tantummodo de herbariis, quæ separatim vendere abnui, nunc se in iisdem litteris pretium 1000 guineas obtulisse contendit. Certus sum te jam litteras illi dedisse; responsi ego, me omnia vendidisse. Fidem et honestatem incolumem servare studui erga omnes; nec mihi conscientia imputat, quod alio quam Sibthorpio collectiones venum dare volui.

Interea tu valeas, amicorum optime, et me ama.

J. G. Acrel.

Triste tibi nuntium mittam, celeberrimum Bergmannum d. 16 hujus mensis diem obiisse supremum, ad fontes soterias Medevicenses, ubi salutis ergo aliquantisper moratus fuit. Apoplexia ab hæmorrhoidibus interceptis occubuit.

Mr. James Edward Smith to his Father.

Honoured Sir, August, 31, 1784.

I have now the pleasure to inform you, that it is unnecessary for us to trouble our friends any further about the custom-house business, for I went there yesterday, and was told that an order had come from the treasury, that every thing except the books should be admitted and delivered to me without duty, or any charges whatever. I was at the same time assured that every attention possible should be shown me, and the greatest care taken that nothing should be injured. I am principally obliged to Sir John Jervis* for this indulgence, and I understand it is almost a singular instance. I shall write to thank him, as he is out of town.

J. E. SMITH.

Dr. Acrel to Mr. James Edward Smith.

Vir nobilissime, Upsaliæ, d. 13 Augusti, 1784.

Quæ ad historiam naturalem spectant, omnia, non minus quam manuscripta, plantæ siccæ et nondum chartis adfixæ, libri, etc. etc. in thecis ligneis (great wooden binns) probe consarcinata, et, ubi opus fuerit, scobibus ligneis (saw-dust) interspersa sunt: insuper sigillo telonii regii munita, ne ea in telonio R. Stockholmensi aperiri necesse sit. Numerus thecarum est 26: et jam ante 9 dies abhinc ad Stockholmiam itinere navali feliciter transportatæ, ibi secure et placide in domo publica (warehouse) navem Londinum pandentem expectaturæ. Cum mercatore quodam Stockholmensi ita peregi, ut primam et securam transmittendi occasionem arripiet; sive id fiat cum navi Anglico vel Suecico.

Inter rariora jure numerantur 97 figuræ ligno

* Afterwards Earl St. Vincent, and at this period one of the representatives of the borough of Great Yarmouth.

This favour Sir James never ceased to recollect with grateful feelings towards the distinguished nobleman, who conferred it on a perfect stranger.

incisæ, quæ olim ad opus botanicum Olavi Rudbeckii patris, *Campi Elysii* dictum, pertinent; et quidem ad tomum 1^{m} cujus non nisi duo exemplaria in rerum natura existit, ideoque inter rarissimos libros habetur. Magnum illud incendium Upsaliense A^{o} 1702 totam editionem consumsit, et solummodo hæ figuræ supersunt. Magni heic æstimantur ob raritatem.

Plura scribere prohibet temporis ratio. Cura ut valeas, mihique fave.

Nobil. nominis tui

Cultor humillimus,

J. G. ACREL.

Mr. James Edward Smith to his Father.

Honoured Sir, Sept. 9, 1784.

On Monday I had a most excellent letter from Dr. Acrel, dated August 13. He says that having now received half the sum, the heirs had consented to forward the collection, and it was sent to Stockholm August the 4th under the care of a trusty mercantile friend, to be put on board the first good ship for England. Captain Browell would not wait for it.

Dr. Acrel says, there is among the books a copy of that very rare book the 1st volume of Rudbeck's *Campi Elysii*, of which there are only two or three copies in the world, almost the whole edition having been consumed, with the whole town of Upsal, in 1702. I have heard Mr. Dryander say, Sir J.

Banks would gladly give 100*l.* for it; he has the 2nd volume, and so shall I. I shall not remove to Chelsea till about Monday.

J. E. SMITH.

Mr. James Edward Smith to his Father.

October 2, 1784.

Upon inquiring this day about the probable time of the arrival of the ship from Sweden, I find it may be here in a day or two. Its name *The Appearance, Captain Axel Daniel Sweder.* The cases are marked J. E. S. No. 1 to 26, and must be very large, as the books, which are near three thousand, take up six of them only, the plants five, minerals four, insects two, shells, fish, and corals occupy three. The freight 80*l.*, and 5*l.* for the captain's fee.

I was at the custom-house to-day, and saw the letter from the treasury, which is very handsome and full. All the expenses that Dr. Acrel has been at on my account amount to 4*l.* 10*s.* which he desires I'll repay him in medical books.

J. E. SMITH.

The ship which was conveying this valuable cargo had just sailed, when the king of Sweden, Gustavus III., who had been absent in France, returned home, and sent a vessel to the Sound, to intercept its voyage; but happily it was too late. At the end of October, 1784, the packages were safely landed at the custom-house.

The whole cost of the collection, including the freight, was 1088*l.* 5*s.*

So nice does Dr. Acrel's conduct appear in the negotiation between himself and the purchaser, rejecting all offers till the first was concluded, that a report of his being bribed with a hundred pounds was circulated in various quarters. In a letter to him, Sir James says, "Dryander immediately contradicted this malicious falsehood; but it gives me much concern that your conduct, which has been so honourable, should have made you enemies. I should be very unhappy if you suffer on my account, and shall always be ready to bear witness to the rectitude of your behaviour, and can at any time produce all the letters that have passed between us, as a proof.

"Between ourselves, it is certainly a disgrace to the university that they suffered such a treasure to leave them: but if those who ought most to have loved and protected the immortal name of Linné failed in their duty, he shall not want a friend or an asylum while I live or have any power, though ever so small, to do him honour."

Sir James's first idea was to deposit his purchase in some spare rooms in the British Museum; but he found some objections to the scheme, and preferred taking a house, that it might be safer, and more accessible to himself and his friends. He therefore hired apartments in Paradise-row, Chelsea, whither it was immediately conveyed; and often has he recurred with great pleasure to the first winter after its arrival, when, with Sir J. Banks and Mr. Dryander, they examined the herbarium minutely, and carefully unpacked and arranged the whole collection.

With no premeditated design of relinquishing physic as a profession, yet from this hour he devoted his time and all the powers of his mind to the object for which he had hazarded so much; nor was there ever a period, in his subsequent life, of misgiving or regret that he had made a wrong choice: neither was his love of botany pursued to the exclusion of other literature or lighter pleasures; but it was the charm of his existence, always at hand ready to take up, always leading the mind forward, and filling his hours with satisfaction.

How well he estimated his own powers, and came up to the expectations formed of him, may be learnt from the opinions of his fellow-labourers in the fields of science.

Mr. Pitchford to Mr. James Edward Smith.

Dear Sir, Norwich, Nov. 26, 1784.

Since I wrote my last short letter, I have had some conversation with Mr. Smith your father about your collection. As he seemed desirous that your friends should communicate to you every thing they thought about the matter, I esteem it an honour he does me, in supposing mine worth your perusal; and as I should be very happy in rendering you any real service, so you must accept the following as written with that intention, to be rejected or embraced as your own good sense and prudence will point out, and offered merely as hints which may or may not be well grounded, as I cannot but have an imperfect knowledge of some parts of the matter.

First then, as the human mind is limited, and your first business is that of the study of physic, give me leave to say, that as your thoughts must unavoidably be much engrossed by the variety and pleasingness of the many objects about you, I would not suffer them to get too much possession of my mind ; but after I had arranged everything to my satisfaction, and taken a complete catalogue, I would let the collection remain as it was, after I had digested the following considerations: viz. whether I should dispose of any part of it at all ? and if I did, whether it would be right to do it directly while the public curiosity was awake, or wait ? —If you mean to give lectures, the minerals cannot be disposed of, no more than the insects or plants. I conclude therefore that you will not dispose of these at present, as you cannot tell whether you may lecture or not.—Secondly, as to the matter of publishing, I am to suppose (without knowing it to be a fact) that you have complete manuscripts of *Syst. Nat.*, *Gen.* and *Sp. Pl.*, *Fl. Suecic.*, and the eighth volume of *Amœn. Acad.*, for an improved edition of each. But first let me ask you, whether your hortus siccus is named by this manuscript edition ?—If it is not, it certainly must, as many alterations may have taken place. If it is, it remains whether you mean to publish these, or not ? At present certainly not, as you have not time ; but

Mr. Rose thinks you should by all means offer them to the booksellers at once, who he thinks would give a very handsome sum (he guesses 500*l.*); whereas you must run a hazard, with the certainty of a great loss of time in the publication, revisal, &c., if you are the publisher. If you cannot sell them to your mind, they can but remain till some future occasion brings them forth.—Thirdly, there is one work which I am sure you have in your power to publish yourself, and you only, provided you could have possession of Ray's plants, viz., a Flora Britannica, *the most correct that can appear in the Linnæan dress;* and such English plants as are not in the Linnæan collection must have new names, which, if you cannot get Ray's plants, must have that author's synonyms as far as it appears certain they are his. Such a work as this I think you might safely publish; but I have no idea that you will have it in your power to do it at present, as your thoughts must be employed on other matters.

Upon the last matter on which I would venture to write my thoughts I own I am entirely at a loss; for, after all the conversation I had with Mr. Smith on the subject, we both, I believe, remained in the same uncertainty, viz., how far you can with propriety show your hortus siccus? I am afraid you must have difficulty in it, as your natural good-nature would be put to a trial to refuse any one a sight they are so desirous of enjoying; and yet with regard to the English plants, if you mean to publish yourself, you would not choose to be anticipated. I hope you do not think me invidious in saying this: I mean it only for your sake, and if you do not publish yourself, there is nothing in it, except the time it would cost you in the exhibition, which I suppose you would regulate so as to make it as little burthensome as possible. I own, for my own particular, that I should be severely mortified if I was in London and could not take a peep.

I have thus offered you a few crude thoughts on subjects which I own I am not sufficient judge of; and, as I said in the beginning, you will accept them for the intention with which they are written, and as induced by Mr. Smith your father; and if you can spare leisure to send me a few lines, I shall receive them with pleasure.

I need not say Mr. Rose and the rest of your botanical friends here desire their best compliments.

Dear Sir,

Your ever obliged and affectionate humble servant,

J. PITCHFORD.

I have lately seen specimens of *Athamanta Oreosel.* of Hudson, gathered by Mr. Relhan within a mile and half of Cambridge, and named by him *Atham. Libanotis,* in which we think him right.

Mr. Hudson has found plenty of *Corrigiola littoralis* on the Devonshire coast.

Professor J. Sibthorp to Mr. James Edward Smith.

Dear Sir, Göttingen, Jan. 1, 1785.

Give me leave to congratulate you upon your

late acquisition of the Linnæan cabinet. The disappointment I feel in not possessing it myself is in great measure alleviated by the kind opportunity you offer me of consulting it upon my return to England. We were competitors from a laudable ambition, and I trust are not worse friends for our competition. You have left me only one wish, that in case you should ever be disposed to part with it, you will give me the first refusal. You very fortunately closed with Acrel a short time before he received a letter from me offering him a thousand guineas; and, as I was told by your banker Wilkieson at Amsterdam, you got it for nine hundred. The Swedes I hear are very angry at Acrel's disposing of it; and indeed Sweden must have very little money, and very little respect for the memory of Linnæus, to suffer his collection to be so soon expatriated.

If I can render you any service in the course of my travels, I beg you will command me. I am now in Göttingen, a German university of much learned labour and sober science, with a library the richest in modern books of any in Europe. The facility of procuring these books, the opportunity of learning the German language, with the civility and literary society of the professors, give me the utmost satisfaction. Should you think of spending some time on the continent, I know of no situation more eligible, or that I should recommend to you more strongly than Göttingen. Pray have you a *printed catalogue* of Linnæus's books? If you have, will you do me the pleasure to send it me? or could you get your manuscript transcribed for me? as I imagine it is not very bulky. Do tell Dryander that my very respectable friend Sir Joseph may procure many of his *desiderata* at Professor Spielman's auction, which will be in Strasbro' in April next. I dare say you will often see my worthy friend Mr. Lightfoot. I beg you will make him my best compliments, and tell him I have got specimens (almost a complete collection) of the land and fresh-water shells of Switzerland, which I only wait for an opportunity to send to the Duchess of Portland. Has our friend Dickson published his lyncean discoveries? Broussonet tells me of strange wonders (a *Rosa unifolia* from Ispahan) in the Paris garden; and I imagine L'Héritier is almost ready with his first Decas of *Plant. Rarior.* Thunberg has lately attempted to throw some light upon the *Bohun Upas*, or Poison-tree, which he imagines to be a species of *Sideroxylon.* Ferber will probably succeed to the chair of Bergman. Schreber is writing a monography on the genus *Aster*, and a little dissertation on the *Boletus suaveolens.*

A Flora and Fauna of Lombardy is expected from Scopoli, and Professor Lachenal and Saussure are collecting the materials for the complete natural history of Switzerland. I am here as a practitioner, making experiments upon the *Caryophyllata*, to chase away intermittents; and as our hospital physician is a pupil of the school of Vienna, the *Cicuta*, *Arnica*, and *Pulsatilla*, are in constant use —*Valeant quantum valere*. . . . .I look on, but with impartial eyes. It is here somewhat colder than in

England; the principal amusement going in sledges, everybody wrapped up in furs.—I shall be very happy to hear from you.

I am, dear Sir, with much respect,
Your sincere Friend,
J. SIBTHORP.

The most devoted friend to natural science, and one most personally attached to Sir James, was the late Bishop of Carlisle. The following is a passage from the very first letter he appears to have written to him, and is dated January, 1785.

"Natural history is to me an object of perpetual pleasure; and whatsoever, therefore, you will be pleased to lay by for me will not be thrown away upon one who is insensible of the trouble (I have *propriâ personâ* collected and dried full 3000 specimens) or value of collecting specimens of plants. Other branches also of natural history have been my study.

"Your *noble purchase of the Linnæan cabinet* most decidedly sets Britain above all other nations in the Botanical Empire; and it were much to be wished that the studies of individuals with respect to the science at large would become so animated and so successful, that she might be induced to *fix* her seat amongst us."

On the 21st of February 1785, Sir James informs his father, "I take the chair this evening in the

Natural History Society for the first time. We meet in a room in Leicester Square. My certificate was given in to the *Royal Society* last Thursday, signed by Drs. Garden * and Combe †, Sir John Cullum ‡, and Mr. Hudson §. Sir Joseph Banks (the President) thinks I have no fear of being rejected; at least, he says, I shall have *him* on my side."

Rev. Mr. Lightfoot to Mr. James Edward Smith.

Dear Sir, Uxbridge, May 5, 1785.

Your magnificent collection afforded the highest gratification to the Duchess of Portland ||, and the little presents you made her obliged her extremely.

* Dr. Alexander Garden, a memoir of whom is given in the Linnæan Correspondence, vol. i. 282.

† Dr. Charles Combe, author of a Catalogue of the coins in Dr. William Hunter's collection, and other works.

‡ Sir John Cullum, author of the History and Antiquities of Hawsted and Hardwick in the county of Suffolk, and elder brother of Sir Thomas Geary Cullum, Bart.—This accomplished writer of one of the best topographical works extant, died in October 1785, aged 51.

§ Mr. William Hudson, one of the earliest Linnæan botanists in England, and author of the *Flora Anglica*, published in 1762, in one volume octavo.

|| Margaret Cavendish Harley, heiress of the Harley and Holles families, married in 1734 to the second Duke of Portland, and long celebrated as the munificent and intelligent patroness of natural history, especially conchology.—Her Grace died the 7th of July 1785, and her fine collection was afterwards sold by public auction in April 1786, and occupied thirty-seven days. The number of lots was no fewer than 4156, and the produce amounted to 11,524*l.* 4*s.*—*Linnæan Correspondence*, and *Catalogue.*

This introduction will I hope make you better acquainted with her, and be productive of a visit to her noble seat of Bulstrode, when I shall hope to have a share of your company there and at Uxbridge.

As your election to become F.R.S. approaches, if you think there will be the least occasion for my presence, be pleased to write me a line, and I will be ready to tender you my services at a day's notice; and will at the same period beg another day or two to go through with the doubtful plants, if you will indulge me so far.

I am, Sir, with great regard and affection,

Your most obliged humble servant,

JOHN LIGHTFOOT.

Please to try if you can make out the *Murex ambiguus* and *despectus* of Linnæus, as also his *M. corneus*, against I see you.

On the 28th of May, Sir James acquaints his father that "he was admitted a Fellow of the Royal Society on Thursday, *without a single black ball;*" and adds, "I paid my money, 32*l.* 11*s.*, and took my seat the same evening: my success was indeed very flattering, and I believe gave my good friend the President great pleasure.

"I received a letter lately from Monsieur L'Héritier, a counsellor at Paris, on a botanical subject; in which he says, he will at any time give me my purchase-money and any other expense that I may have been at for my collection; but not expecting I should listen to this offer, he earnestly desires to buy the duplicates of plants, if any, at whatever price shall be judged reasonable.

"I have lately employed a few hours in translating into English a little treatise of Linnæus's recommending the study of nature: it begins by showing that this study leads to a proper knowledge of the Deity, then takes a comprehensive view of the œconomy of nature, and ends with a number of curious and striking facts in the history of a number of animals.—It is the Introduction to a large work of his, which few people here have seen; I think this preface would be very acceptable to the public*.

"I inclose you a preface to it of my own, and wish it may meet your approbation. I do not mean to put my name to it; all my friends will know it is mine; it is not of consequence enough to make me known, nor is my name of importance enough to make it sell."

Rev. Henry Bryant to Mr. James Edward Smith.

Dear Sir, Heydon, May 28, 1785.

Yours of the 19th came safe to hand, the contents of which gave me infinite satisfaction, as they informed me of your good state of health, and your laudable endeavours to acquire knowledge and literary fame; and whatever may be in my power to forward such pursuits, you may at all times com-

* Published in June 1785, under the title of "Reflections on the Study of Nature."

mand. You are now in possession of the greatest part of the natural wealth of almost the habitable world; and therefore the public must look up to you for information, which I trust you will be ready to give, as they will at all times be thankful to receive from you.

I make no doubt but we have been in numberless errors, both in the Cryptogamian and other classes; which does not show that words are inadequate to convey our ideas to one another, but only that the first attempts of the Linnæan method of doing it in a certain number of words, is not always sufficient for the purpose; and therefore, where more are necessary, more should be employed.—I know no thing of the preface of the *Musæum Regis Adolphi Frederici*, and shall be very glad to see it when it comes forth. My essay on Branded Wheat was partly intended for the same purpose you mention.

I have lately found the *Bryum aureum* and *flexuosum*, and a new one for which I want a name. Likewise the *Riccia natans*, which I suspect is not in your collection; and many other good things, any of which shall be at your service.

HENRY BRYANT.

Mr. Smith to Mr. James Edward Smith.

Dear Son, Norwich, June 2, 1785.

I received yours of the 19th and 23rd of May, by Mrs. Kindersley, and on Sunday that of the 28th with the very agreeable news of your being admitted F.R.S., of which I give you joy most cordially, and especially on the unanimity with which you came in, which is a great addition to the honour, and must be to your satisfaction. I believe 'tis looked upon here as a very extraordinary thing, and does you great honour in the opinion of the world.

As for your care of, and the use to make of your collection, 'tis at your own discretion; I would only wish to guard you against either the flatteries or circumventions of pretended friends, and men who may be interested in misleading your conduct with respect to it. I know the purity of your own heart, which being incapable of any mean or base designs, will hardly permit you to suppose others can: experience will certainly convince you of the contrary, and there can be no harm in beginning to be on your guard in good time in matters of importance.

I thought Mr. Pitchford's the letter of a true friend; and I thought your answer considerate and manly. I will not conjecture myself into uneasiness; but as you well know the interest I take in your prosperity and happiness, which is much dearer to me than my own, I cannot avoid communicating to you my most secret thoughts, and to the best of my abilities (I wish to God they were greater) be your watchman against any evil that may approach you. I had rather say twenty useless things than omit one piece of advice that may be of the least service to you; and I have that opinion of your affection for me, that you will not put that down to the score of impertinence, which proceeds from the purest sources of parental love and esteem.

I have read *Le Médecin de Soi-meme* once with great pleasure, and am reading it again with more: there are some most excellent things in it I am certain ; and tho' 'tis a severe satire upon interested and dishonest practitioners, I think such upright ones as yourself will applaud the good sense and honest advice of the author: besides, I am mistaken if the wisest may not gather some more wisdom from it. It seems to me Linnæus had read it, and thought it worth while to mark some passages with his pen. I will return it as soon as Francis and I have read it, and I beg you will never part with it.

I am reading Milton with great reverence and pleasure: 'tis immensely sensible and often very entertaining, tho' the language is uncouth. 'Tis curious to observe him when he is upon a subject that is not to be explained or supported, how he labours and struggles, you may see without satisfying even himself; how then can he his readers? But when he is supporting truth and liberty, he carries it with a high hand indeed. I never met so nervous and triumphant an opposer of temporal and spiritual tyranny, as far as I have yet gone in the books. They are curious and valuable for the sake of the publisher, and his having marked with his pen the passages that most forcibly struck him.— This work is an invaluable gem in your library*.

The annexed observations upon this fine edition of Milton, are from Mr. Smith's common-place book.

" *The Works of John Milton, Historical, Political, and Miscellaneous. Now more correctly printed from the Originals than*

As for the people of England, what with factions, plundering and being plundered, and luxury, they seem to be dead to their true interests, nay to their safety,—and, I fear I may add, to their existence as a

in any former edition, and many passages restored, which have been hitherto omitted. To which is prefixed an Account of his Life and Writings. In 11 volumes, quarto, 1753.—This very fine and beautiful edition of Milton's prose works was published by that very eminent friend of liberty and the rights of mankind, Mr. Hollis, who bestowed the most part of the income of a very large estate in promoting them to the utmost of his power. For this end, to perpetuate the writings, and disperse the doctrines of the authors who have been the most conspicuous friends to civil and religious liberty, he reprinted several of their works in very good editions, and presented fine sets of them to almost every university, learned society, and eminent man, in Europe, perhaps the world. Among the rest, Milton, holding the first rank upon many considerations, did not escape this honour, as well as Toland's Life of Milton.

"The books I read were presented by Mr. Hollis to the late celebrated Linnæus, and came with his library and collection into the possession of my son, James Edward Smith. They are addressed to Linnæus, in the first leaf, in Mr. Hollis's own hand, though not signed with his name, and there are many notes and marks in manuscript by the same hand throughout both volumes.

"Milton's prose style is very faulty ; 'tis intricate, stiff, and the periods immensely long, formed in the manner of the Latin, some tracts more so than others ; but the strong sense and forcible expressions that abound in them, make full amends for the pain of reading much that is uncouth and awkward. The most liberal and free notions and principles both in church and state, are maintained without reserve, and they will be a perpetual monument of the vigour of the minds of our ancestors in the age of Milton."

It is remarkable that Mr. Smith uses the epithet *uncouth* in

people. I know you and my good friend Mr. Clay will smile at my grave politics; and you in the midst of your library and kingdom of nature look down on the miserable mortals who are busy in

speaking of Milton's periods; yet Milton employs the same word while speaking of himself, in the finest elegiac poem that was ever written, for "the force of language can no further go."

> "Thus sang the *uncouth* swain to th' okes and rills,
> While the still morn went out with sandals gray;
> He touch'd the tender stops of various quills,
> With eager thought warbling his *Doric* lay."

How would the excellent father of Sir James Smith have delighted in the interchange of thoughts and ideas with Dr. Channing, upon Milton alone! Speaking of the intellectual qualities of our great poet, in his "Remarks on the Character and Writings of Milton," Dr. C. has observed, "that the very splendour of his poetic fame has tended to obscure and conceal the extent of his mind and the variety of its energies and attainments.—Of all God's gifts of intellect, Milton esteemed poetical genius the most transcendent. He esteemed it in himself as a kind of inspiration, and wrote his great works with something of the conscious dignity of a prophet. It seems to us the divinest of all arts; for it is the breathing or expression of that principle or sentiment which is deepest and sublimest in human nature,—we mean of that thirst or aspiration to which no mind is wholly a stranger, for something purer and lovelier, something more powerful, lofty, and thrilling, than ordinary and real life affords.

"But we rejoice," continues Dr. Channing, "that the dust is beginning to be wiped from Milton's *prose* writings, and that the public are now learning, what the initiated have long known, that these contain passages hardly inferior to his best poetry, and that they are throughout marked with the same vigorous mind which gave us Paradise Lost. We recommend them to all who can enjoy great beauties in the neighbourhood of faults, and who would learn the compass, energy, and richness of our

working out their own misfortune in corruption and injustice. I hope, my dear James, your useful, pleasing pursuits will never be interrupted by any thing that may happen to kingdoms and states:

language; and still more do we recommend them to those who desire to nourish in their breasts magnanimity of sentiment and an unquenchable love of freedom. They bear the impress of that seal, by which genius distinguishes its productions from works of learning and taste. The great and decisive test of genius is, that it calls forth *power* in the souls of others. It not merely gives knowledge, but breathes energy. There are authors, and among these Milton holds the highest rank, in approaching whom we are conscious of an access of intellectual strength. A 'virtue goes out' from them. We discern more clearly, not merely because a new light is thrown over objects, but because our own vision is strengthened. Sometimes a single word, spoken by the voice of genius, goes far into the heart. A hint, a suggestion, an undefined delicacy of expression, teaches more than we gather from volumes of less gifted men.

"His moral character was as strongly marked as his intellectual, and it may be expressed in one word, *magnanimity*. It was in harmony with his poetry. He had a passionate love of the higher, more commanding, and majestic virtues, and fed his youthful mind with meditations on the perfection of a human being. We have this vivid picture of his aspirations after virtue:—'What God may have determined for me I know not; but this I know, that if he ever instilled an intense love of moral beauty into the breast of any man, he has instilled it into mine. Ceres in the fable pursued not her daughter with a greater keenness of inquiry, than I, day and night, the idea of perfection. Hence whenever I find a man despising the false estimates of the vulgar, and daring to aspire in sentiment, language and conduct, to what the highest wisdom, through every age, has taught us as most excellent, to him I unite myself by a sort of necessary attachment.'

"He reverenced moral purity and elevation, not only for its own sake, but as the inspirer of intellect, and especially of the

and I trust Providence will ever protect such as are studying to do good to mankind with such upright intentions, and so good a heart, as it is one of the chief causes of the happiness I do enjoy to know you possess.

We join in the tenderest love.

I remain, Dear James, ever yours,

JAMES SMITH.

Mr. Smith to Mr. James Edward Smith.

My dear Son, Norwich, Nov. 28, 1785.

I wrote to you on Wednesday last, to propose to your consideration the delaying your journey to Leyden till the spring, that you might have milder weather, more favourable to your health as well as pleasure and comfort, which are all so dear to your parents and family; and I advanced as an argument, the safety of your museum and library in the long winter nights. To all which I shall be glad to hear your opinion; to which we shall I hope concede with firmness, as we are satisfied your prudence and good judgement will fix your resolves upon what

higher efforts of poetry.—'I was confirmed,' he says, in his usual noble style, 'I was confirmed in this opinion, that he who would not be frustrate of his hope to write well hereafter in laudable things, *ought himself to be a true poem*; that is, a composition and pattern of the best and honourablest things: not presuming to sing of high praises of heroic men, or famous cities, unless he have in *himself* the experience and the practice of all that which is praiseworthy.'"

is best for you to do; as you know all circumstances better than ourselves: and we trust that kind over-ruling Providence that has so distinguishedly conducted and protected you so far on the journey of life, will continue your friend and guide. To God Almighty then we commend you in our daily prayers; and we rely on your virtue and piety that you will not lose his favour and regard.

Mr. Martineau thinks the spring, when the waters are set at liberty in the marshes of the Low Countries, is more unfavourable to intermittents than the depth of winter; but I am more afraid of the severe cold for your constitution than the fogs and exhalations, which I think you may more easily avoid. Indeed, my dear James, I believe my own health and happiness depend in a very great measure upon yours.

There is no doubt but you will make the best use of your stay at Chelsea, and I trust that your publications will bring you both honour and profit: whilst you are thus employed, I dare say you will not neglect the study of medicine, which, if not the entire foundation, I must look to as the superstructure of your fortune, and I forebode of your fame too.

I have finished Milton. There is so strong a beam of good sense and profound judgement, that I was highly pleased with his prose works, and could read them again if I had leisure.

Pray have you read Sonnerat's account of the Medicine, and especially the Mythology, of the East Indians? Perhaps you have not, though you may the Natural History. I have got to the end of the Mythology, to which there are some curious notes indeed, which I shall not make my observations upon till you have read that part of the work: it entertains and informs me very much, and is written, in my opinion, in the best manner I ever met with any account of countries. I judge the author to be very liberal, and a sound deep philosopher.

Your ever affectionate,

JAMES SMITH.

Dr. Hope to Mr. James Edward Smith, F.R.S.

Dear Sir, Edinburgh, April 24, 1786.

A few days ago I had the pleasure of receiving your kind letter and acceptable present. I thank you for a copy in English dress of a Dissertation I long wished to see. I thank you for the honour you do me in dedicating it to me*; and I thank you very much for the many kind expressions in your letter.

The more I thought of publishing on a large scale the entire plant of *Assa fœtida*, the more difficulties I perceived in the execution of such a design; till finally I saw clearly I had been mistaken, and that Sir J. Banks had done the thing in the best way for me, and for which I am much indebted to him.

I hear with great pleasure of your intention, upon your return to Britain, to publish a new edition of the *Species Plantarum*. I most heartily wish you success. You have so happy a genius for botany, and so much unremitting zeal for the improvement of the science, that I heartily wish His Majesty, by

* Dissertation on the Sexes of Plants.

a good pension, would induce you to give up physic. My son attends with great vigour every branch of physic;—had he leisure for it, he should make a botanical tour through Scotland this summer.

Believe me, my dear Sir, with much sincere regard,

Your most obedient humble Servant,

JOHN HOPE.

The concluding letter from his father was written immediately before Sir James set out upon his continental tour, as will appear by the following chapter.

Mr. Smith to Mr. James Edward Smith.

Dear Son, Norwich, June 4, 1786.

Every thing has gone on as it should do in my absence, and except the prospect of parting with you for a time, we are as happy as we can expect to be in this world: and as *hope* is the most comfortable food the mind can partake of, we look forward with pleasing anxiety to your happy return. In the mean time we recommend you to the Divine protection, which can carry you safe *where thickest dangers run*, to restore you to us again in the fullness of time.

I am dull and weary: you must excuse my adding more, than sincere good wishes for you on your journey.

I am, my dear James, ever affectionately yours,

JAMES SMITH.

CHAPTER III.

Sir J. E. Smith begins his travels.—His correspondence with various friends.—Publication of his "Sketch of a Tour on the Continent."—His taste for the Fine Arts.—Introduced by Viscountess Cremorne to the late Queen Charlotte.—Instructs Her Majesty and the Princesses in Natural History.—Loses the Queen's favour by some expressions in his Tour.—Rousseau.—Approbation of his Tour by Dr. Pulteney and Mr. Pennant.—Origin of his acquaintance with Colonel Johnes.—Visit to Hafod.—Miss Johnes.—Letters of Mr. Johnes.—Letter from Mrs. Watt, the only daughter of Ellis; and other friends.—Introductory Lecture at the Royal Institution in 1808.—List of Foreign Professors.

In somewhat less than two years after Sir James became possessed of the collections of Linnæus, he began a tour through Holland, France, Italy and Switzerland, on the 16th of June, 1786. The immediate object was to obtain a medical degree at Leyden.

The name of Linnæus, he tells us, opened every door and cabinet to him, though he disclaims the weakness of assuming to himself the honours which were paid to that name.

The following letters to and from the traveller will throw some light upon what was passing in natural history in England during his absence, as well as describe the objects and pleasures of this delightful tour. How frequently soever these countries have been visited, descriptions of them seem never unwelcome to the reader. A traveller, who preceded Sir James, observed of Italy, that it is "a fine well-known academy figure, from which all sit down to make drawings, according as the light falls, and their own seat affords opportunity." So, like the pencil or the chisel, the pen also portrays, not merely the scene, but the character of the mind which directs it, making either a beautiful, a grand, or a mean representation.

The friendly letter of introduction to Professor Allamand, which the traveller carried with him, from Sir J. Banks, may, without seeming too arrogant, precede the other correspondence; and is given as a proof of the high esteem in which, at so early a period, Sir James was held by that eminent man.

Letter of Introduction from Sir Joseph Banks to Professor Allamand, Leyden.

Sir, London, June 16, 1786.

Give me leave to recommend to your notice the bearer of this, Mr. Smith: he is an enthusiast in natural history, and I really hope will one day become one of the chief supports of that science.

He has purchased at a very liberal price the herbarium, library and other collections of our great master Linnæus, and has for some time had them in his possession. You will find him well informed in every branch of natural history, but particularly so in botany: he intends to take his degrees in Leyden, in the physical faculty; and I trust you will find him able to stand a good examination in that also. Allow me, good Sir, to request that you will introduce him to Professor Van Royen, and claim in my name the Professor's civilities to him: he brings a parcel of American ferns as a present from me to Dr. Van Royen's collection. I am called upon suddenly for this letter, and pressed by business have not time to make it more particular; but you shall hear from me again, and I shall have the pleasure also of answering Mr. Van Royen's obliging letter at my first leisure.

Believe me, Sir,

Your most obedient humble Servant,

JOSEPH BANKS.

J. E. Smith to his Mother.

Honoured Madam, Leyden, June 26, 1786.

I wrote to my father from Rotterdam, and hope he received my letter. I had a very pleasant voyage thence along a canal to this place. On Friday morning I called on Professor Sandiforte, to whom I had letters, and that afternoon was fixed for my examination; it lasted about forty minutes in Latin, and gave me, and I hope the Professor, much satisfaction. Next day I was examined by the College, and had two aphorisms of Hippocrates given me to write on, which I gave in this day, and am now entitled to my degree when I please: I only wait for the printing of my thesis. I doubt not you will rejoice with me at my having got through the trial, about which I felt no small anxiety; although I think scarcely so much as I did before my election to the Royal Society. I thought the examination very proper and sensible; nothing but what was useful was asked, although some questions were very minute. Mr. Vaughan, one of my companions, passed his examination at the same time as myself. The students here are treated like gentlemen by everybody; indeed the people are extremely polite, rather troublesomely so. When one goes along the street every well-dressed person pulls off his hat in so slow and formal a manner, that one had need have nothing else to do but to return their compliments; and all the students bow to each other in the same way.

Yesterday I was at a party at Professor Allamand's, the oldest professor in the university, to whom I was particularly recommended by Sir Joseph Banks: there was a large party, and very polite one; their manners quite French, and nothing but French was talked: there were four or five *card-tables,* although it was Sunday; and before I was aware of it, a whist party was made on my account, being an Englishman; so I could not avoid playing, although it was a desperate undertaking for me, who do not pretend to be a whist player, nor ever played at cards in French before: however, I was successful, and was at least as good a player as the rest; for 'tis not a game any people shine in but the English. Nothing could exceed the politeness of the company, and of Monsieur and Madame Allamand: today he carried me to the Botanical Professor, who walked with me two hours in the garden this evening. I found my friend Van Meurs at Leyden; he is gone to Amsterdam, where I am to meet him on Thursday nex to see that town. My two companions are very good kind of young men, and a great comfort to me; we have taken lodgings together;—one very large and good dining-room, and two chambers, for about two guineas per month, which divided between three makes it come cheap to us. All the people here speak French; every thing is contrived in the best manner to provide for the convenience of students, as they are the chief support of the town. The principal street in which we live is truly noble, full of very handsome houses and fine public buildings; the houses have platforms of grey marble before them, which are washed clean every day; the middles of the streets are paved with stones like the London pavement (not flag stones), and the sides with bricks; no gutters in the streets; and the whole is so clean that you might safely sit down in any part. The insides of the houses are neat in proportion, and superbly ornamented with marble; the rooms very lofty and large; every part so wonderfully cleaned and polished that you would be charmed with it: 'tis the same with our beds, and every thing we have to use. The houses on the outside resemble some good old houses in Yarmouth more than any others that I have seen;

but every house in the town looks as if it had been cleaned and painted within this fortnight.

Adieu, my dear Madam;
I am your dutiful Son,
J. E. SMITH.

Rev. Dr. Goodenough to J. E. Smith, Leyden.

My dear Sir, Ealing, July 3, 1786.

I give you much joy upon your succeeding with such éclat at Leyden, and am much obliged to you for your kind remembrance of me; you cannot repeat it too often: there are very few people with whom I correspond for pleasure, because I can find nobody, scarcely, who loves to live with his eyes open, and has philosophy enough to talk of any thing but common occurrences. I pity your sea-sickness, which horrid consideration curbs my roving spirit, and bids me think that I am very well off in staying at home. How delightful must your hours pass with Allamand and Van Royen! Your account of the *Chamærops humilis**, planted by Clusius, is truly wonderful. Is it likely still to live; or does it bear marks of weakness and decay? The age of vegetables is a curious and useful subject;—you talk of getting specimens from Van Royen—.(Just at this moment, which is ten o'clock at night, I have caught the *Cimex personatus*, which settled on my paper.—N.B. This is the second of this species which I have taken this year, both of them at night,

* See Tour, i. p. 11.

whilst I was writing.)—I say, you talk of getting specimens : Pray think of my poor list if you have it with you, or if not, pack up for me any thing curious (particularly English) which may offer.

I never saw so many of the *Musca bombylans, mystacea* and *pellucens*, as this year. I observe that some years particular insects are excessively common; and in others are very rarely to be met with. I remember about eight years ago the *Phalæna Geometra sambucaria* passed over my garden night after night, like flakes of snow almost: perhaps I speak a little too hyperbolically, but I am confident I have taken fifty on a night; but since that time I have never seen twenty, that I know of. If you pass by any places where the *Sium* or *Phellandrium* is in plenty, do look for the *Curculio paraplecticus*, and take as many of them as you can. I am told it is not uncommon abroad. The *Ptinus elytris striatis* is probably the *Ptinus mollis* of Linné. Is it reddish brown?

I am glad to hear that you had your talk out with Sir Joseph. I wrote to Marsham last night, and told him what you said. See how our minds move! Marsham was at that instant writing to me upon the same subject. He says he has had a conversation with Forsythe, and that we may have him if we please. But Forsythe thinks that we might form a party in this Society, which he says wants *weeding* very much. But Marsham adds, "he told me many things which serve to convince me that that is impossible." I think of going to the next meeting of our Society, July 11th, and then shall see Marsham, and you shall hear what is likely to be done. I shall long for your return, as by that time I reckon that I shall have ten thousand questions to ask you. You will therefore be apprized of some little troublesomeness on my part. In the mean time, believe that you have the very best wishes of your very sincere friend,

SAMUEL GOODENOUGH.

J. E. Smith to T. J. Woodward, Esq.

Dear Sir, Leyden, July 14, 1786.

I heard of your having received my last with the Portland Catalogue; and now write to let you know whereabouts I am, in hopes of being favoured with a line at your leisure.

I have gone through my examinations, and shall take my degree tomorrow. I am very well pleased with this place, and with the Professors. Van Royen and Allamand are as friendly and communicative as possible. The garden is by far inferior to many in England; but there are some things in it which we have not; some of which I am promised.

The garden at Amsterdam is rather neater, and has perhaps more plants in it, than the Leyden one; but Burman shelters his ignorance under his professional dignity, and is very difficult of access. I could not get a sight of his herbarium; nor did he seem to be acquainted with some very well known botanical facts. How different is Van Royen! I spend almost every morning with him in looking over original herbariums of Herman! Boccone, and Rauwolf, as well as his own and his uncle's collections.

Conium Royeni proves to be nothing else than *Caucalis daucoides*.

This gentleman has just resigned the botanical chair, having reserved to himself the use of the garden: and a young man, of small skill, is appointed, against the approbation of Van Royen, who wished to have had Thunberg to succeed him; but this is not publicly known. I have seen no wild plant worth gathering except *Eryngium campestre*, which was not in flower, and *Menyanthes nymphæoides*, which was out of my reach. The former is very plentiful on the sides of the road from Helvoet to Rotterdam and elsewhere.

My inaugural dissertation is printed; 'tis very trifling, but all my own or nearly so. You shall have it as soon as I can send it; but you must consider it merely as an exercise and a sketch.

Amsterdam is a fine town, and so is Leyden; the former is all bustle, the latter stately and silent. The streets wonderfully neat, and the houses very elegant. The Dutch exceed us much in expense in fitting up their houses. Every hall and kitchen almost is paved with blue and white Italian marble.

I cannot meet with a single copy of Leers's Flora. I have many commissions for it. Charity begins at home, but you are next in my list.

Yours &c.

J. E. Smith.

J. E. Smith to Mrs. Howorth.

Dear Madam, Leyden, July 15, 1786.

I hope I need not make any apology for not having sooner availed myself of the permission you gave me at parting; for you will do me the justice to believe that I should not easily neglect an opportunity of enjoying your conversation, or even the shadow of it: in fact, I have of late been very busy indeed, nor have I had a moment at my command, till this day, for some time past. This morning finished my academical business, to my great joy. Well has Miss Lane imagined that the great wigs must have a formidable appearance; there were near twenty of them assembled; some to bait me, and others to watch my behaviour. However, thank God! I came safe out of their hands, and have now nothing to receive from them but congratulations. I go hence on Monday for the Hague, and thence to Antwerp, Brussels, &c. nor will I forget to pay my devotions at the tomb of the good Archbishop of Cambray, as I have already done at that of Boerhaave.

I feel some reluctance at leaving Holland, and particularly Leyden. I like the people better and better, and have made some agreeable acquaintances here. There is something very delightful in the recollection of being in a country of universal toleration and unbounded liberty;—the first country that afforded an asylum for the Protestant reformers who were driven from their native soil, and long without a resting-place, till Amsterdam wisely received and protected them. You may perhaps have read of the memorable siege of Leyden, which happened at the time when this country was about to shake off the Spanish yoke. The people were reduced to eat the leaves of trees, as well as horses, dogs, every morsel of leather, and other animal substances in the town; which at length were all exhausted. A pestilence came on which carried off more than half the inhabitants; and in this dreadful exigency the besiegers demanding the townsmen to surrender, the latter appeared on the walls, and vowed that they would first each cut off his left arm for food, and fight with his right. The governor wrote to the Prince of Orange, that without help from him, or from Heaven, they could not resist two days longer.

At this crisis, providentially surely, the wind changed, and blew in such a manner that the Spanish army, fearing a flood, made a precipitate retreat. They were no sooner gone, than the wind returned to the same point as before, the waters retired, and there was an easy access to the town for the people with provisions, who flocked in on every side. The churches were crowded with the famished wretches, who, just saved from the jaws of death, one moment greedily devoured the welcome food, and another, with sobs and inarticulate exclamations, returned Heaven thanks for their deliverance; insomuch that no regular or methodical service could be performed (surely never was said a more sincere or a more acceptable grace !). And here a new distress occurred; the poor creatures, who were too eager and incautious in gratifying their craving appetites, many of them fell down dead on the spot, from having fasted so long; so that the magistrates were obliged to regulate the quantity of provision for each person. The day after this signal deliverance, the Prince of Orange went to Leyden to express his admiration of the inhabitants' behaviour, and gave them their choice, whether to be for a time exempt from certain taxes, or to have an University founded in their town: they wisely chose the latter, and have derived much profit from it.—May not one be proud to belong to such an University? I look with reverence at the houses, which are of an earlier date than the period I have described; and I contemplate with pleasure the portraits of the great founder of the College, and his no less illustrious son, Prince Maurice, which are in the Public Library. I could tell you some very interesting anecdotes of persons who have lived in this town; but at present must omit them.

I assure you, I am, &c.

J. E. Smith.

J. E. Smith to his Mother.

Honoured Madam, Rotterdam, July 20, 1786.

This day I came through Delft and saw the magnificent monument of William I., the great Prince of Orange, not without great veneration. I put my fingers into two holes in the wall of the house where he was murdered, which were made by the pistol balls after they had passed through his body.

He had just dined and was coming down stairs, when near the bottom the assassin pretending to have some business with him, shot him almost instantly dead. There is an inscription on the wall to commemorate it.

What a great man was Maurice Prince of Saxony; to whom Germany owed its liberty, and whose daughter married the great Prince of Orange, by whom she had Prince Maurice! What a constellation!

I am sure you would be charmed with the Dutch neatness, as I am. Their beds here are made in the form of those in the pictures to our History of the Bible. I should have liked very well, if I had had nothing else to do, to have stayed longer at Leyden, for I began to form some very agreeable acquaintances there. It is the custom in this country for strangers coming to a place to make the first visit: if I had known that at first, I should have made acquaintances sooner. I was very often at Professor Allamand's, and learnt many little particulars of etiquette, which it is not amiss to know. There are many very delightful walks about Leyden, but all flat. The Hague is the pleasantest place I have seen: on one side of it is a delightful wood, which extends three miles, in which is the Prince of Orange's summer palace; on the other side is a fine avenue two miles long, which leads to the sea, and at the end is a village: here are even some little elevations of ground, with woods and thickets of birch.

Your affectionate and obedient Son,

J. E. Smith.

Mr. Woodward to J. E. Smith, Paris.

Dear Sir, Bungay, August 13, 1786.

I was very much flattered by your early attention to my request of letting me hear from you. It seems very strange that our *Caucalis daucoides* should be inserted a second time under the name of *Conium Royeni:* Linnæus certainly did not see the specimen, and named it after the description, or possibly merely on the authority of Van Royen, without any description. Did not Burman put you in mind of Hudson? I do not mean any reflexion on the latter, whose abilities, when unclouded by arrogance and self-sufficiency, I admire: but that difficulty of access to his herbarium and professorial dignity which you complain of, seem strongly to mark a similitude of character between the Amsterdam professor and our author. I must congratulate you on finishing the business of the degree, and the being now completely Dr. Smith; though I do not apprehend the business had anything very terrific in it. I shall be much flattered with the sight of your inaugural dissertation: you speak of it in very modest terms; nevertheless I am sure I shall find something to admire in the style and manner of the composition. This and the other *primitiæ** of your pen will be laid by as choice memorials of our early friendship; and should it please God to

* These were, Reflexions on the Study of Nature; and A Dissertation on the Sexes of Plants; published in 1785 and 1786. They will be enumerated among Sir James's works in a subsequent chapter.

give us a length of years, how delightful will it be to read them over together some years hence, when the business and hurry of your profession, and works of greater consequence, in both the lines of physic and natural history, may have almost made you forget your juvenile performances! There is however a freedom of style and spiritedness of composition which appears in both the translations, that I would wish you to cultivate: they have been much admired by everybody who has seen the translations, and they give to them an appearance of originality which makes me very desirous that you should cultivate the same style in original compositions.

Believe me, dear Sir,
Yours most affectionately and sincerely,
THOS. WOODWARD.

J. E. Smith to his Father.

Honoured Sir, Paris, August 21st, 1786.

On Sunday, August 6th, I went with my friend Broussonet to Versailles, which I need not describe to you at present: I shall only say it is more superb than I had an idea of; but it is tiresome, and not pleasing (I mean the garden). Saw the King and most of the family, but the Queen was in bed. The daubing of the ladies' cheeks is beyond conception; nature is quite out of the question: old hags, ugly beyond what you can conceive, (for we have very inadequate ideas of what an ugly woman is in England,) are dressed like girls, in the most tawdry colours, and have on each cheek a broad dab of the highest pink crayon, or something like it. The King is a pretty good person, rather fat, his countenance agreeable: he had some prodigiously fine diamonds. In the evening, after making two or three visits, seeing the menagerie, &c., we went to St. Germain-en-laye, and slept at the country-house of the Marechal de Nouailles, a fine old gentleman who was a great favourite of the late King, as he is of the present; he contributed chiefly to give the late King a taste for gardening and botany, and was a correspondent of Linnæus; he received us very politely, but had a large party of his family with him, so we had little conversation. With him lives Mr. Le Breton, a young man of genius whom he patronizes, and who was in England with Broussonet this spring. It is he who is translating my two pamphlets into French. When the preface to the last one was read to the marechal, it drew tears from his eyes, and he expressed the highest approbation of it. You have heard of the Chateau de St. Germain, built by Francis I.: its situation is very fine, but Louis XIV. did not like it, because from it the spires of St. Denis (where he was to be buried) appear in view; so he built Versailles, in a situation by no means comparable to it.

After an early dinner, hearing that the King was coming to St. Germain to shoot, the marechal sent Broussonet and myself in his chariot, and himself and Le Breton rode on horseback to the place. The game had been all driven together into some fields and thickets, around which the people were kept at a distance by soldiers. The King came about three o'clock, alighted from his coach, stripped off his coat, ribbands, &c., and appeared in a linen jacket and breeches, with leather spatterdashes. He was attended by eight pages in almost the same kind of dress, each of whom carried a gun, and one of these guns was always ready charged for the King; as soon as that was discharged, another, and so on: next to these were ten or twelve Swiss guards, all (as well as the King) on foot; about were some of his Majesty's principal officers, whose business it was to attend, with a physician, surgeon, &c. on horseback, and a few persons of distinction, as the Marechal de Nouailles, and their friends, of which number were Broussonet and I, for it was a great favour to be allowed to follow the King. His Majesty went several times up and down the fields, killing almost every thing he aimed at. Hearing there were some Englishmen in his train, (there were one or two beside me,) he desired the Marechal de Nouailles to ask us if we had heard any thing of the attempt on our King's life, and bade him tell us that he himself had had a full account of the affair, and that the King was safe. This was a very polite piece of condescension. Since that I have seen the whole story, for all the English papers are at Paris. In the evening Broussonet and I returned to Paris. You will easily imagine here are things innumerable to be seen, and I must postpone particulars to future conversations. Some are fine, and some paltry. Traces of Henry IV. and Louis XIV. appear everywhere. Here are many fine buildings, but mostly left unfinished, or now in the act of finishing. In a little shabby apartment in the Benedictins Anglois, lies poor James II., under a rusty black pall and tattered escutcheons, waiting to be carried back to England! So very deplorable a spectacle softened my contempt into pity. I am quite well, and in constant entertainment.

Your ever dutiful son,

J. E. Smith.

Mr. Dryander to J. E. Smith, Paris.*

Dear Sir, London, Sept. 5, 1786.

I am very much obliged to you for both your letters. All the parcels which came with the letter from Leyden were sent as directed.

Far be it from us to encumber the library with

* "Mr. Jonas Dryander, a Swedish naturalist of eminent talents, and a distinguished pupil of the great Linnæus, was born in 1748. He was domesticated under the roof of Sir Joseph Banks as his librarian in 1782, and continued in that situation as long as he lived. Mr. Dryander also held the offices of librarian to the Royal and Linnæan Societies. He was one of the first founders of the latter in 1788, and took a principal interest in all its concerns, especially in drawing up its laws and regulations, when this Society was incorporated by charter in 1802. He moreover fulfilled the duties of a very active vice-president till the time of his decease in October 1810, in the 63rd year of his age.

"The study in which this most acute and correct man found ample scope for the exercise of his talents, was bibliography. His *Catalogus Bibliothecæ Historico-Naturalis Josephi Banks* is a model for all future writers in this line; but a model rather calculated to check than to excite imitation. A work so ingenious in design and so perfect in execution can scarcely be produced in any science." J. E. S.—*Supplement to the Encyclopædia Pritannica.*

such an enormous quantity of nonsense as the collection of the works of Albertus Magnus, which you mention. Of all dull books, the most dull are those of scholastic writers; and only the desire of having all books on natural history, good, bad, and indifferent, could induce me to take up Albertus Magnus's book on animals, among the wants. If the edition of Rome 1478, or Mantua 1479, should ever occur for a couple of guineas, I suppose we must buy it, though it is throwing away money on trash.

We have at last received Hedwig's answer to the prize question at Petersburg, with 37 coloured plates, and also the two first fasciculi of his *Stirpes Cryptogamicæ.* I have had his account of the fructification of *Filices* and *Algæ,* and have no objection to his account of them, except as far as relates to *Equisetum,* which does not seem convincing to me. *Lycopodium* he acknowledges not to have been able to find the male parts of. I have not yet had time to read what he says about *Fungi.*

I am, dear Sir,

Your very faithful friend and servant,

J. DRYANDER.

J. E. Smith to his Father.

Honoured Sir, Paris, Sept. 13, 1786.

Paris is wonderfully like Edinburgh in many respects; nor should I at all like to live here. Some places here are very fine and beautiful, as the Tuilleries, to which we have nothing comparable. The statues, which are in such profusion in all the gardens and squares hereabouts, are extremely noble; and such is the management of the Government that nothing is injured by the populace: there are marble statues in the gardens of Versailles and the Tuilleries exposed to all kinds of people, and yet unhurt, which deserve to be kept in glass cases. I have seen no really beautiful and perfectly pleasing morsel of architecture except the portico of St. Genevieve now building, and the celebrated colonnade du Louvre. The paintings are all removed from the gallery of the Louvre, and packed up; but I hope to see the Luxembourg gallery soon. I have visited with great veneration the bedchamber of Henry IV. which is now one of the apartments of the Académie des Sciences, who meet in the Louvre: it is a small ill-lighted room, wonderfully richly carved and gilt; his bust stands in the place of the bed. Here I suppose he was brought bleeding, and left for the people to look at.

The monument of the Valois family, and those of Francis I. and Louis XII., all at St. Denis, are exquisite indeed; but the design of them is odd, for the kings and queens are represented almost naked, in the agonies of death, or just dead; their limbs, features, and hair, in ghastly disorder; and the bodies as if having *been opened* and sewn up again, for the stitches are as curiously done as any part. All the church and every monument are kept most scrupulously neat, which is a great advantage.

The Duke of Orleans has a fine collection of pictures, which I have seen: among them is one exactly like your picture of nymphs and satyrs, which hangs on the stairs, if I mistake not; but 'tis lighter and in better preservation than yours: the master's name on the frame is C. Polembergh,—'tis not a good picture.

I hope you are not displeased with the criticism of my work in the Critical Review for June. I am much flattered by it. Should be more proud of the critic's praise if he would but allow poor Linnæus any judgement.

Your dutiful son,

J. E. Smith.

Mr. Smith to his Son, Paris.

My dear Son, Norwich, Sept. 20, 1786.

I received your most pleasing acceptable letters dated August 21st and September 13th. The news of your health and safety give us all great joy; but the pleasure is much increased by the entertaining account of what you see and do. You are laying up a treasure of knowledge that will serve for a high entertainment through your whole life. I would not have you be discouraged by my last letter from pursuing your travels from an idea that it will be disagreeable to me, or that I may repine at the expense. I am too well acquainted with your prudence and your virtue to entertain any apprehensions that you will waste your time, your health, or your money, in such pursuits as are pernicious or reprehensible, because you are beyond the reach of the observations of your friends and neighbours whom you respect, and whose opinion you esteem, as is too often the case with young people at a distance from home. I know your objects and inclinations are of the most noble kind, that as your mind is possessed with the purest principles of truth and virtue, so your genius and understanding, excellent as they are from nature, are, the one aiming at, and the other acquiring, the most valuable treasures of human knowledge, upon which to establish a character and reputation that will be of advantage to yourself and all your connexions, as well as an honour to your family and country. I would not, then, check you in your progress. I hope it will please God to enable me to provide for your expenses without injuring the rest of you. For the risk of life and health I trust in Providence and your prudence for your preservation, to which I hope my unworthy prayers may contribute something.

I cannot object to your visit to the Marquis Durazzo anything but so long a journey in short days, in which the passage of the Alps, if you go that way, may be disagreeable, if not dangerous. But I should think there may be a way along the coast from Provence to avoid the Alps. If you go to Turin, you will call on my correspondents, who no doubt will show you every civility.

As for the election, we never had one upon which there is so much to say: it is so difficult to do it justice, that I would not have you expect it from me, who am neither practised in descriptive writing, nor have abilities for it. The canvass for the Honourable Henry Hobart began two months ago, and the day of election was last Friday, the 10th instant. The intermediate time was filled up with carousings of each party three times a week in various quarters of the town, when there was drunkenness and noise more than enough. The committees met very frequently; Hobart's at the King's Head and Johnson's Coffee-house; Beevor's at Tuck's and the Angel. As the dispute was not upon the ground of political principles, for both candidates professed the same, that is Whiggism and an attachment to the present ministry, I wondered to see them so eager; but as it was for power and interest, and which of the two factions should rule, I ought to have known that the corruption of the present age would be as zealous as the principles of the last. Those who could use a pen and tag verses * were set to work:

* From the herd of mere verse-taggers one exception must be made,--the author of the following eclogue, an intimate and highly esteemed friend both of Mr. Smith and his son. Whoever recollects the late Mr. John Taylor of Norwich will recognize in these lines the good temper, the pleasantry, and wit, which at all times enlivened his conversation, and were the emanations of a good heart as alive to mirth and enjoyment as it was void of malice and detraction. The Mr. Hampp who figures in the eclogue was a cordial friend of the author, a German by birth, and whose broken English is happily imitated.

CITY ECLOGUES. ECLOGUE THE FIRST.

Scene—A Club. *Time—Evening.*

The clock struck seven,—the cheerful sun retires,
And only gilds our castle and our spires.
The market walk now fills from every street;
There jarring parties, various interests meet;

the press teemed daily with poems, songs, epigrams, on both sides, vilifying the characters and ridiculing the persons of each party who were at all conspicuous. The day that was to terminate the dispute proved good weather, and every room in the market

---

Each candidate resumes his wonted ground,
And all his friends and followers throng around.
Now hope inspires, now gloomy fears succeed,
And show what thorny paths to honour lead:
Now still and silent is the vacant loom,
And hot and noisy is the ale-house room;
For hither, thirsting after news and nog[a],
And loving, if not Hobart's cause, his prog,
Freemen and freemen's wives and friends repair,
And pay due reverence to the leathern chair;
For there presides, with face of Belgic stamp,
That son of Liberty—Bavarian Hampp.
He, at the sheriff's uncontroul'd command,
Amongst the friends of Hobart takes his stand:
He knows each wise contrivance to a hair,
Which brought his master Thurlow to the chair;
And boasts to know, however you may doubt,
The gibes and jolts o' th' day which threw him out;
And therefore, as a manager right able,
He claims attention at the council table.
Now, rising from his chair, his cane he waves,
As who should say, "Be silent, English knaves!"
Silence ensues; our hero strives to speak,
And tortures English ears with German Greek:
Tired with his eloquence, the clamorous rabble
Drown his oration with their deafening gabble;
Till hearing something said about *the Diet*,
They thought the supper coming, and were quiet:—
"Te Diet, sers, I mean te Parlament
To vich dis Mr. Hobart sall be sent;

[a] A kind of ale.

was filled with well-dressed ladies, fluttering their white handkerchiefs out of the windows with a favour at the corner, of the colours of the candidate whose interest they espoused,—Hobart's deep blue and orange, Beevor's pale blue and white; they made a pretty show. The area of the square was crowded with stavesmen and spectators: the candidates rode as usual.* The contest was very strong, but all was carried on with very little violence; so little that the Countess of Buckinghamshire with her two

---

Dere must he take te care of all te laws,
And make more to dem if he find te cause;
And if te king of money sall fall short,
Why, he must to hem come, and ask hem for 't;
For he vill have te string of all your purses,
And must look sharp to vat te king disburses.
Vell, sers, all dis can Mr. Hobart do,
For he can read and write as well as you;
He knows quite vell de Engelsch constitution,
And is so great as me at elocution:
I know myselve te interest of dis city,
And Hobart is te man, I know, to fit ye.
As for dat Beevor, which some people talk of,
Let me alone, *I'll* make dat fellow walk off:
Who dares to speak fon wort of Beevor here,
Te schondrel sall be scalp'd from ear to ear;
Forth from dese club my friends sall kick him out,
And I will eat his share of beef and krout.
Are fagabons to say who sall be chos'd,
And gentlemen of blood to be oppos'd?
Donder and *blixem!* 'tis a thought so vile,
As makes te hairs upon myn head recoil;
Sooner den have te lot on Beevor fall,
Got! ye sall have no Parlament at all."

* "At the elections for the county of Norfolk, for Norwich, Yarmouth, &c., this custom is invariably observed. In the city ust named, the candidate, standing erect on a platform, is car-

daughters and three other ladies, who were in Mr. Schuldham's small room when Mr. Hobart in one of his chairings was coming his third round, issued out to meet him, near the Hall, and then proceeded with him to the King's Head, his quarters.

Perhaps you don't know what a fine person Lady B. is,—tall, handsome, and extremely elegant: most of the ladies besides were handsome, and all young. To see such in the centre of a throng of stavesmen, rending the air with their shouts, was a novel sight, and struck everybody with surprise and pleasure, except some who might envy what they had no opportunity of rivalling. It pleased the people so much, that they requested her ladyship to gratify them once more with the pleasing honour, to which she condescended, and the ladies took a tour round Mr. Hobart's ground.

I beg your pardon for the suspicion I entertained too hastily; 'tis the foible natural to age. I am very glad to hear I am mistaken, and rejoice at every instance of the integrity of human nature.

I remain, dear James, ever yours,

JAMES SMITH.

---

ried on men's shoulders three times round the place of election, and is frequently tost by them into the air. Those who have seen this ceremony will not fail to be struck with the words, already referred to, of a Roman historian, (Tacit. Hist. iv. 15.) 'impositus scuto et sustinentium humeris *vibratus*;' the exact agreement of which with the yet remaining practice will scarcely allow us to doubt that the *elevation* of kings, here, as well as in the other countries above mentioned, was the original mode of their inauguration."—*Taylor's Glory of Regality*, p. 29.

Mr. W. Jones to J. E. Smith, Paris.*

Dear Sir, London, Sept. 20, 1786.

I have not felt my mind for a long time under such pleasing sensations as this day, from your attentive favour. I therefore take pen immediately in hand, lest every day deferred I might grow more indifferent to thank you for it, and to assure you that you possess a place in my memory and esteem as frequent and favourable as I can have in yours; and to tell you *si tantus amor casus cognoscere nostros*. I read your letter with great patience until I came to the Prince of Orange's cabinet of insects, when I found an unaccountable fidgetting about me, very restless in my seat, until I had taken two or three turns across the room to compose myself; and just as I was recovered, the double Tuberoses occasioned a relapse, that 'twas some time before I was able to proceed. 'Tis a flower I have a great partiality for, but their culture in England is difficult without a hot-house,—*quo fata ferant ubi sistere detur*. May health attend you, and every pleasure your journey can give! I am sensible that cannot be little, when you have access to every thing that is rare and beautiful, and more especially in those things particularly adapted to your taste.

* An excellent entomologist, well known in the scientific world, though, like other men of superior genius, modest and retiring from the observation he so well deserved.

Mr. M——, as you say, I find very anxious about the success of the new society.

I am persuaded from that open, honest simplicity that prevails in every part of your conduct, that you have no views but what are disinterested; but that may not be the case with every one that joins you. I have been united in societies of various kinds, and have been heartily vexed and dissatisfied. Omit it a few years: at the end of the first weigh every objection, and demur a little, and I think that afterwards the spirit of procrastination will increase; yet I would not have the thought wholly laid aside. We may enjoy every satisfaction from each other's information and company, as we might have done in society, and this by a breakfast to our select friends, once a quarter, either by you, or alternately as agreed. By this will be discoverable the probable success of such a society; and if it ever matures, let it come forth. Even the idea of associating should be foreign to our first purposes, and only meet as select friends. Under this view we raise no envious spirit against us from the present existing society; and prevent in future that acrimony which, I am sorry to say, exists too much amongst ingenious people. In short, I am horribly afraid of a wasp's nest. St. Paul's advice to Timothy was a good one, even in common life: "Lay hands suddenly on no man."

If, amidst the various avocations that surround you, one hour can be spared, employ it by scribbling a few lines to, dear Sir,

Your very sincere friend,

W. Jones.

J. E. Smith to Mrs. Howorth.

My dear Madam, Paris, Sept. 24, 1786.

I abhor the thought of staying any part of the winter at Paris: the streets are so dirty, and the houses so cold, with nothing but brick floors and marble tables, that the idea chills one.

I have lately had a most agreeable jaunt, in which I wished for nothing more than your company, to visit the tomb of Rousseau. My companion was an Englishman of great taste and sensibility, who enjoyed the expedition no less than I did. We went first to Chantilly, about twenty miles from Paris, where the Prince of Condé has a noble seat, with very fine gardens, perpetual fountains, and every decoration that art can furnish, but all in the old style; 'tis however worth seeing, as being one of the finest things in its way, and we have nothing like it in England. Here we slept, and next day had a very romantic ride of eight miles through the forest to Ermenonville; we arrived about dusk, and put up at a little inn, where the present Emperor, and the King of Sweden had been accommodated before us. The landlord knew Rousseau, and spoke of him with the greatest esteem. The day of his death this man saw him about seven o'clock botanizing; he complained of having had a sleepless night, from the headach. Before ten he was dead. Water was found collected in his head. Our landlord preserves his snuff-box, and the shoes in which he died; they have wooden soles and straw tops. One of his admirers has written something on the box; and another has written on the shoes, that he was proud to inscribe his name "*sur la simple chaussure d'un homme qui ne marchoit jamais que dans le sentier de la vertu.*"

The next morning being very fine we rose at six, and had a most enchanting ramble through the gardens of Monsieur le Marquis de Girardin, which form a striking contrast with those of Chantilly, being laid out in the most romantic style, what the French call *à l'Angloise.* They consist of about eight hundred acres, a great part of which are wild woods, and rocky hills and dales as wild as the highlands of Scotland. We first passed a beautiful cascade, and went along a winding path through a wood by the side of the lake, from time to time meeting with inscriptions disposed with great judgement. We took a boat to go to the Island of Poplars, honoured with the ashes of Rousseau. His tomb is elegantly simple, of white stone; on one side is a piece of sculpture representing a mother of a family reading Emilius, with other emblems; on the other is inscribed, "*L'homme de la Nature et de la Vérité.*" He desired to be buried in the garden, and the Marquis chose this spot. I shall not attempt to describe to you what I felt on seeing and touching this tomb. I brought away some moss from its top for you.

In another island near it is a lesser monument, over a German who taught the Marquis's children drawing; and being a Protestant could not be buried in consecrated ground. Hence we passed by some inscriptions in honour of Virgil, Thomson, Shenstone, and some others, to the Temple of Modern Philosophy, an unfinished building; on each of the pillars already erected is inscribed the name of some great man, with a word expressive of what he excelled in: thus to Voltaire is given, *ridicule;* to Rousseau, *nature;* to Priestley, *air;* to Franklin, *thunder*, &c., &c. On an unfinished column is written in Latin, "Who will complete this?" This temple overlooks the lake; near it is an hermitage embosomed in a wood. From this spot we went to some simple wooden buildings, where every Sunday the Marquis and his lady amuse themselves with having the neighbouring peasants dance, &c., on the plan described in the Nouvelle Heloise. The woods around them are very fine; and after passing through them we came to a solitary elm-tree, on which the Marquis has written, "*Le voici cet orme heureux où ma Louise a reçu ma foi.*" From hence is an immense prospect, finely varied with fields, woods, and water. Descending the hill among heath and juniper, we came to two charming Italian inscriptions by the Marquis, which lead to a rock on which Rousseau has engraven with his own knife, "*Julie.*" I have some moss for you from this very rock. Ascending another hill we came to the House of Rousseau, a little hut so called, in which he wrote several verses; for he often used to visit it during the short time of his residing here, which was only six weeks before his death, although he often used to come to Ermenonville with the

Marquis's family before. Of his dwelling-house I shall speak hereafter. Within this hut is written, "*Jean Jacques est immortel.*" From it is another fine view ; it stands among craggy rocks.

Descending into another valley, we went by the water side through groves and across a meadow to the tower of *la belle Gabrielle d'Estrées,* who was mistress to Henry IV. Tradition says this garden was their first place of rendezvous, which occasioned the Marquis to build this tower ; it is in the Gothic style, and ornamented with trophies and verses. Among the rest is the very armour which belonged to a faithful follower of Henry IV., whose name I forget, and who passing through the street where that prince was murdered, a few days after that event, fell down in an agony of grief, and died the next day.

Passing by a pretty grotto by the side of a bubbling fountain of the finest water I ever saw, we at length arrived at Rousseau's garden, one of the sweetest spots I ever beheld, quite sequestered, and planted in the most romantic style ; it chiefly consists of an irregular lawn surrounded with a variety of trees and shrubs, and ornamented with flowers, but apparently all in a state of nature ; nor is the hand of art to be traced at all, except in the beautiful velvet of the turf. On a tree is an inscription, signifying that there Jean Jacques used often to retire, to admire the works of nature, to feed his favourite birds, and play with the Marquis's children. Near this spot is a house intended for his dwelling, but he died before it was finished ; 'tis a comfortable cottage, with a little garden of flowers before it, and is embosomed in apple-trees, vines, &c. In a small arched building near it, the Marquis at first intended to have buried Rousseau, but changed his mind. From this place we soon reached the front of the house opposite to that whence we set out, and our delightful tour was at an end.

I think you will not be displeased at my giving you so particular an account of it, so I make no apology for the length of my letter; but I have more to tell you.

Hearing that the widow of Rousseau was living at a place not far out of our road to Paris, and that many strangers visited her, we felt a strong desire to do the same; but had some fears lest we should discover something in her which might excite disagreeable sensations, and even perhaps lessen our veneration for her husband; for we heard that she had been his servant, and after having lived with him in that capacity ten years, he said to her, "*Ma bonne amie*, I am satisfied with your fidelity, and wish I could make you an adequate return. I have nothing to give you but my hand. If you think that worth having, it is yours." They were married; and lived together sixteen years afterwards very comfortably. She was several years younger than her husband.—At last curiosity prevailed, and we went to see her. She received us with the greatest politeness, and appeared much pleased with our visit; spoke in the most becoming manner of her husband, and readily answered every question I put to her. What I principally learned from her was as follows:—The character of Julia was drawn from Madame Bois de la Tour of Lyons, a lady still living, with whom Mr. and Mrs. Rousseau often spent a great deal of time: she has a large family, and is the admiration of all who know her. The story of Julia has not however any connection with hers How far that is founded in truth, Mrs. Rousseau said was only known to its author. The idea that Ermenonville was the scene of it, or that the real father of Julia lived there, is without foundation. She assured me that the Confessions of Rousseau were really all of his own writing. She confided the manuscript to the Marquis de Girardin, who expunged several names and anecdotes relating to people still living, but against her consent; for she thought the whole ought to have been published as the author left it. I think more ought to have been expunged, at least the name of Madame de Warens ought to have been kept secret.

We asked her which was the best portrait of Rousseau. She showed us a plaster bust, which was cast from his face a few hours after death, and which she said resembled him exactly. The expression of the face, as well as its form, is vastly superior to that of any likeness of him I ever saw. There is great serenity in the countenance, and much sensibility. The mouth is uncommonly beautiful.

I saw at Chantilly a wax bust, which was cast from the face of Henry IV. four hours after his death; it has the same features which appear in the portraits of him; but such a melancholy gloom is diffused over the countenance, that it is quite a pitiable object. I account for this difference from the death of Rousseau being less violent, and his mind more serene at the time, than the King's could have been, considering all circumstances, besides the mechanical cause of the great loss of blood in the latter.

We returned to Paris extremely well pleased with our jaunt, and particularly so with having seen Madame Rousseau. I learned at Ermenonville that the King of England allows her fifty pounds a year, which I never knew before, but which ought to be known. *Le grand Monarque* allows her nothing.

I am acquainted with a gentleman at Paris, who knew Rousseau intimately, and often used to botanize with him (you see how I delight to put you in mind that he was a botanist). He describes him as the most unaffected and unassuming of men, free from all airs or petulancy in conversation, and even very sociable latterly, at least since he knew him, which was many years before his death. He was always warmly attached to those who loved natural history, especially to the pupils of the Linnæan school, and he adored Linnæus. What would I have given to have seen him!

Adieu, my dear Madam.

Believe me faithfully yours,

J. E. Smith.

Rev. Dr. Goodenough to J. E. Smith, Paris.

Dear Sir, Ealing, Sept. 25, 1786.

Your letter dated Paris, Sept. 8th, has just reached me, and given me prodigious pleasure, not only from the very favourable expressions towards me with which it abounds, but from my having such an actual proof that you are alive and well. To speak in modern political terms, with respect to the regard which you are pleased to testify towards me, "*the reciprocity is not all on one side*."

I had a very pleasant excursion with Curtis to Maldon, then along the Essex coast to Mersey Island, and thence to Harwich. At this last place I was laid up with a terrible boil, which threw me quite into a fever for a week, and imbittered the latter part of my time. However, before this unfortunate circumstance, we worked well from seven in the morning till eleven at night, with only the interruptions of breakfast and dinner, and those, short repasts. I had no conception we should have found so many insects. But we happened to find a pond overrun with *Typha major* and *Festuca fluitans*. It was incredible what a number of curious things we found; *Sphex fissipes*, and your little *Cantharis miniata*. But as if fortune designed us to be niggardly, she would not let us take more than *two* of each; so there was one for each of us, and not one for a friend. I looked over every stalk and blade for two days together, five or six hours at a time, to find more, but in vain! A great variety of *Coccinellas*, of *Muscæ* also, particularly of the spotted-wing sort, *Cardui* and *Cerasi*, &c. *Ichneumons* a great many, particularly *I. sarcitorius.* Bees, several new ones, one very specious indeed, with a red thorax and blue abdomen, fasciated with white. *Sphexes* not a few. A new *Chrysis* or two. Some new *Carabi*, besides some of the Linnæan beauties. All the new ones which we took I described, or rather as many as time would permit,—between fifty and a hundred. Some were drawn on the spot by Curtis's draughtsman.

Of plants, we found in many places *Lepidium ruderale*, and particularly all along the Essex coast, on the sea marshes, Hudson's *Dactylis cynosuroides.* Hudson's *Poa loliacea*, his *Lolium bromoides*, *Salsola fruticosa*, *Brassica campestris*, *Atriplex laciniata* and *serrata*, *Chara flexilis*, *Ruppia maritima*, *Ægilops incurva*, *Bunias Cakile*;—*cum multis aliis.* I have not been able to see Sir Joseph Banks yet, although I have called several times. I think of our New Society with pleasure, and long for your return on that account, as well as others; but, as you say, the longer you stay the more you will be worth at your return. A thousand thanks for your *Agrostis minima.* I shall be much obliged to you to get me any of my *desiderata.* *Gnaphalium luteo-album* is *inter mea desideratissima.* Pray look sharp after the *Origanums*, and if you can see either *O. ægyptiacum* or *syriacum*, do be so good as to have *very* correct drawings made of both or either of them for me.

I am preparing a complete list of the Linnæan

nomenclature, through all the classes from *Mammalia* to *Lapides*, which will be ready for the press by your return. I should have liked to have given it to the public "By a Fellow of the Linnæan Society." Whatever you do, take care of your health, and use your time with courage and activity. Above all, believe me ever your most sincere friend,

SAMUEL GOODENOUGH.

I depend upon hearing of your motions to the southward, and your successes. When you come towards the Mediterranean, perhaps you may find it worth your while to think of conchology. At Leghorn there may be an Ægyptian correspondence carried on; a step towards getting *Origanum ægyptiacum.*

Rev. Dr. Goodenough to J. E. Smith, at the Marquis Hippolito Durazzo's, Genoa.

My dear Sir, Ealing, Nov. 3, 1786.

I am so far glad at finding that the prodigious loss which we have lately sustained has made its way to you, as it saves me from the pain of mentioning it to you. A better child, I verily believe, never lived: my very heart doated upon her, not from any weak or irrational motive, but from her goodness, her love of improvement, her duty to us, and her actual attainments in every thing which we wished her to apply to. Our grief was highly enriched by observing the whole neighbourhood in tears for her. It was hard parting with her at her time of life.

I thank you most sincerely for your very entertaining letter of the 21st of October. Your account of the morals of Paris is just what every one who lives with his eyes open and *will* see, has given me; so that that does not tempt me to cross the water.

But I could give one of my eyes almost to be with you, and join you in turning over Tournefort's rich herbarium. I am sorry to find that you leave any behind to examine at your return: I am always fearful of something or other happening or intervening which will draw off the attention from pursuing a labour, once deserted, to the end; and here I would say, that I hope you will find time to go *regularly* through Vaillant's herbarium. What if it should detain you a week or two longer; it will be better to seize the opportunity which now offers, and perhaps will never return, than after you have returned to England, to sit down all your life regretting that you had not completed your view. On this account also I hope you will describe every thing in Tournefort's herbarium that *appears* new; it can be easily thrown away if you should find yourself in a mistake: and for this purpose I should again wish you not to be impatient of any loss of time; or rather I should say of a *longer employment* of time. In Italy, if you can make it convenient, you would certainly be repaid in examining Allioni's collections. Do you mean to come through Switzerland? If you do, try for a view of Haller's cabinet.

The Earl of Northampton and his son Lord Compton (a pupil of mine formerly) reside somewhere near Lausanne in the Canton de Berne. Show Lord Compton my handwriting, and I am sure he would do any thing for you that you could wish. He is a collector, as is his sister. Before I call you back to England, where we wish you, once more let me say, do not mind the trespassing upon a little time or money, in securing any intellectual attainment. You can easily make up every thing afterwards. And now for matters at home. Your letter of the 21st instant did not reach me till after the anniversary meeting of our Society. It appears from the business of the day, that you might have saved yourself your *shudderings*. I could not attend; but from Marsham's account, every thing was settled by the party before the business was proceeded upon. A most curious committee (instead of that farrago of committees of last year) was chosen. J. Hunter, Home, Curtis, Swainson, Lee, Marsham, Mitchel, and Drury. What can they do in the publishing way, or indeed in any thing great? It is a joke to think of it. You may easily then conceive how much I wish for your return. A due share of *activity* and *firmness* in a few intelligent working people will do more than our present unwieldy body, with all their *members*, and all their *wealth* (for they have voted the purchase of 100*l*. stock in the 3 per cent. consols.) This might seem to cut a dash, could knowledge and credit be bought.

Your translation of the Linnæan treatise on the Sexes of Plants is *very* well spoken of in the

Monthly Review,—the only Review which I see. Pray write to me as often as you have leisure. Do not for this one grand excursion (as you must ever look upon it) grudge either *time* or *money* to *satisfy* your mind in all scientific matters. Commit every thing of science to writing, *etiam tritissima* as says Linné. It is scarcely worth while to burthen the mind with remembrances of houses and churches ; all books are full of them, and will call every thing to your remembrance. When Howard set out upon his prison-visitation, he refused seeing one of these lesser *spectacula*, alleging that one pursuit was sufficient at a time. I pray God to bless you ; and am ever yours,

S. GOODENOUGH.

Pray take one thing into consideration. *Members* and *wealth* are so far real necessaries to a society, as it enables them to carry matters into effect ;—to purchase, reward, publish. A society at Bruxelles is stopped at present for want of money to enable a publication going forward. I say this only to remind you, that along with diligence, the *primum mobile*, we may think of increasing the number of the society as far as may be *safe*. This occurs to me at the moment upon thinking of the society at Bruxelles.—Adieu.

J. E. Smith to his Father.

Honoured Sir, Montpellier, Nov. 18, 1786.

At Lyons, a letter from Broussonet to his cousin Monsieur Lajard procured us the greatest civility from him, as well as from Dr. Brun, Dr. Frossard the Protestant minister, and several others. We were charmed with the hospitality we experienced here, and with the manners of the people: I could have spent a month with them gladly. The ladies too begin to improve on us in every respect, as we go south. After seeing the hospital (which is more praised than it deserves), the public library, &c., &c., which took us four days, on the 10th we took a voiture to carry us to Montpellier. We are now in a fine temperate climate; like a fine English September. The most striking objects we saw in our way were the *Pont St. Esprit*, an old stone bridge of twenty-two arches, over the stately Rhone; the *Pont du Gard* near Nismes, a most noble Roman aqueduct; and the Amphitheatre at Nismes; the country for the last three days covered with lavender, thyme, box, evergreen oaks, and many fine plants still in flower; olive-trees laden with fruit in greater abundance than has been known these hundred years; fig-, and white mulberry-trees. Ripe olives are purple like a damson, but of the most abominable bitter taste. At Montpellier we have met with the kindest reception from Professor Broussonet, father of my Paris friend; we dined with him today: he has introduced us to several interesting acquaintances. We could gladly stay here a month or two had we not greater objects in view.

This town is situated on a hill very pleasantly: its streets are narrow, crooked, and paved like those of Norwich; so is Lyons. But of all that I have yet seen, the *Place de Perou* is the finest thing by far; it is a vast platform, out of one of the gates of the town, surrounded with a stone balustrade. From it you have the command of a most beautiful country, covered with olives, studded with villas, and bounded by blue mountains; the Mediterranean Sea to the south, and beyond it in fine weather may be seen the Pyrenees, and on the other side the Alps; no finer situation can be conceived.

The Place is disposed in grass plats; in the centre is a fine equestrian statue of Louis XIV., in bronze, and at the side opposite the entrance an open temple (over a basin of water), which with the flights of steps leading to it are in so fine a taste, and so striking, that it is like a design in an opera, a fairy palace, or a dream of some great genius rarely seen on paper, and scarcely ever in reality. To this temple water is brought by a vast aqueduct, worthy of the Romans; and hence it is distributed into various fine marble fountains in the town. This Place I had often heard of, but had no conception of it till I saw it.

We shall stay here about a week longer; then go to Avignon, Marseilles, &c., to Genoa.

I am extremely obliged to you for your liberal provision of money.

Believe me, honoured Sir,

Your ever affectionate and obedient Son,

J. E. SMITH.

J. E. Smith to his Father.

Honoured Sir, Marseilles, Dec. 5, 1786.

My last was from Montpellier, Nov. 18th. We were so well pleased with that place and its inhabitants, that we stayed there till the 27th, and then left it with regret, and went in a coach with an intelligent and liberal-minded superior of Cordeliers, to Nismes: next day a violent rain prevented our seeing much of the celebrated antiquities of this place. We saw, however, the *Maison Carrée,* a fine Roman temple, very entire (see Thicknesse's Travels). The Amphitheatre we had seen in our way to Montpellier. Mr. Granier, a friend of Broussonet, showed us the library and collection of the Academy, left them by Mr. Seguier, a celebrated naturalist and antiquary. After dinner we pursued our journey, and next day reached Avignon, where the most striking objects are the vast walls of the city, and the great old palace of the Popes. The town is lifeless and unpleasant. Next day we went in a chaise to the fountain of Vaucluse, so celebrated in the history of Petrarch; but in my opinion far more interesting in itself than for all that has been said or sung about it. It is a river at least as large as ours at Norwich, and ten times as rapid, which rises at once from an unfathomable rocky basin at the foot of a rock many hundred yards high, which hangs over it. The water, which is as clear as crystal is *supposed* to be, but looks sea-green as it runs, falls from the edge of the basin over fragments of rock, forming cascades, compared with which all the paltry squirts of Versailles are mere baubles ; and then winds along a most romantic valley, overhung with rocks of every varied form, which are here and there clothed with fine evergreen shrubs, and sometimes stuck with cottages, which seem equally in danger from the smallness of their foundations, and the vast crags above them.

Every Frenchman that comes here thinks it incumbent on him to set his brains a-jingling, and scribble something about love and Petrarch and Laura. We were tired to death with a man who travelled with us to Montpellier, who repeated to us some verses of his own and others of his friends, all about Vaucluse: they excited my indignation so much, that I thought it almost profanation of nature and feeling to indulge any similar ideas at so prostituted a spot ; and indeed even the passion of Petrarch for Laura has always seemed to me too artificial and affected to be touching. Notwithstanding all this, I could spend a solitary month at this charming spot, in the most delightful meditations ; but I should rarely think of Petrarch. We slept at a comfortable inn, at a town not far distant, and returned next day to Avignon. The following morning set out for Marseilles, where we arrived in two days. Passed through Aix, a pleasant town, in a most delightful country, covered with olive-trees with vines between them, and here and there a towering cypress. Aix is the only place I have seen (except Marseilles perhaps) to which I should ever think of sending a consumptive patient, considering situation, climate, country, &c., and more especially as there is a fine walk which passes through the middle of the town, in which are three noble fountains constantly running,—the middle one of warm water, like that at Bristol. This is the country for fine water: every village in the South of France has its perpetual fountains.

The approach to Marseilles the night before last struck us with great admiration. From a high hill in the way we had a complete view of the town, its harbour, and a very rich extensive valley studded with villas, and clothed with olives. This is by far the finest town I have seen in France, Paris in general not excepted. The streets are broad, straight, and finely built, the pavement good and clean, and the people more so than in any other place we have been at. It is as busy a place as London or Amsterdam, crowded with people of all nations, and with every kind of merchandise. The markets loaded with vast variety of fish unknown to us, and with pomegranates, melons, dates, flowers of every season, and the finest grapes at one penny per pound. Here we seem at length to have overtaken summer. The sky yesterday was as bright as possible; today we have had rain, but the air is so warm that a fire is only necessary in the evenings. We meet with great civilities here, as we have done everywhere. I think the inhabitants of the South of France the most engaging people I ever saw; there is little grimace, and much real urbanity and hospitality. The women are pretty, and very interesting; they do not in general use paint: at Montpellier that custom is as infamous as with us. The lower kind of people are much more civil here than in England; but they speak *patois*, the ancient language of the Troubadours, which we find it difficult to understand, except by the help of Italian, which it much resembles.

The Abbé Raynal resides here, but we have not yet been fortunate enough to see him: he is safe here, as this place is endowed with many privileges.

Believe me, dear Sir,
Your dutiful Son,
J. E. SMITH.

J. E. Smith to his Father.

Honoured Sir, Genoa, Dec. 30, 1786.

November 7th, we left Marseilles in an excellent voiture, which we hired by the day to convey us to Nice; travelled through a most romantic country (like Switzerland, as I conceive), and in two days reached Toulon. Foreigners are not permitted to see the arsenal; but we saw a very decent hospital and little botanic garden.

November 9th, got to Hyeres to dinner. Here we first saw gardens or rather woods of orange-trees, loaded with fruit; myrtle and aloes in the fields, and had much botanical entertainment. Hyeres is a shabby little town, but in one of the finest situations possible: the Isles d'Hyeres are much celebrated for their beauty, but we could not visit them. Next day a very bad road led us to Brigancienes, most beautifully situated in a valley; and after dinner we rode through one of the most picturesque countries I ever saw: woods of pines, thickets of myrtle, aloes, cistus, rosemary, noble cascades of every varied form, all combined to delight and surprise us. The following day dined at Brignolle, famous for the *Prunes de Brignolle*, which we have corrupted into Prunellas; and the same night arrived at Cottignac, where we went on purpose to see Monsieur Gerard, one of the most famous European botanists of the Linnæan school, author of the *Flora Gallo-Provincialis*. He received us in the most polite manner, and devoted the next day entirely to us; gave me copies of two letters of Linnæus to him, with leave to publish them, as well as his to Linnæus. I learnt a great many things from him, and we settled a plan of future correspondence. November 13th, slept at Draguinan, and next day at Frejus, where are remains of a Roman amphitheatre, and of a very noble aqueduct. 15th, our road lay over a very high mountain, from the top of which we had the finest and most extensive prospect I ever beheld. Frejus, at our feet, was almost imperceptible; on one side was the beautiful Mediterranean quite calm; on the other, hills rising o'er hills, clothed with myrtle, juniper, and pines, (not nasty black Scotch pines); and beyond all, the Alps of Piedmont covered with snow, and glittering in the sun. Towards evening we descended to Cannes, a little sea-port close to the water's edge.

16th. A fine bright and rather frosty morning. Our road lay by the sea-side, through the most beautiful shrubberies of myrtle, with its blue berries and some flowers; fine heaths, which with us are kept in greenhouses; most stupendous clusters of the great aloes (improperly called American), and other fine plants. We passed by Antibes, had a fine view of Nice and its bay, and after fording the Var (a very unpleasant business), with the Duchess dowager of Leinster and her train, we arrived at Nice. This town is finely built, and well secured from the cold winds. It is full of our country people, and indeed exactly in the style of an English watering-place. The natives fawn upon, laugh at, and cheat the strangers, who come here from all parts of England, Scotland, and Ireland, to get rid of their consumptions and their money. Here I first learned with great concern the death of Dr. Hope. I rejoice, however, in having paid him the compliment I did, and particularly in having written a letter, which I sent with the dedication copy of my book, and which seemed to give him particular pleasure.

The road from Nice to Genoa is bad beyond description, being for the most part along a path on the brink of precipices hanging over the sea, and so rocky and steep that every comparison of flights of broken steps, Norwich pavement, &c., &c., is very insufficient to give an idea of it: but the mules never make a false step, though they have often all their four feet together on a loose slippery stone, and sometimes climb up places so steep that 'tis difficult to keep from sliding off their rumps. The weather was cold and windy; for though we rode through myrtle thickets, and among trees which even at Montpellier are kept in greenhouses, and had groves of orange-, lemon-, and palm-trees about us, yet there was ice an inch thick in the road, and snow on the hills at no great distance. The weather is, however, now very fine, rather cold but bright.

On my arrival at Genoa, I called on my friend the Marquis Durazzo, and was a little struck at the magnificence of the house (notwithstanding what I had heard of the Genoese nobility), the ranks of servants, &c.; so that I began to dread some formality and reserve. I was, however, most agreeably relieved from all such ideas the moment I saw him; and I found him the same cheerful, easy, unassuming man as when I knew him in London. He made me most heartily welcome; insists upon all ceremony being laid aside; that he will make us acquainted with all his friends, and procure us access to every thing worthy our notice. Last night he introduced me to a family party; everybody received me in the same agreeable manner, and talked English or French (both equally well), out of politeness to me; while I blundered out some Italian. His family are persons of the first consequence here; their palaces far beyond what we have, except perhaps Chatsworth; and their collections of pictures celebrated in books of travels. A marble staircase in the Marquis's house is just finished, at the expense of about 5000*l.* sterling.

We shall not stop at Rome for more than a day or two in our way to Naples. We have letters to Mr. Walsby, secretary to the Duke of Gloucester, now at Rome.

Your affectionate Son,

J. E. SMITH.

Mr. Smith to J. E. Smith, Genoa.

My dear Son, Norwich, Jan. 4, 1787.

Pray make no apology for writing a small character that your letters may contain more; they are so entertaining, independent of the interest we take in every thing you do, and in all that takes your attention, that we are never tired, but read them over and over again.—I have looked into Thicknesse for the *Maison Carrée,* into *Le Voyage de France* for Aix, Nismes, the *Pont du Gard,* Marseilles, &c.; and much the more enjoyed your and their descriptions. For the climate, the fruits, the good cheer, the frugality, and though last, not least, the fountains of Aix, I amuse myself with the delights of living there. Could we but find civil and religious liberty, besides the above blessings, I could wish myself and my family resident there.

I was not sorry you were not bit with poetic phrensy at your visit to Vaucluse. 'Tis by no means a vein to be wished. No muse rewards her votaries so ill: to say how many she has ruined would be endless. Whether Garth, Akenside, and others in your line reaped any advantages in glory or fortune from courting her, I know not. I would lay they did not come near a Sloane, a Lee, a Heberden, a Jebb, in one or the other; but dare say if we looked about, we might find many such ingenious men as Goldsmith, who would have lived happier and obtained more renown, as well as riches, in being more ardently fond of their own sciences than of a rhyming muse.

Your judgement of Petrarch and Laura is just what I formed when I read Petrarch's life; yet I confess there was a sort of fascinating pleasure attended the perusal. One might style it a sentimental romance, as Buffier would. Too many delusions are pleasing.

I am sorry to tell you ———, so lately married in an extraordinary manner, very soon fell into bad health, and this week she died,—a very serious lesson to those who set a step of the utmost importance against prudence and against duty. 'Tis most likely she is a sacrifice to disappointment and repentance. Alas! too amiable, too valuable a victim!

We join in wishing you the continuance of health, pleasure and safety; which, whilst you take care of yourselves, we trust you will be blessed with by Heaven.

I am, dear James, ever yours,

JAMES SMITH.

J. E. Smith to his Mother.

Honoured Madam, Pisa, Jan. 22, 1787.

We staid at Genoa till the 18th instant, being very highly entertained with the fine buildings, pictures, &c. of that superb city, and extremely pleased with the perfect ease and familiarity with which we were received by the Durazzo family, who all seemed to strive to amuse and make us welcome. We found their attention of great use to us in procuring us that of everybody else; for the Genoese are rather stately: but this family having undertaken to introduce science into the country, and particularly natural history, found some amusement themselves in our company. I must postpone my descriptions of Genoa to our future conversations, which I often think of. I rejoice that I have been there on many accounts, independent of the pleasure I received at the time.

The road towards Pisa being very mountainous, my companion preferred the sea; so he had a very pleasant voyage to Sestri, thirty miles, where he landed to sleep the first night, and where I met him the same evening after a rather perilous ride on horseback; however, I dared not ride down the hills, but got off to walk whenever I saw danger. The road beyond Sestri being so extremely mountainous and bad, that everybody told me I could not go without great danger, I was induced (notwithstanding my promise, which I hope you and my father will pardon) to go by sea. We could not leave

Sestri next morning, as it rained; but the day after being very fine, not a breath of wind, and the sea smooth as possible without a wave, we were rowed thirty miles to Lerici, all the way within a few yards of the shore, which is most sublimely beautiful; nothing could be finer than this voyage, nor was it worth sixpence to be insured. From Lerici we came yesterday post to Pisa. This day we have been highly entertained with the fine old cathedral, the hanging tower, &c. of this place; the town is very pleasant, streets wide, clean and airy, the river very fine, but the inhabitants too few for so large a place. We have determined to go straight to Florence to-morrow.

Your ever dutiful Son,

J. E. Smith.

We are abundantly furnished with letters to every place we go to, from the Durazzos, Dr. Batt, Broussonet, &c.; so that we shall see the principal literary people in Italy.

The Marquis Durazzo (who knows Spallanzani well) assures me that he is now in prison at Vienna, under a charge of embezzling some things intrusted to his care by the Emperor. His friends hope he is innocent; so do I, for the honour of philosophers.

J. E. Smith to his Father.

Honoured Sir, Rome, Feb. 12, 1787.

My last was addressed to my dear mother, from Pisa, Jan. 22nd; and I now sit down to continue my narrative of our journey thus far. We met with a very good voiture with two mules, which conveyed us to Florence, and the same afterwards brought us to Rome, and had a very safe and pleasant journey; except that the country inns were very bad, and some of the country little better than the borders of Scotland. We had a very sharp wind all the way to Florence, so that we had but little enjoyment of the famous vale of the Arno, through which we passed. It is highly cultivated; the fields bordered with elms or mulberry-trees, each of which supports one or more fine vines, which hang in festoons from one tree to the other; but at present they are quite without leaves, except here and there in a warm corner, where the old leaves remain. We were quite in raptures with Florence, where we spent eight days, the greatest part of which was passed in the gallery. Nothing can be more handsomely contrived than this place is for the convenience of strangers. The Grand Duke (a worthy brother of the Emperor) gives the most positive orders that no money should be taken on any pretence for showing the gallery, the library, museum or any thing, except only one of his palaces. The civility and extreme attention of the guides deserves no less praise than their sovereign's liberality. The gallery is open every day, except Sundays, from 9 till 1, and from 4 till 5. Here we contemplated at leisure the Venus de Medicis (of which I had a very inadequate idea before), the Venus of Titian, the Wrestlers, the Slave whetting his Knife, the Apollo Venator, with many other master-pieces of sculpture and painting, all in one room; among the rest St. John in the Wilderness, by Raphael, I could never cease to admire. We studied the countenances of the Roman Emperors, and various great men of antiquity, in their original busts. I would give one of my ears for the bust of Marcus Aurelius. The portraits of the painters interested me much. I was sorry to see Sir Joshua Reynolds cut so very poor a figure among them as he does. His picture is one of the worst of his works I ever saw,—a bad likeness, and faded all away; it hangs too in so low a situation that it appears a mere mass of daubing. Some of the best portraits in this collection are by painters of whom we know little or nothing. The inlaid tables are above all praise. You have read so many descriptions of this justly celebrated collection, that I need not attempt to describe it at present. I shall only say that it more than answered my expectations.

The collection of the portraits of illustrious men I think unworthy of the rest; most of them are only bad copies, not originals.

The chapel built for the sepulchre of the Medici family, but never finished, is very large, and entirely lined with the most precious jaspers, agates, lapis-lazuli and various gems, in the finest taste possible: stones of which we generally see only small bits in snuff-boxes or rings, are here used for panels or cornices. There are innumerable fine statues and pictures to be seen about the town, to mention which would take several sheets. Our afternoons were generally spent with the celebrated Abbe Fontana, whose civility to us was very particular, and we learned much from him. Our inn was the most comfortable and elegant one I ever was at, so that on the whole we had no alloy to our happiness at Florence. We stopped a day at Sienna (a fine old town in a very fine high situation) on purpose to see Dr. Mascagni, a very ingenious anatomist, whose discoveries in the lymphatic system have gone far beyond those of Monro or Hunter, and who is going to publish a fine work on the subject. He is one of the most modest and unaffected of men. We arrived at Rome full of expectation and admiration at every step when we came within its walls, and were much struck with the entrance, which you have, figured in one of Piranesi's plates, and which gives a very just idea of it, notwithstanding what we have been told to the contrary. We got here Wednesday the 7th instant, and are lodged extremely well in part of a house hired by the Duke of Gloucester (who is now at Naples). A Portuguese abbé*, whom I knew at Paris, is extremely useful to us; and as we have several letters, we shall I hope see every thing worth seeing, as well as several distinguished persons. Thursday was almost entirely spent in St. Peter's, which, although with respect to architecture perhaps scarcely superior to St. Paul's, and even in some parts inferior, is so infinitely beyond it in internal decorations, as well as in cleanliness both within and without, that a comparison excites one's pity. We have seen several antiquities likewise; Trajan's and Anthony's pillars, the Coliseum,

* M. Correa de Serra.

the arches of Titus, Severus, and Constantine, the castle of St. Angelo, as well as many churches. Tomorrow we go to the Vatican. Saturday carnival began, and I need not say we were highly diverted. After it is over (in about ten days) we go to Naples, stay there two or three weeks, and then return here for three more; thus we shall be here in the holy-week, the best time possible.

I have now learned all the particulars about Spallanzani. Scopoli *, who is a man of the first cha-

* In a letter to his father in 1791, Sir James tells him, in speaking of his publications: "I hope you have seen the Critical Review for January, in which the second fasciculus of my *Icones* (uncoloured) is so very handsomely reviewed. I believe I never translated to you what I have said of poor Scopoli, which the reviewer quotes with approbation. It is nearly as follows:—'T. A. Scopoli, universally celebrated for his *Flora* and *Entomologia Carniolica*, after various labours in metallurgy, zoology, and botany, at length, by favour of the Emperor, became the public botanical and chymical professor at Pavia. This indefatigable man devoted his leisure hours to the collecting all such new or ill understood natural productions as fell in his way; whence arose that splendid work, *Deliciæ Floræ Faunæque Insubricæ*, his last production, and brought forth in trouble. Although devoted to the most inoffensive pursuits, although dear to all good men, and esteemed by all Europe, how much, alas! did he suffer from the arts and malice of the malevolent! When he would have exculpated himself from censures unjustly cast upon him, his excellent sovereign, seduced by the arts of the same men (for sovereigns are generally obliged to judge but superficially), absolutely commanded him to be silent. He submitted to his fate, but his indignant soul sought for liberty and justice at a higher and more impartial tribunal, May 8, 1788.'"

In speaking of Professor Scopoli in his Tour on the Continent, Sir James adds: "It is needless here to enter into the particulars

racter, and beloved by everybody, has found out how he has been used, and is dispersing the true story all over Europe. My idea of the rivalship between these two men proves exactly true; indeed I knew it could not be otherwise. Spallanzani is in but little credit among real philosophers, who examine for themselves; for many of his experiments are found false, and he is considered as a mere random theorist.

I am, dear Sir, yours,

J. E. Smith.

---

of the celebrated *fracas* which happened at Pavia not long since, relative to some articles missing in the public museum. Nothing is more painful than to find blemishes in a distinguished character; nor should I mention the matter at all, but for the purpose of justifying the innocent. All the above-mentioned professors were unanimous in their account. The government was also well informed of the truth, and the emperor Joseph II. took care to have his sentiments known. But, unwilling to lose a person whose scientific abilities were of importance to the university, the accusers were somewhat harshly condemned to silence. Poor Scopoli, one of the most concerned in the business, died of grief the following year; and his only justification was a printed circular letter, sent to the principal literati of Europe, in which the real cause of his death was mentioned, and which authority itself did not suppress. May my honest though feeble endeavours help to revive his blasted laurels, and protect a name which ought to be dear to every good man, and doubly so to every naturalist! It is the privilege of an Englishman, thank Heaven! safely to assert the cause of justice whether in his own case or that of another; a privilege which, under even a mild and beneficent sovereign, cannot, we find, be always obtained in an unlimited government. May happy Britain long most jealously watch and preserve this inestimable blessing! May she be cautious, too, of

Sir Joseph Banks to J. E. Smith.

Dear Sir, Revesly Abbey, Sept. 24, 1786.

Enclosed you will receive two letters for Naples, one to Sir W. Hamilton*, the other to Greater the gardener, whom I sent over to the Queen of Naples; you will find him a knowing man, and I expect you will receive much real assistance from him. I shall have great pleasure in hearing that your journey is pleasant and successful. I know of no news to send you from this country. Dryander writes me word that Swartz, the Danish botanist, who has been all about the West Indies, arrived in London a few days ago, and stays till spring.

Believe me, most faithfully and truly,

Your humble Servant,

JOSEPH BANKS.

hazarding it in pursuit of other advantages, however flattering, remembering how much she has to lose, and how little to gain."— Vol. iii. p. 72.

* *Sir Joseph Banks to Sir William Hamilton.*

My dear Sir William, Revesly Abbey, Sept. 24, 1786.

As the bearer of this, Dr. Smith, is one of my most intimate friends, I take the liberty to request as particular an indulgence in his favour as your occupations will allow. As his wishes are merely on the subject of science, he will repay, I am sure, any civilities you are pleased to show him.

Botany is his chief pursuit, and in that he has made very great progress.

Believe me,

Most faithfully and affectionately yours,

JOSEPH BANKS.

J. E. Smith to his Father.

Honoured Sir, Naples, March 5, 1787.

I was much disappointed at not finding a letter from you at any of the post-offices (for there are several) at Rome, nor any here.

We staid at Rome till Feb. 25th, and spent our time very happily indeed. We saw all the Carnival, a description of which (as you have seen so many descriptions of it) I must defer to those future opportunities so often referred to, and thought of by me with great pleasure.

We saw the Pretender every day: he is a heavy sickly-looking man, very much like what Mr. Bacon of Earlham was latterly. He drinks very hard. The principal things which we have seen at Rome are the Coliseum, the arches of Severus, Titus, and Constantine, columns of Trajan and Antoninus, the Pantheon, the Museum Clementinum (where are the most celebrated statues, as the divine Apollo Belvidere, the Antinous, Laocoon, and a thousand other inestimable things), the paintings by Raphael in the Vatican, the Moses of Michael Angelo, which is the first modern statue, as the Transfiguration of Raphael, is the finest picture, in the world. We have also seen the Palace Farnese, and several other palaces and villas, with churches innumerable; yet we shall have enough to employ us very closely for the three weeks which we have allowed ourselves to stay on our return, and shall then leave Rome with more regret than ever I left any place. I think it best to see the principal things well, and leave the rest slightly seen, or unseen altogether; but I think nothing of the kind will ever interest me after I leave Rome. We were three days and a half in coming to Naples in a voiture; the inns were bad, but the road equal to any in England; the weather very fine. Another voiture with excellent company was with us, and we liked our journey very well, especially as the country through which we passed was so very interesting. We visited the villa of Cicero, where he was killed, and also his tomb. We arrived at Naples February 28th. Sir W. Hamilton is most particularly civil, and his acquaintance will be of great use to us. Mr. Walsby, who is here with the Duke of Gloucester and family, is also very polite. On Saturday evening we were presented to the Duke and Duchess, who received us very courteously, especially the latter, who talked with me for near half an hour with the greatest affability, and so did the young prince and princess. The Duke was engaged almost all the evening in playing on the bass-viol with the celebrated Giardini, who is in his train; there were only two gentlemen and ourselves besides the family, so that it was quite like a private visit. Lady Almeria Carpenter is always with them, and we had some conversation with her. She is the most angelic creature that I have seen since I left England,—I might almost say the only really handsome woman. The Duchess too is a very fine woman. We are now at liberty to go any evening we please, as they are generally at home; we shall see them again at Rome.

This morning, in consequence of an invitation from the French ambassador, we went to hear mass performed for the Abbé de Bourbon, natural son of the late King of France, who died here of the small-pox, aged 24, the day we arrived. It was the most superb thing of the kind I ever saw. The church was hung with black, and an astonishing profusion of cloth of gold and ermine, and lighted with about four thousand large wax tapers. In the centre was a most magnificent Ionic temple of wood, painted like gray marble and gilt, in which was a sarcophagus of purple velvet embroidered with gold, at each of the corners of which stood a skeleton holding a great torch.

The body was buried on Saturday. Mass was performed by the Pope's nuncio; the music was quite heavenly; there were a hundred instrumental and fifty vocal performers, and two organs. The company consisted of the first people of both sexes, all in mourning. The abbé is much regretted, particularly by the English; and the French say he was an Englishman in his heart.

Naples is a very long and extensive town, amazingly populous; the people very dirty, and the most barefaced villains that ever I met with.

Here are some rich churches, and many tolerable pictures; but none very fine. The architecture in general, as well as the taste of the ornaments of the town are very bad; some of them like the taste of our James the First's time; others like the flouncing, fluttering, scrawling style of the Dutch burgomasters, or London and Norwich aldermen. How different from Rome, Florence, and even Genoa! Mount Vesuvius is in full view of our window, and this evening has thrown out flame several times.

March 6th. This morning I have seen the Chartreux, a very rich monastery, where is a capital Nativity by Guido, which has been engraved, and some other good pictures by Lanfranc, Spagnoletto, &c. The treasury is amazingly rich. From a castle above this convent is one of the most celebrated views in Europe, and the day being very fine we saw it to great advantage.

I forgot to tell you, that we got into the Pope's chapel on Ash Wednesday, and saw him put ashes on the heads of all the clergy. He is a very handsome man; and although said to be ashamed of this ceremony, yet he went through it with great dignity.

We find Italy much dearer than France; but I hope to finish my tour without being immoderately extravagant.

Your affectionate and obedient Son,

J. E. SMITH.

Rev. Dr. Goodenough to J. E. Smith, Milan.

Dear Sir, March 11, 1787.

Your last letter from Genoa, dated January 13, reached me the beginning of last month, but not before it had been anxiously wished for. The account you give me of your travels through groves of oranges and beds of myrtle, puts me in mind of the enchanting scenes of the Arabian Nights Entertainments, or Milton's

——— Spicy gales
From Araby the blest.

I do not know what I would not give to be with you. When you return home I think I shall never have done asking questions about something or other. As it now seems determined that we shall have a commercial treaty with France, and restore the golden age, which I foolishly imagined was only in Utopia, I hope a few exports and imports in natural history may be both allowed and obtained, *while it lasts;* for I own myself one *qui timeo Danaos et dona ferentes,* and cannot bring myself to think that there is much sincerity on either side in this business. Perhaps if mutual interest can be made to take place, it may exist a while. Yet between two rival nations, I do not see how there can be a *mutual* interest. If the gain be *equal* on both sides, then they are as they were before *relatively* considered, and there is no bond of union. If one side have the greater gain, the other will appear foolish, and in the end try to break the bargain. In this train of thinking, I am not at all sanguine about this measure, which I fear I ought to call the bantling of mercantile interest, favoured in prejudice to the true offspring of Britain, national valour and sound policy. Now I am upon politics, just let me tell you the exact state of things at present. The Houses of Lords and Commons have both pledged themselves to adopt the treaty *generally;*

that is, reserving a right of modifying particular matters of inferior moment. Mr. Pitt carried it in both houses by a majority of two to one. The whole body of traders are full of ideas of present profit; and as yet, though a slight grumbling is beginning in some quarters, not a single petition has been presented against it. Lord Lansdown says that the scheme was his, and therefore, though a secret enemy to Pitt, is for it. Four charges have been urged against Mr. Hastings (for maladministration) in the House of Commons. He has been adjudged liable to an impeachment upon *three* of them, to the no small displeasure of a certain great personage. Some people go so far as to say, that Mr. Pitt, who voted for Hastings's impeachment, will lose his situation for his behaviour. Lord Mansfield, angry that ministry will appoint Sir Lloyd Kenyon to succeed him, refuses to resign. He wishes Buller to succeed him. A pamphlet, entitled, "A short Review of the Political State of Great Britain," affecting a great impartiality and knowledge of present affairs, has lately made much noise, and run through several editions. It by no means deserves the character which has been given it (it has been translated into all foreign languages), for it is evidently partial to the court side, and is not a true prophet. Witness the case of Hastings, whom this writer trumpets to the skies. I look upon this pamphlet as little more than a jumble of the best sort of coffee-house conversation. Though you are not a zealot in politics, nor I myself, yet I thought, at your distance from England, you might not dislike to hear so much of our affairs. Portugal is to be considered as of very inferior moment: but ministry are very much perplexed about Irish affairs; for the treaty says, the French are to be *the most favoured nation;* so the blood of St. Patrick thinks that very strange.

As to natural history, which I busy myself about every day, more or less, I am in the first place amusing myself with the idea of our New Society, which must take place of our present *gross* body, the instant it starts. The present society goes on in the usual way, of having a fossil or a plant go round the table: nothing is or can be said upon it. It is referred to a committee to consider of it; the committee call it by some name, and send it back to the society. The society desire the committee to reconsider it: the committee desire the society to reconsider it. In the mean time nothing is done; indeed it does not appear to me that any of them can do any thing. I have had a short conversation with Sir Joseph, who repeated to me what he said to you. Were he not President of the Royal Society, I am sure he would join us.

My *Systema Accentuatum* is ready for the press: but I shall not take that step till you return.

I am much obliged to you for your letters, and beg that you would continue to write to me, wherever you are. Your letters are my first-rate amusement. In the mean time believe me,

Your very affectionate friend,

SAMUEL GOODENOUGH.

Marsham is here with me, and has just showed me your letter to him, which was wonderfully entertaining, particularly your account of the famous gallery, and your laying down your heart at the foot of the Venus de Medicis.

I hope you will contrive to see Allioni: I look upon him as one of the few who search into nature, and see with their own eyes. Besides, where can you see the Alpine plants in such plenty, or such perfection as with them?

Do keep in your mind *Curculio paraplecticus*,—my desiderata in general. I should be glad of a little more *Agrostis minima*. Above all, write to me often, again and again.

Ever yours, with best wishes, S. G.

Thomas Jenkinson Woodward, Esq. to J. E. Smith, Milan.

Dear Sir, Bungay, March 16, 1787.

I have now before me your two very friendly and highly entertaining letters, the first from Paris, October 2nd, the other from Rome, February 12th. It gives me great pleasure to find that in such distant journeys, and amongst the literati of all countries, and such interesting objects, your old Norwich friends claim a share in your recollection and regard; and I do assure you the distinction flatters me highly. I am very much obliged to you for Bulliard, which, with your Parisian letter, I received very safe about Christmas. I shall certainly thank you for the promised copy of Leers, at the eighteen livres. You will also be so kind as to procure for me on your return to Paris, such other Fungi as Bulliard may have published since you bought them, together with the *Histoire des Champignons,* and such other letter-press as may relate to that subject.

Your thesis I shall hope to receive on your return. Retzius is, I take it, a Swedish professor, who has published three fasciculi of observations in botany, in 1779, 1781, and 1783. The second is dedicated to Sir Joseph Banks. There are a few plates, particularly the real *Astragalus arenarius,* and another called *Hippuris lanceolata,* which I apprehend to be the *tetraphylla;* some good observations, and I apprehend some error. I think I saw the book in your library. I thank you for *Gnaphalium luteo-album,* and *Agrostis minima,* both of which are treasures.

I need not say how often I have wished myself with you, for it has happened every step you record of your tour, and every library and herbarium you have examined. The names of the old botanists seem to carry one into enchanted ground; and the travelling through the herbaria of Tournefort and Vaillant, &c. is like taking a journey to Jupiter or Saturn, to verify the number of their satellites, and periods of their revolutions.

Your *Senecio acanthifolius* must be a magnificent plant, and the rest I long much to see. If you return through Switzerland as you propose, possibly you may have an opportunity of inspecting Haller's herbarium, which would, I should think, be of great advantage to an accurate *Flora Anglica,* as so very many of the Swiss plants are also natives of England. I have travelled with you in your second letter, from Paris to Rome, and enjoyed the beautiful and sublime scenery you describe, and the sight of so many plants growing wild, which we either immure in greenhouses, or look upon as valuable ornaments to the shrubbery or flower-garden. I hope you drank out of the fountain of Vaucluse, to the memory of Petrarch and his beautiful Laura.

The two days you spent with Gerard must have been delightful, and particularly advantageous to your proposed reform of the *Syst. Veget.* Your time at Genoa must have passed delightfully with your friend the Marquess and his family. And the time spent in examining the gallery at Florence was charmingly employed. The *veteris vestigia flammæ* however, always breaking forth, makes me regret the impossibility of examining to purpose the collection of Micheli, as it might have been of consequence in ascertaining the synonyms of many English *Cryptogamia.* I was ill-natured enough to be highly pleased with your account of Spallanzani's disgrace, and that principally on account of the strictures it pleased the reviewers to pass on your second tract, because they judged it improper to attack so great a personage. To people any way qualified to judge of the subject, there could be no doubt of the propriety of your strictures on Spallanzani; but there are so many readers of the Reviews, who pin their faith upon them, and though entirely ignorant of the subject of natural history, choosing to retail their opinions on that as well as other subjects, that it gave me great pleasure to have, what one might call, a practical defence of your assertions. In consequence I have mentioned to some friends the anecdote.

You expect an account of some discoveries; but whilst you are daily treading new ground, how am I to supply you with any thing new? I have not met with any thing new in the *Cryptogamia* this winter, and in the other classes one has nothing to expect. Dickson's journey to Scotland has however produced enough, at least according to Crowe's account, to surprise you on your return. He says that he believes he goes not more than half way, when he declares that Dickson has discovered one hundred and fifty new species of Mosses, Jungermanniæ, &c., and Lichens, and some of them so completely unlike any thing discovered before in those genera, as to be wonderful; but whether they will excite your wonder as much as ours is rather doubtful. For my own part, not having been at London this winter, I have seen none of them; and Crowe's portmanteau, in which were many of these treasures, was stolen on its way to the coach, by which they, as well as his wardrobe, were irreparably lost. I was at Norwich last Saturday, when I saw Mrs. Smith, your father being out of town; and she was so kind as to communicate to me great part of a letter from you of the same date as mine; but they were so different, that each of us received much pleasure from the exchange of intelligence. Pitchford and his wife desire their affectionate compliments. I showed them your letter, with which they were charmed.

Norwich has just been in another uproar. Yesterday se'nnight the committee of the House of Commons, after a long sitting on Sir Thomas Beevor's petition, declared the last election void, from bribery proved on both sides; and on Saturday a writ came down. Yesterday was the election, when, after a hard struggle, Hobart succeeded by a majority of eighty and upwards, as I have heard; but a scrutiny is talked of.

We have experienced a severe loss in the death of my worthy and much respected father-in-law and friend Mr. Manning, which happened on January 15th. His kindness to me was such that I can never cease to regret his loss, and must ever highly respect his memory.

I am, dear Sir,

T. WOODWARD.

J. E. Smith to his Father.

Honoured Sir, Rome, March 26, 1787.

Your letter of February the 12th, which I found here, March 23rd, on my return from Naples, was as grateful to me as "cold water to a thirsty soul;" and I take the first opportunity of thanking you for it.

I now proceed with my own history.—The 9th instant we were most delightfully occupied in seeing the antiquities on the west side of Naples; the grotto of Pausilippo, Puzzuola, Baiæ, the Elysian

Fields, temple of Venus, (from which I brought some sprigs of myrtle, as I did of bay from Virgil's tomb,) baths of Nero, grotto of the Sibyl, lake of Avernus ; in short all that you will find in books, and which I need not describe. They are highly interesting, and I could have spent a month about them instead of a day. I had them all, however, already so well in my mind, that a slight view was sufficient to stamp a lasting impression.

Two days after, we visited the palace of Capo di Monte, where are some good pictures, but bad ones innumerable ; the best thing there is an onyx cup, found in Adrian's tomb.

The catacombs are wonderful for their extent, and worth seeing. We afterwards visited the Grotto del Cane, Virgil's tomb, and some other things thereabouts.

Next day was devoted to Vesuvius. We left Naples at six in the morning, and in an hour and a half got to Portici, where we took mules, and ascended as far as we could in that manner, but were obliged to walk about a mile over the loose ashes to the top, which is very fatiguing ; but by going very slow we accomplished it well. We walked round the inside of the old crater, in the centre of which is arisen a new mountain within these twelve months, to the top of which we ascended and looked into the present crater ; but it was less striking than I expected, as its inside is not perpendicular, and the smoke, although very moderate, prevented us from seeing far. I can give you no idea of the grotesque appearance of iron cinders, encrusted with salts of the most vivid green, red, and yellow colours; the chasms from which issue smoke and heated vapours, with snow at three feet distance from them. All this I had no conception of, much less of the lava, of which we saw a fine stream, about a yard in breadth, issuing from one side of the hill: we dipped our sticks into it, and it immediately cooled into a cinder around them;—the melted iron running from a furnace gives the best idea of it. We descended much faster than we came up, after having gratified our curiosity, and enjoyed the view, which is beautiful and extensive beyond description. As to the hazard we ran, I believe it to have been none, however formidable in the description; for there was no rumbling in the mountain, as there always is before any change, and all the top is an iron crust firmer than any other rock in the world, and very safe to walk on.

We descended into the theatre of Herculaneum, which is all that is left open, the rest having been filled again for security, as Portici stands upon it. Next day went to Pompeii, many of the houses of which are quite laid open, with their beautiful stuccos and mosaics in full view; the streets, gardens, &c. are quite perfect: nothing can be more interesting, and they daily discover more. The museum at Portici contains most of the things that have been found: vases, bronzes, statues, paintings, implements, even dates, figs and bread, with many other such things, in their perfect form, although discoloured. At the porcelain manufactory at Naples we saw a superb set of china, intended as a present to our king. I received great civility from the Danish ambassador, who is a botanist, and who was earnest with me to correspond with him. Saturday last, March 17th, we left Naples for Caserta, where the court resides. We spent that day and the next with Sir William Hamilton and the queen's gardener, who was sent over by Sir Joseph Banks; and on Monday a very bad road brought us to Monte Cassino, the chief of the Benedictine convents, to which we were well recommended, and where we were most hospitably and elegantly received: 'tis well worth seeing, but I must postpone a description of it for the present. We spent Tuesday there; and in three days more a very bad road, through a most beautiful country, brought us to Rome. We passed through Frescati (Tusculum), and saw the ruins of Cicero's Villa, and the Villa Aldobrandini, with some other gardens. Yesterday and today we have seen the antiquities of the Capitol, and today ascended to the top of Trajan's noble column. We stay here a month.

I am, honoured Sir, your dutiful Son,

J. E. Smith.

Mr. Smith to J. E. Smith, Milan.

My dear Son, Norwich, March 29, 1787.

Your letters, my dear James, give us the most exquisite pleasure: but although the narrations are so pleasing, and the reception you meet with makes your parents' hearts exult with joy; though we contemplate the advantages you may derive from your journey in regard to information and connections, with a pleasing prospect of future happiness and interest;—all these agreeable subjects for reflection do not interest us near so much as what concerns your health and safety, the confirmation of which in every letter carries my soul, loaded with gratitude, up to the throne of our gracious God, displaying your letter as the testimony of his great goodness in preserving and protecting you, who are the joy and comfort of those to whom you owe your birth, and who are never ceasing to offer up vows for your happy return, when you have completed your purposed tour. Excuse, my beloved son! these sallies of the passions,—my heart is full.

Your brother Francis is pursuing astronomy now with the earnestness he undertakes every thing, and he is making astronomers of the girls and me. We are ever popping out to peep at the stars, and are intimate with Orion, the Ursæ, the Lion, and Bull, &c. Sir J. Banks told me he had received a letter from you. I shall write to you to Turin, and hope you will let us hear from you as soon as you arrive at Venice, at Milan and Turin, if you have only time to say you are well.

Yours for ever,

JAMES SMITH.

I am told the road from Milan to Turin is in danger of banditti. I beg of you omit no precautions, nor spare expense, for your safety. I write now for the caution I give you, which I don't let your mother see.

Mr. William Jones to J. E. Smith, Milan.

Dear Sir, London, April 9, 1787.

I thank you for your favour from Rome, and particularly for your apology for not writing sooner; because it justified my jealousy, when you wrote to Mr. Marsham, that you had either forgot me, or was offended that I did not embrace your scheme with the same ardour yourself and some others possessed for it. I own, not one tittle of the plan expressed in your favour can be objected to by such as are qualified for it; but am certain 'tis your partiality only can rank me in that number, for I feel in myself too much ignorance when with the ingenious and scientific, to conceive myself entitled to it. This, and an indolence of mind that frequently accompanies me, seem to throw obstructions where otherwise my ambition might tempt me. To cultivate your esteem, I shall be always proud of communicating the little I know, and I am afraid so little, though all, that I shall hardly merit it; nevertheless, I may venture to assure you it shall not be charged to my account, if our friendship ceases sooner than our lives.

Rome, more than any city upon earth, would please me to visit; I conceive it an inexhaustible source of delight to a curious mind. 'Twas the sentiment of an ingenious traveller, that of all the places he had seen or should see, it was by far the most delightful;—that a voyage to Italy might properly enough be compared to the common stages and journeys of life. At our first setting out through

France, the pleasures that we find, like those of our youth, are of the gay fluttering kind, which grow by degrees as we advance towards Italy, more solid, manly, and rational, but attain not their full perfection till we reach Rome; from which point we no sooner turn homewards, than they begin again gradually to decline; and though sustained for a while in some degree of vigour through the other stages and cities of Italy, yet dwindle at last into weariness and fatigue, and a desire to be at home, where the traveller finishes his course, as the old man does his days, with the usual privilege of being tiresome to his friends by a perpetual repetition of past adventures. This last is only to finish the climax, and by no means applicable to you. You have a fair opportunity of adding your suffrage to the above, or giving your dissent; and as this remark was made more than fifty years since, you will observe how far manners may have changed with time. I must give you another quotation. Brydone, in his Travels through Sicily, speaks of large aloes making a magnificent appearance, and I believe thought them indigenous, and enumerated sundry plants, which, when I read him, struck me with surprise; but you assign a very probable reason why more southern plants should flourish in that country. In the west of England, near the sea, 'tis no unusual sight to observe myrtles in the open air bear the winter well; and I have noticed some springs, when near London the severity of weather has destroyed the blossom of the wall-fruits, near the sea in Sussex they have not perished.

Mr. Latham will be much obliged to you if you can procure a good description, &c. of a bird called Courier, described by several, but only seen by Aldrovandus, from whom the others took their account. The description given is too short and imperfect to give a just idea of it.

I saw Dr. Goodenough and Mr. Marsham this day. They and theirs are well, and are anxious to see you; but not more than, dear Sir,

Yours most sincerely,

W. Jones.

J. E. Smith to his Father.

Honoured Sir, Rome, April 11, 1787.

Yesterday I received from Genoa yours of January 4th, inclosed in one from M. Durazzo, which informed me of the death of his father, a fine old man of eighty-five, whom every body loved, and I was very much pleased with. I am not partial to the bread of Italy, 'tis all made with leaven, and so sour that nothing but custom can reconcile one to it, nor can I ever eat it without something to conceal the taste. Since our return from Naples we have been very busy in seeing houses, pictures, and statues, and with the ceremonies of the Holy Week. The palace Borghese contains I think the finest collection of pictures I ever saw; there are above seventeen hundred originals, scarcely any of which are not good, and some exquisitely fine. You will find accounts of them in books. There is a Venus of Titian, one of the two which contends with that at Florence for originality; several things by Leonardo da Vinci, a charming painter, of whose works I have seen but few; many by Raphael, Julio Romano, Titian, Guido, Paul Veronese (especially Cupid caressing Adonis), St. Cecilia by Domenichino, a most divine performance, &c., &c. I have learned one thing, which I hope will be agreeable intelligence for you, which is, that your Virgin with the Infant Jesus and St. John, which now hangs disgracefully on the staircase, is by no other hand than Raphael, in his second manner,—you know he had three very different ones. The first, in which he imitated his master, was a bad and stiff one. In the second, he improved very much in drawing and colouring, and finished very highly (as yours is), so that some of his pictures in this manner are very highly valued, especially as they are rare. His third manner is quite original, although various in itself, and of various merit; of this are all his most celebrated pieces, and no one has yet equalled him in design and drawing, although his colouring has been excelled. I suspected your picture to be by his hand when I was ta Genoa, from one which I saw there; and since, I have made so many observations on the pictures painted by him and his scholars as well as others, that I am without a doubt on the subject. What settled my opinion was a very much admired picture at the palace Borghese, which is extremely like yours in the thought and execution.

I have seen some of Andrea del Sarto's, which in

some measure resemble it, but still are sufficiently different. Leonardo da Vinci and his scholars are not in the least like it, and I wonder Worlidge thought it was one of theirs; but perhaps he was never in Italy, where, only, Raphael's pictures can be seen in any number. What is characteristic in your picture is a certain sublime tranquillity in the countenances, especially I think in St. John's; the little gold wiry hair of the children; the Virgin's very thin veil; the colours of her drapery; the colour and design of the back-ground; and the softness of the finishing, with some errors in the outline of the faces here and there, but no defect in grace or delicacy. I have so often contemplated the picture, that I know it perfectly, and have taken no small pains to make observations on the subject; but I'll tell you better when I see it again. The villa Borghese, which is without the city, is the finest about Rome, and as rich in sculptures as the palace is in paintings; besides that the house is extremely elegant and neat, abounding with the most precious porphyry and marbles. Of all the splendid sights that can be seen, I think the Holy Week exceeds them; indeed no prince who pretends only to temporal authority can assume the dignity that the Pope does, nor can any other have such a place as St. Peter's to exhibit in. The present Pope conducts himself with admirable grace and dignity. Nothing can be finer than his blessing the people from the middle window of the great front of St. Peter's, the great bell ringing, guns of St. Angelo firing, and the soldiers, who are drawn up in the area, trumpeting at the same time. Immediately after he has pronounced the blessing, with his arms extended, all the people kneel to receive it. His washing the pilgrims' feet, and serving them at dinner, are sights altogether novel; but the finest thing of all is his saying mass on Easter Sunday. The altar of St. Peter's was set out with candlesticks and statues of pure gold, each about four feet high. Behind it, at some distance, was a most magnificent throne for the Pope, who came in a chair on men's shoulders, blessing the people as he advanced. Behind him were carried two immense fans of white peacocks' feathers, and before him about eight different mitres and tiaras, for the most part covered with the finest pearls and gems, besides one which he wore on his head. The ceremonies of the preceding days were chiefly in the Sixtine chapel; there we heard such music, all vocal, as I never heard elsewhere, especially the *Miserere*, which is so very famous. I could have no idea before of such sounds, yet I can conceive that music might be carried further. We dressed in black, full dress, as is the fashion, and were easily admitted everywhere, as the Romans make a point of showing attention to strangers. There were a great many foreigners of the first distinction; the Duke of Gloucester and family (whom we have visited here since they came from Naples), the Duke and Duchess of Buccleugh, Lord and Lady Gower, &c. The fireworks at St. Angelo are inferior to what I have seen, except the vast fountain of fire, which Wright has painted. Two evenings in the week the inside of St. Peter's was lighted only by a huge cross of copper, studded on both sides with lamps, and hung in the middle of the church, which was quite a promenade, like Ranelagh. We leave Rome in twelve days for Loretto, Bologna, Venice, and Milan.

Your dutiful Son,

J. E. Smith.

J. E. Smith to his Father.

Honoured Sir, Rome, April 24, 1787.

We could not get off this morning, but shall leave Rome this afternoon or tomorrow. I have got a letter from Broussonet of Paris, informing me that I am chosen a member of their Royal Agricultural Society. We have been busily employed in seeing all we could, but must still leave some things undone. The church of St. Paul without-the-walls pleased me very much, on account of its great number of noble antique pillars of marble, forty of which came from Adrian's tomb, and are most of them of a single block of Parian marble near twenty-five feet high. A few days ago a very elegant new monument was opened in the church of the Twelve Apostles, for the late Pope Ganganelli; 'tis one of the finest I ever saw: 'tis of the usual form; a large base, on which is a sarcophagus, at one end of which is a figure of Humility, and at the other Temperance, both quite in the antique style, and far superior to Bernini's heavy works.

Above is a figure of the Pope giving his benediction. The artist is to be paid about 3000*l*. He is a Venetian. There is a perpetual crowd about this monument, and nobody talks of anything else. The people have begun to kiss a part of the drapery of one of the figures, which hangs low; for the late Pope is adored by all kinds of people except those who helped to take him off. The people likewise kiss the monument of Innocent XI. at St. Peter's. I have not met with the "Voyage de l'Abbé Richard," but with that of "De la Lande," which is an excellent one. I have also just been reading Addison's Travels, and a poor book it is. Smollet's gives a very just account of the manners of the people, but he always gives the dark side.

I wish I had Lady Miller's Letters; from what I remember of them they are very just, and have afforded me many useful hints.

Your dutiful Son,

J. E. Smith.

J. E. Smith to his Mother.

Honoured Madam, Bologna, May 7, 1787.

Although I wrote to my father so lately as April 24th, yet having made a long step since, I think you will have no objection to hear we are got thus much nearer home. On the 25th after dinner we left Rome, casting many a longing lingering look at the fine things we passed by. I have left no place with half so much regret. The last object we lost sight of was the dome of St. Peter's, from a hill about six miles off. We engaged with the same voiturin who had brought us from Florence to Rome,—a man of character, and with whom we were well pleased,—to carry us to Bologna by Loretto (366 miles) for 10 sequins (that is about 5*l.* each), and provisions and every thing included. We arrived here quite safe and well May 5th at noon; and had a very agreeable journey, except that the weather was part of the way very cold, and we had much rain. One half day we were obliged to lie by on account of a storm of rain and wind in our faces as we went by the sea side from Savignano toward Rimini. Friday, April 27th, we visited the famous cascade of Terni, said to be one of the finest in the world. It is a very considerable river, which falls above 460 feet perpendicular, as it is said, but I can scarcely conceive it to be quite so much: it is almost all changed into foam in its fall, and is dashed up an amazing height from the rocks below; it fills the neighbouring valleys, which are very romantic, with a perpetual mist, and when the sun shines there is always a rainbow. We slept at Spoleto, where there is a fine old aqueduct. Next day our road lay through one of the finest and richest countries I ever saw; it was along a very extensive valley, cultivated in the highest degree with corn and vines, and bounded by hills clad with olive-trees, and stuck with towns, white convents, villas and cottages, to their very tops. The hedges were full of what we call Italian May, in flower, and Venus's Looking-glass grew on the banks. At

Foligno we saw a very fine statue of massy silver, as big as life, representing St. Felix, one of their bishops; it is in a chair of silver, and is carried about the town once a year. After dinner we began to ascend the Apennines, and slept at a poor little place called Serravalle, in a very deep rocky valley, high among the mountains, where every-thing wore the appearance of February; the trees just budding, and the spring flowers now in bloom. This is one of the lowest passes of these mountains; there are others which are now covered with snow. The road is commodious enough. Next day we descended into a milder climate, and I found several fine plants. Passed through a rich and beautiful country (which continued all the way to Bologna) to Macerate, a very fine situation, and next day dined at Loretto, where we spent the rest of the day in seeing the celebrated curiosities of that *holy place.* They are very civil to strangers here, and delight to show them every thing; no kneeling or any ceremony of that kind is required. The place swarms with poor labouring people, who come on pilgrimages here from every part of the world: they go round the holy house on their knees, singing or saying their prayers. The holy house is converted into a chapel; the outside is cased with marble; it stands under the dome of a very large church: the inside is divided into two parts by an altar and grate, all of massy silver; the part behind the altar is lined entirely with silver, and here in a niche of gold stands the holy image of the Virgin, in a robe covered with above six thousand diamonds, some of them of immense value. About the niche hang great numbers of offerings made by the greatest princes in the world; among others, a figure of a child in gold, weighing twenty-four pounds, given by Louis XIII. of France, on the birth of his son Louis XIV., and many others equally valuable. Here are kept what they call the Virgin's gown, dishes which she used, the cup out of which our Saviour took pap, &c., all in massy silver cupboards. But the most stupendous thing of all is the treasury, a large room where most of the offerings that have been made are disposed in glass cases all around; that is, all the choicest of gold and jewels, for silver can scarcely find a place. There are some most beautiful as well as precious things, and many interesting in an historical light; as the fine crown and sceptre of Christina queen of Sweden; a gold heart set with diamonds, given by our James the Second's queen, and many others equally curious. There are garments for the priests, covered with millions perhaps of pearls and precious stones, in embroidery and other forms. Every thing is kept in the most exquisite order, and shown freely to the public twice a day. One thing which struck me extremely was a huge mass of emeralds, each as big as the handle of a table-knife, sticking in their natural bed; the whole must weigh one hundred and fifty pounds; it was given by Philip IV. of Spain. Here is a Virgin and Child by Raphael, for which they say Lord Spencer offered about 3700*l.*: but this church is not in want of money, as you will easily believe; their revenues are in proportion to their other riches. We laid in a stock of rosaries and crucifixes of various kinds, which are made at Loretto, and which we had touched with the holy house and porringer, to make them (as Lady Miller says) as efficacious as possible, to present as curiosities to our friends in England. Their cheapness is as remarkable as any thing, and some of them are really very pretty things; as you shall see, I hope, one day. We were obliged to buy a little silver Madonna, and some flaring artificial flowers to stick in the front of our voiturin's hat; that was indispensably necessary, and if it had been neglected, nobody knows what ills might have befallen us. Next day we proceeded amid troops of pilgrims to Ancona, a pretty sea-port, and thence to Sinigaglia. We saw nothing very remarkable all the way to Bologna, but passed through some very pretty neat little cities, viz. Rimini (which last week felt another shock of an earthquake), Forli, Faenza, Imola, and Cesene. Most of the towns in this country are built like Bologna, of brick; the houses stand on stone pillars, and form a clean broad covered way for foot-passengers on each side of most of the streets, which is very convenient. We have been busy since our arrival here, in seeing churches and pictures, with which this place abounds; some of the latter prodigiously fine: but no churches will do after seeing Rome. Tomorrow evening we set off in a barge, to go by canals and rivers to Venice, where we are to arrive in thirty-six hours, and where we expect to meet several of our friends. We shall probably stay ten days, or at most a fortnight, to see the place, and the celebrated ceremonies at the Ascension. Direct to me as before, at Genoa: the letters go post free to M. Durazzo's; and he can send them after me to Paris, where I shall likewise get them post free. I may perhaps call again at Genoa, as it is very little out of my way, and I am warmly invited.

Your ever dutiful Son,

J. E. Smith.

Sir J. Banks to J. E. Smith, Milan.

Dear Doctor, Soho Square, May 11, 1787.

Many thanks for your letter, and the news it contains. I am very glad to find your travels produce so much amusement to you.

I have nearly finished Swartz's herbarium,—a most valuable one it is; he is certainly the best botanist I have seen since poor Solander's death, so that the addition I have received from him is immense. I am in hopes to procure his being sent out by the East India Company next year to Bengal. The Royal Society flourishes much. Herschel sent an account there the other day of three volcanos* which he saw burning in the unenlightened part of the moon, the largest of which appears twice as large as Jupiter's third satellite, and he

* In a letter from Dr. Younge, dated Milan, June 22nd, he tells his friend Smith, "Oriani thinks he has found out what Herschel calls volcanos; he is at present in doubt whether they be really so."

conceived to be three miles in diameter. We expect from him a dissertation on the subject, which is not new; as Hevelius saw a similar phænomenon, though he was not able to account for it. Herschel had before discovered two new satellites to his Georgian planet.

Mr. Walker, apothecary to the Infirmary at Oxford, has produced a frigorific mixture with salts and acids only, by which he has frozen mercury last week when the thermometer was at forty-five. His cheap process, which produces forty-six degrees of cold, may be useful to cool your wine, &c. It is oil of vitriol mixed with an equal quantity of water, cooled by putting into it Glauber salts finely powdered. These I think are our three things, the newest this season, for the Philosophical Transactions. We begin to look out for your return, which I suppose will be in autumn; and have a long list of queries for comparison with your herbarium. Sibthorp is expected here shortly.

Yours most faithfully,

Jos. Banks.

Marquis Ippolito Durazzo to J. E. Smith, Milan.

Dear Sir, Genoa, June 8, 1787.

Your letter of the 23rd of May came safe to my hands, and afforded me some notices of you, which I wanted of. I thank you for the part you take on the loss of my father, which has been very much felt by all our family. It seems, by the history of the Capuchin, you were astonished to hear I was no more a bachelor. I cannot tell you how this changement of thinking took place in my mind so suddenly. In few days every thing was proposed and concluded.

The box shall remain with the other till you come, and I shall pay what is necessary,—I say, till you come, as I hope you will do it. Why, sir, do you take the liberty to ask me if your coming to Genoa is convenient or not to me, after so many petitions and prayers which I made to you for this object? I expect you consequently without any doubt, and will prepare some interesting things for you. You must, therefore, write to me some lines, that I can know the time of your coming; and shall be very happy to meet you again, and present you to my wife before that I can see you again in England.

I wrote to Mr. Broussonet in the month of April, and received no answer at all from him, according to his custom. Believe he will be always the same for me, and consequently I shall be obliged to look for another correspondent in Paris, in the scientifical line.

Present my compliments to Mr. Younge. I send you those of Mr. Cattaneo and my sister; and am with the greatest regard and consideration,

Your most humble and obedient servant,

Ippolito Durazzo.

J. E. Smith to his Father.

Honoured Sir, Milan, June 8, 1787.

Nothing could exceed my joy and satisfaction on finding here your very affectionate letter of March 29th, together with ten others from various friends. I always feel (thanks to your good instructions) so exactly the same sentiments towards Heaven on receiving your letters that you express on receiving mine, that I need not repeat them even if I could express them better; and whenever I approach the place where I expect news from home, I endeavour to prepare my mind in such a manner, that if it were unfavourable I hope I could bear it, and if favourable, as thank God it always has been, such a disposition of mind doubles my pleasure.

My last was to my mother, from Bologna, May 7th. I believe we shall fly down to Genoa for a few days, to congratulate my friend the marquis on his marriage, which is now in hand, and to strengthen still more my connection with his family, and especially with his elder brother, who is now the chief of the family, and who was beginning to be much attached to me: his son will soon come to England. From Genoa we go by the *low* road to Turin, in which there is nothing to fear. Our journey from Bologna here has been very prosperous and agreeable, and I now proceed to give you a sketch of it. To Venice we went by the Courier by water, pretty much as in Holland, in two days and two nights; the country we passed through was execrable, like the worst fens of Holland or

Lincolnshire. We dined excellently at Ferrara, a vast lifeless town, or to use Ganganelli's words, "*une belle et vaste solitude, presqu'aussi silentieuse que le tombeau de l'Arioste qui y repose.*" We had not time to visit Ariosto's tomb. I regretted more that we did not pass through Cento, where are some celebrated paintings of Guercino's, one of my favourites. I have made ample notes on the pictures of Bologna, which I hope will amuse you at my return, so I say nothing about them now. I shall tire even you with what I have to say about pictures and statues. The approach to Venice struck me, but the place on the whole disappointed me; it has formerly been one of the finest towns in Europe, but others have so far surpassed it, that it is now only the most singular one. We spent seventeen days there busily enough; saw with due attention most of the pictures of Paul Veronese, Titian, and Tintoret, which abound in Venice; and I think I understand these painters' works tolerably; their principal merit you know is in their colouring. I was amazed to find every thing in Venice prodigiously dirty. St. Mark's church, one of the richest in the world in oriental marbles, porphyries, jaspers, &c., as well as mosaics, is so blacked and dirtied in every possible manner, that one thing can hardly be distinguished from another; the taste of it is very bad Gothic; pillars intrinsically worth some hundred pounds each, even to cut in pieces, are piled on one another in all parts without any judgement. We saw Venice in all its glory, it being Ascension time, when everybody resorts there; and we paid for it pretty well, although not so much as I expected, for the price of every thing is nearly doubled at that time. The Doge's marriage with the Sea is a most noble and singular sight; we went in a gondola among thousands perhaps of other vessels accompanying the Bucentaur, which is a splendid ugly thing, not much better than our Lord Mayor's barge, and heard mass with the Doge, &c. at an island about two miles from St. Mark's-place, called Lido. We returned in the Bucentaur itself, as all genteel foreigners were permitted to do, with the Doge, who sat in a throne at the upper end, with the Pope's Nuncio (a shrewd-looking fellow) at his right hand, and the nobles, in crimson gowns and great wigs, about him. St. Mark's fair began on Ascension-day; it lasts three weeks, and is held in an elegant temporary building in the back part of the Place of St. Mark (not the part that appears from the sea); there were all kinds of shops, coffee-houses, &c. under a circular colonnade, and all the world walk there, many in dominos and masks, especially in the evening and all night till day-break; there is a tedious sameness in the amusement, and I never had patience to stay very late. We picked up much curious *Materia Medica* here; and, in searching for the various articles, were amazed at the state of superstition, ignorance and folly in which physic seems to be at Venice. Indeed all science is at a low ebb there; we found nothing like a man of literature, except those apes of them collectors of old useless books. We saw at Venice (what I had long wished for) the making a nun, or the *profession* as it is called, of a young lady of noble family: there was a great deal of company, and more diamonds on the ladies' heads than ever I saw at one time, except at Loretto; the music was very fine: the Italians always treat you with good music on every occasion. The young lady seemed in high spirits, but I should fear a flatness might follow when all the splendour and pomp were over. At the end of a year she is to come out of the monastery for three days, and may then, if she pleases, recant; but if she persists in taking the veil, as is most probable, it is irrevocable. "*Voilà* (said a Venetian near me) *encore une victime des préjugés*."

We left Venice May 28th, and went in the barge (as in Holland) to Padua, which is like a vast rambling village more than a city. The church of St. Anthony is very rich; the church of St. Justinian by Palladio is very fine, and contains a good picture by Paul Veronese. Here is Mr. Ardouino, a celebrated naturalist, with whom I was much pleased.

May 30th got to Vicenza by one o'clock, and saw most of the fine buildings of Palladio's, for which this town is famous, especially the theatre, built after the style of the ancients, and reckoned one of the best pieces of modern architecture.

Next day travelled through a most sweet country to Verona; the road was flat, but Mount Baldus, with its innumerable snow-capped summits, was always on our right hand. Verona is a large handsome town; the amphitheatre is one of the most entire which remain; there are many other curious pieces of architecture, some antiquities, and a few pictures in Verona.

June 2nd we went through Mantua (chiefly remarkable for being the birth-place of Virgil) to Guastalla, where we slept, and next day (Sunday) dined in Parma, which is an extremely handsome well-built town and very neat; it abounds in officers and abbés. The Duke, cousin to the King of Spain, always resides here: we saw the Duchess, a fine stately woman. Correggio's paintings (which are only to be seen here in perfection), although few, are well worth going a great way to see; one or two of them, I think, are among the best pictures of Italy: he excels in grace and in colouring. Here is a noble public library, established by the Duke; the first librarian is father Affò, a capuchin, known by many biographical and historical works. We had a letter to him from the elder Mr. Durazzo, and he has been very civil. The royal printing-office is one of the greatest curiosities in Parma, as that art is here carried infinitely beyond what has ever been done before, or perhaps ever will again; for the genius, the taste, and the application of the director are immense: he gave us some specimens of his work, the most exquisitely elegant things I ever saw. This printing-press is now only beginning to be known in England. The famous theatre, built by the Farnese family when sovereigns here, is the largest in the world; the pit used to be filled with water to represent sea-fights: it is now disused on account of its size and the great expense of light-

never make a false step, though they have often all their four feet together on a loose slippery stone, and sometimes climb up places so steep that 'tis difficult to keep from sliding off their rumps. The weather was cold and windy; for though we rode through myrtle thickets, and among trees which even at Montpellier are kept in greenhouses, and had groves of orange-, lemon-, and palm-trees about us, yet there was ice an inch thick in the road, and snow on the hills at no great distance. The weather is, however, now very fine, rather cold but bright.

On my arrival at Genoa, I called on my friend the Marquis Durazzo, and was a little struck at the magnificence of the house (notwithstanding what I had heard of the Genoese nobility), the ranks of servants, &c.; so that I began to dread some formality and reserve. I was, however, most agreeably relieved from all such ideas the moment I saw him; and I found him the same cheerful, easy, unassuming man as when I knew him in London. He made me most heartily welcome; insists upon all ceremony being laid aside; that he will make us acquainted with all his friends, and procure us access to every thing worthy our notice. Last night he introduced me to a family party; everybody received me in the same agreeable manner, and talked English or French (both equally well), out of politeness to me; while I blundered out some Italian. His family are persons of the first consequence here; their palaces far beyond what we have, except perhaps Chatsworth; and their collections of pictures celebrated in books of travels. A

to me could not be exceeded: hearing I proposed to publish some works of Linnæus, he voluntarily offered all his notes, which will be a great acquisition.

Mr. Ippolito Durazzo is married since I was here, to a very agreeable woman. His father left him and his three brothers above 7000*l.* sterling a-year each, in clear cash, besides the estates to his eldest son.

I have now spent a month at Genoa in cultivating the friendship of many worthy people, and especially the Durazzos. Mr. Ippolito wants me to spend a whole summer here, and takes delight in showing me an apartment in his new house, which he calls mine, and tells me how it shall be fitted up for me; but I assure him this cannot be. He means in a year or two to revisit England with his wife; but she is to learn English first. His nephew Mr. Marcellino, who will be the chief of the family, is also to come to England one day; he is a most charming young man. I really meant this day to have left Genoa, but am now obliged to stay another week, which I hope you will pardon when you hear the reasons. First, Signor Jacomo, Mr. Ippolito's eldest brother, is just returned from a long journey, and I could not with any propriety leave him so soon after. Next, he has brought home great numbers of natural curiosities from France and Switzerland, which he wants me to see. But the chief reason of my stay is, that there being a vacation of public business next week, he and Mrs. Theresina his wife, with Mr. Marcellino and Mr.

Ippolito, go for four or five days to Cornegliano, their country-seat, a magnificent and delightful palace by the sea-side, four miles from Genoa, where their museum is; and they all so earnestly entreated me to go there with them to assist in arranging and naming the various things, that I could not refuse it without being rude and ungrateful. Indeed they merit every attention from me; their friendship is valuable and sincere; their acquaintance is in the highest degree agreeable to me; and people of more elegant, cultivated minds are nowhere to be found.

I remain, dear Sir, &c.,

J. E. Smith.

Dr. Younge being on many accounts in a hurry to get home, we agreed to part at Pavia; and he joined a friend who was going through Switzerland.

Dr. Younge to J. E. Smith.

My dear Sir, Geneva, July 6, 1787.

You will probably be much surprised to learn that I have already passed the Alps, and find myself much at my ease on the borders of the Lake of Geneva.

On the 25th ult. I left Milan, without having had the pleasure of seeing Locatelli. The 26th was entirely employed upon the Lago Maggiore, which I confess did not answer my expectations. We entered from the great lake into the smaller one of Margozza: on the banks of this last I found a plant of the class Gynandria, a *Lobelia*? for I have no botanical guide as heretofore; and the (O my treacherous memory!) *Ros solis*, heretofore called, growing in a micaceous soil; not that this last plant is rare at all, but I was very glad to find it in a particular state, which secures to it a rank among the irritable, if not the sensible plants. I have got three specimens, all of which have small flies inclosed within the leaf as it were, and hurt.

We slept at the inn on the banks of the lake; and the next day, being informed that there was a pass that way, by which the post always went, by Milan to Geneva, we set out on horseback to encounter all the difficulties which might present themselves. Our journey during the first day lay along the plain betwixt the mountains, under trellises covered with vines, whose flowers had strongly the smell of mignonette: slept at Domo D'ossola in the Sardinian dominions. Leaving this early on the second day, we soon began to ascend the hills, rising on each side to a very considerable height above us, and a rapid stream roaring among the fragments of rocks in the valley beneath. The common accidents in mountainous countries was our lot; for a heavy shower coming on about noon almost wetted us through. It cleared up soon, and as we proceeded we found the road almost dry. We entered into the country of the Valais, where we heard no language but a sort of German, to me utterly unintelligible. The higher we ascended, the more difficult we found the passage, which I compared to the passage along the rocks from Nice to Genoa, with this difference, that instead of a sea below, we had a rapid river. The scenes of rude grandeur, and the beauty of the natural cascades tumbling sometimes among woods of pines, and then bursting into day, over the rugged and bare rocks, were inexpressibly pleasing. One cascade in particular, about a league before we arrived at Simpelen, is as fine, I had almost said, as Terni; but the stream is not above half as large, though the fall is higher. We were now at such a height, that the Alpine plants began to present themselves. I found *Alchemilla alpina* in abundance, a *Pinguicula* (an *alpina*?), a *Pedicularis*, and *Geranium*, I think *pyrenaicum*. At Simpelen we supped on excellent milk; and after a breakfast of the same we again set out to mount still higher. Mount Simpelen takes its name I believe from the village at which we slept. The ascent is not sudden, but long. Found a *Gnaphalium*, two *Gentians*, *Primula*, *Pinguicula* in abundance, &c., &c. But what I looked for most was the *Rhododendron*, which at length I found in great profusion: I think it is the *ferrugineum*, if I may judge from the under surface of the leaves.

At the top we found a sort of swampy plain, about a mile and a half over; this being passed, we began to descend rapidly among forests of pine, and by some of the most dangerous passes which we met with, for they were not only narrow, but a shower of rain, which continued for two hours, and wetted us through, had rendered them very slippery. About three o'clock in the afternoon of the third day we arrived at Bryg, situated in the valley at the foot of the mountains, and near the banks of the Rhone. We were taught to expect that carriages might be had here, but we were again obliged to set out on horseback. This day produced nothing in addition to my collection, except a few insects. The road lay chiefly along the banks of the Rhone, whose waters were considerably swelled with the rains; so that we went a little out of our way through the fields, and did not cross the river till near eleven o'clock at night, by a wooden bridge just by the city of Syon, where we slept, and were well accommodated. *Quis tumidum guttur miratur in Alpibus?* is an old exclamation; yet notwithstanding, I could not but wonder at the immensity of the size, and the frequency of the complaint, of the *goitres* in this country: scarcely a woman above twenty years of age in the streets who had not more or less of this deformity, I would call it, though habit may probably have rendered it a beautiful protuberance. The men are by no means exempted from it, though in them it is not so common; but they are subject to what may acknowledge probably a similar origin, viz. swellings of the *parotid* and *submaxillary* glands: those I particularly noticed were on the right side. *Solamen miseris socios habuisse doloris;* so common is it, that no one looks upon the people thus deformed with any sort of idle curiosity, nor are they anxious to prevent or diminish its growth when once it begins to make its appearance. What is the remote cause.

We proceeded from Syon to Martigny, from thence to Villeneuve on the banks of the Lake of Geneva.

The next morning, 3rd of July, took a barque from thence to Vevay, where we breakfasted. Vevay is almost directly opposite to Meillerie;—see Rousseau. We found some difficulty in proceeding by water to Lausanne, all the boats being engaged by the market-people of Vevay. To lose no time we took a voiture, and after a very warm ride among the vineyards, which almost cover the hill sides in the Pays de Vaud, we arrived at Lausanne about dinner-time. Here there is nothing to see except the prospects; and Mr. Tissot, whom I visited on the 4th in the morning, introducing myself as a stranger and a physician: he returned my visit at the inn, the *Lion d'Or*. I have promised for you, who will I think be pleased with him, though he seems to be affected with the complaints of *sedentary people*.

If you are not displeased with me for writing so long a letter, write as much in return.

I am, &c.,

W. YOUNGE.

J. E. Smith to Mrs. Howorth.

My dear Madam, Genoa, July 7, 1787.

You desire me to write to you, and I am sure I neither ought, nor do I wish, to withhold from you that satisfaction. Of poor Mary I never had great hopes: her friends should consider her as one more treasure laid up for them in heaven; a beloved object secured from all possibility of trouble, doubt, or fear; possibly a guardian angel, blest with the means of directing their steps, and smoothing their path to that abode where minds like theirs shall be fully gratified with unembittered love and joy.

You do me the honour to consult me about mental as well as bodily ailments (in gratitude for which I only wish you a better adviser). I shall venture to encourage you about my dear young friend Douglas. Believe me, you have little to fear about him; I say *little*, because a good mother can never be without fear, but I know him perhaps as well as you do. His feeling affectionate heart, and his piety to you, so different from a blind fondness, are the best things you could wish in him: firmness and resolution will come in due time. "Filial piety is of more value than all the incense which Persia offers to the sun;" it is intended by Providence to prepare the mind for religion, which is only the same feeling ripened, and directed to an higher object, which a good parent ought carefully and very judiciously to point out. And here I cannot but observe, what I believe you do not want to have pointed out, that you are in your children one of the happiest people I know: and what troubles will not such comforts alleviate?

I know not what to say to you about Italian poetry: without doubt there is plenty of it, but I feel myself unequal to recommend any. You guess rightly, that I have learned to like Italian music, which I could not *relish* till I had *heard.* As to pictures and statues, I shall tire you to death about them; and with Rome I am quite enchanted. Would I could live two or three years there! I am sorry to say I must disappoint you about views in Switzerland: 'tis too cruel of you to require of me what I cannot do, because I wish never to decline anything which may give you pleasure; but I find myself quite unequal to sketch views myself, or to admire what other people draw, now I have seen so much of nature.

Your reflections on Rousseau and Socrates are excellent, and perfectly just. I think, and so thought Dr. Johnson (who was certainly a good man, although he had his foibles), that a partial concealment is less hurtful than open profligacy, or an unnecessary avowal of weaknesses and transient faults. We have delicious weather, although rather hot, with terrible tempests. A few days ago a steeple was knocked down; but no people are killed here, because they cross themselves at every flash.

I am, dear Madam, &c.

J. E. SMITH.

J E. Smith to Mr. William Jones.

Dear Sir, Genoa, July 7, 1787.

Perhaps you may wonder at my not having sooner answered your last favour; perhaps too you may have done me the honour to be a little displeased, or jealous if you please,—an honour I value more than you think. I am proud to find myself capable of exciting your jealousy, and my reason for saying so is, that I am no stranger to that feeling myself; but there are very few people that I honour with it. Indeed I rather wish never to feel it again, for it is in me connected with such a degree of esteem and affection as scarcely any human being can merit or return; yet the sufferings it occasions are those of a martyr, and bring with them in a great measure their own reward. This is far different from that jealousy which is founded in pride, accompanied (however paradoxical it may seem) with a sense of meanness, which is so common among foolish people, and so general with married persons, who feel themselves unworthy of their partners. I admire above all things your comparison of the tour of Italy to a journey through life,—'tis just what I have felt. I long to know whence you took it, as you seem to say 'tis not your own.

Leaving Rome we came to Loretto by Bologna, which we saw without any diminution of our enjoyment. But Venice disappointed us; its singularity will always be striking, but nothing there is in a good taste, riches are squandered injudiciously, and dirt deforms every thing. The ceremony of the Doge's marriage with the Sea pleased us by its novelty rather than by its pomp. We had scarcely any relish for the fair of St. Mark, which is held in a temporary building in the square, and lasts three weeks; it is lifeless and uniformly tedious. There is a good library at Venice, and indeed all through

Italy we have met with very fine ones. Padua, Verona, Parma, and Milan, although we found ourselves *going down the hill*, were not quite destitute of amusement for us, even with respect to works of art. The theatre of Parma is the finest in the world, so large and magnificent is it as to be useless; it costs too much to light it up, and it is only studied by architects and admired by travellers. I have seen no collections of natural history worth mentioning, compared to what we have in England. At Bologna are some things, and at Pavia more; but the latter has been so plundered by Spallanzani as to be diminished one-third of its value. We spent four days at Pavia with Scopoli, whose civility was very great. We heard Spallanzani lecture; the composition was admirable, but his manner supercilious and affected. Here my friend Younge, in whose company I had been perfectly happy all through my tour, left me; and I, wishing to enjoy all the summer abroad, continued my journey alone to Genoa, to spend a little time with my friends here, for whom I have a great regard. *These* are pleasures which never cloy: and I hope to enjoy them in their full extent at the end of this journey, as I rely on them to make old age comfortable at the conclusion of the journey of life.

I am always, with sincerest regard,

Dear Sir, most faithfully yours,

J. E. Smith.

Dr. Younge to J. E. Smith, Turin.

My dear Sir, Basle, July 17, 1787.

The truth is, that I have been so long used to a participation of the pleasures of travelling with you, that even in my present situation I meet with nothing new or pleasing which I do not wish you to see as well as myself; and though you may probably get information similar to my own from other persons, yet with regard to the particulars of which I shall speak, you will find none more authentic.

After travelling during the whole night along a road rendered very rough with repeated rains, I arrived at Basle on the 15th instant in the morning. The situation of the city, upon the banks of the Rhine, which is about as large and rapid as the Rhone at Lyons, is charming. The inn where I now am, the *Trois Rois*, is built upon the very edge of it; so that the river runs directly under the window. I recommend this inn to you for good bed and good dinner; but bad tea and coffee. As Basle was the residence of the justly celebrated Erasmus (whose writings I have always esteemed from the days that his innocent jokes alleviated my tasks at school), and of Holbein the painter, whom he patronized, my first business was to inquire after them. I knew that there existed in the library some paintings by Holbein, and some manuscripts by Erasmus;—but how to see the library? It was necessary either to signify my desire in writing or in person to Professor Falkner: I chose the latter, and introducing myself as a stranger of curiosity, the professor politely appointed the hour at which I should come, and met me at the library. This is a very good one, having been founded upwards of three hundred years ago; but I slightly regarded every thing except what related to my two favourite objects. There is a room full of paintings and sketches, chiefly by Holbein, who here appears in a much higher line as a painter than I had before placed him (for I never saw his picture of Sir Thomas More's family). The Passion of our Saviour in eight separate histories appears to have been painted for the doors of a small organ, but are nevertheless in as fine preservation as if they had been painted yesterday. The countenances have not that great stiffness observable in most paintings of the year 1520, about which time this work was executed, and the colouring is fine. But I most admire the Last Supper, in which the figures are nearly as large as life, the characters well expressed, and the colouring inferior only to Raphael's.

There is a capital miniature of Erasmus by Holbein, and another picture of him writing, (a profile,) which still seems to think and write: I question whether as true representations of nature as these two last portraits could be outdone. In another room are preserved manuscript letters of Erasmus. The *Elogium Stultitiæ*, with the drawings made by Holbein in the margins. Erasmus's will in his own hand-writing. Manuscripts of the Council of Basle for the suspension of the Pope's authority, &c. Can you read of these things and not wish to see

see them? I forgot a copy of Columna's Phytobasanos, in the same room with the manuscripts of Erasmus. I inquired at the library from Professor Falkner about the family of the Bauhins, and the herbarium, which he told me had been purchased by Doctor de la Chenal, who is director of the botanical garden here. I paid my next visit to him, and was received by him with great courtesy. He purchased for a very trifle Bauhin's herbarium, which had lain neglected in the lumber-room of the house, and from which some parts had been taken; and he has since incorporated the remains of it with his own. He is extremely fond of botany, and gave considerable assistance to Haller in his great botanical work. His library is one of the best private ones I ever saw: witness Dillenii Hist. Muscor. 4to; Columnæ Ecphrasis and Phytobasanos; Raii Synopsis Plant. 1724. Sloane's Jamaica. Gerard; Jacquin; Linnæus (Flor. Lapp., Crit. Botan., &c.)

Jacquin has called a plant *Lachenalia tricolor* after this gentleman. He has given me a coloured plate of it. I talked to him about *you*, and engaged to inform you of his existence, that you might become acquainted with him. He offered to give you all assistance in his power with respect to Bauhin's plants, about several of which Linnæus has mistaken him.

I must not forget one thing. When I talked of Columna's Phytobasanos as rare, he said that he had a duplicate of it. I did not urge the matter further; but I think it rather odd that he should have two copies, and the successor of Linnæus not one.

Could not one be procured by honest means without seeming to beg it? You may see a fine collection of prints, English, French, German, &c., in the house of Mr. Mechel, a very celebrated engraver here;—paintings likewise. The Dance of Death in the churchyard here is much in the style of Holbein, and will bear examination. No botanical books to be got in the shops here. One old crusty fellow wondered I should come to ask for old books at Basil; they had no such things!

I leave this place tomorrow, and expect to sleep at Strasburg. I shall be impatient until I hear from you.

Yours ever,

W. YOUNGE.

J. E. Smith to Dr. Younge, Paris.

My dear Sir, Genoa, July 21, 1787.

I thank you for your long and entertaining letter of July 6th; and to show my apprehension of not writing an answer sufficiently long to satisfy you, I begin as near the top of my sheet as possible; but I am not in the way of writing so eventful an epistle as yours. I have often wanted to see *Drosera* in the state you describe. Your *Lobelia* is probably *L. Dortmanna*, and your *Rhododendron* certainly *ferrugineum*. I will not fail to visit Tissot. Here are many people with goitres in the mountainous country of Genoa. I have been with Dr. Pratolongo, with whom I am much pleased, to a little country-house of his in the mountains, about eight miles off. We staid there two days to botanize; found *Hieracium cymosum* and *Auricula, Linum viscosum*, and many other good things, *at your service*, except the *Linum*, of which I have not a duplicate; Linnæus had it not. We travelled in sedan chairs. Dined in our way with the Senator Durazzo (master of the great house in Strada Balbi*) at his villa of Pino; he is a very courteous, affable old man, and his brother the abbé quite charmed me. I have

* "The palace of the Durazzo family was erected by the celebrated Fontana; the length and elevation of its immense front astonish the spectator, who perhaps can scarce find in his memory a similar edifice of equal magnitude. Besides the rustic ground-floor, it has two grand stories, with *mezzanini*, and over the middle part, consisting of eleven windows, an attic. The portal, of four massive Doric pillars with its entablature, rises as high as the balcony of the second story. The *mezzanini* windows, with the continuation of the rustic work up to the cornice, break this magnificent front into too many petty parts, and not a little diminish the effect of a double line of two-and-twenty noble windows. The portico, which is wide and spacious, conducts to a staircase, each step of which is formed of a single block of Carrara marble. A large antechamber then leads to ten saloons, either opening into one another, or communicating by spacious galleries. These saloons are all on a grand scale in all their proportions, adorned with pictures and busts, and fitted up with prodigious richness, both in decorations and furniture. One of them surpasses in the splendour of its gildings anything of the kind I believe in Europe. These apartments open on a terrace, which commands an extensive view of the bay, with its moles and light-house, and the rough coast that borders it on one side. In this palace the Emperor Joseph was lodged during his short visit to Genoa, and is reported to have acknowledged that it far surpassed any that he was master of. The merit of this compliment is, that it is strictly true."—*Eustace's Classical Tour*, vol. ii.

now completed my *allotted month*, and really conscientiously meant to have gone away this day; but Signor Jacomo returned only on Tuesday last, and he has brought so many shells, corals, &c., that I cannot go without looking them over a little; besides, I did not choose to leave Genoa so soon after his arrival. But this is not all. Next week there is a vacation of five or six days in the public business, and these are to be spent by the family at their country-house; they pressed me so much to pass this interval with them in studying shells, insects, &c. in the museum, and especially the request of Mrs. Theresina had such an effect, that I could not resist. We go to Cornigliano on Monday afternoon, to stay till Saturday, after which I shall set off for Turin. Indeed I am more and more charmed with this amiable family. Signor Jacomo is the most good-natured cheerful man that can be. It is a constant joke against him that he cannot speak my name, but calls me *Smit*, or *Smish*, though his wife takes great pains to tutor him. I spent Saturday and Sunday last at Castagna, a country lodging of Mr. Caffarena, whose wife is an English woman, *worthy of her country:* his sons are very amiable young men; I am quite at home with them. These are all my acquaintances. Mr. Ippolito Durazzo wants me to spend a whole summer here for botanical purposes; to come in February, and stay till October. He takes delight in showing me a room in his new house, which he calls mine, and tells me how it shall be fitted up to receive me; but this scheme is next to impracticable. Pratolongo is very industrious about botany, and knows a great deal of the matter; we are often together. Batt is coming back to Genoa. We have made a catalogue of the plants at the villetta, which amount to about 500; but I hope to increase them much. Yesterday afternoon I spent there very agreeably with only Mr. Cattaneo; he is a most amiable man. Mrs. Lomellina, the senator's wife, has a pretty villa just out of town, by the Lanthorn; I was there the other evening, and never saw the succulent plants so fine, they are really stupendous. From it is the best view of Genoa I know; I mean for a picture of the town; for the view from Mr. Durazzo's villetta is on the whole vastly superior.

Count Durazzo has promised me a copy of his catalogue of prints, with remarks, lately superbly printed by the man at Parma *; I have not yet seen it. Cattaneo promises to reprint his poems, &c., in one volume.

Adieu, my dear Friend,

J. E. Smith.

Mr. Wm. Jones to J. E. Smith, Paris.

Dear Sir, London, August, 1787.

From some circumstances that attended your favour from Genoa, I have been puzzled to determine whether you have thereby more gratified my feelings, or flattered my pride. I cannot assert that I am insensible to the latter, nor will I believe any one who affects the reverse: the only misfortune

* Bodoni.

that can attend that quality in the mind, is a deficiency in the understanding, to direct it to a proper end. I read more of your heart in your last, to engage my esteem, than you ever yet discovered; nevertheless, I had seen sufficient to claim a partiality before. There are many that talk well, that write well, that fill an amiable character in life, and yet destitute of those tender sensations that vibrate through kindred souls, awake to every touch. There is a pride, which I account a folly, that prostitutes itself, and spreads its sails for every breath of applause; so that it can but catch the gale, it cares not from what quarter the breeze springs, favourable or unfavourable, it never examines, and thereby misses the point, its intended goal. I conceive that, like myself, you have been searching through life for a friend,—'tis a word much hackney'd, I mean in its strictest import,—and sometimes thought myself successful; but soon deceived, wearied and tired, I gave over the pursuit, determined to give no one more admittance to my heart, but with indifference view the world, and pass along. I have since found myself more at my ease, but still there remains an aching void that wants a filling, yet better and easier to be borne than the excruciating pangs of ingratitude and disappointment.

I am particularly sensible of your early attention. Mr. Marsham wonders why you should write to me, who did not press you for another letter, and omit him and Dr. Goodenough, who solicited it most anxiously.

The extract I gave you relative to the comparison of the tour of Italy to a journey through life, was from Dr. Conyers Middleton's Letters from Rome. I will lend you the volume; 'twill merit your attention. I travel with you, though with less fatigue, as my elbow-chair is the more easy vehicle; yet shall be happy to hear of those things which you will be able to give a better account of by having seen. You say you hope to enjoy your friends at the end of this journey, and rely on them to make old age comfortable at the conclusion of the journey of life. Now as I reckon myself among the number of your friends, how am I and some others, who may be twenty years older than you, to make your old age comfortable?—many a one would here cry out, a Bull, an Iricism. They know nothing of the matter; you are perfectly in order; for, having I suppose in Italy picked up, by way of novelty, a certain heretical book, you have noticed that length of days do not consist of a number of years, but that "Wisdom is the gray hair unto men, and an *unspotted life, old age.*" But I fear I am here encroaching upon Dr. Goodenough's province, writing a sermon;—be it so, 'tis in the sincerity of my heart. I am sorry to add, Dr. Goodenough (though I love and honour him) has agreed with Mr. Marsham, if I will not unite in their Society, to banish me to Coventry,—a species of severity I know you would not be guilty of; nevertheless I hope to get through this affair with a little address. I have been collecting some Lepidoptera this summer, which I design as a free-will offering to that gentleman. Free-will offerings are what the clergy are fond of, and by that means hope to make my peace with him. Mr. Marsham I will fight in another manner. He has a book in hand*; and I have matter in my head. At Coventry there is no conversation, you know, consequently no communication: I shall therefore soon settle matters with him.

Fabricius is in London, and much wishes to see you, but will certainly leave us before your return; he is going through my drawings, to correct, amend, and add to a Mantissa that he has now in hand; yet I have more than he will be able to accomplish in the time he has limited to stay. I am sorry you are from home for your own sake; he is a man that must please;—open, free, easy, candid, unaffected; in short *I* like him, and think *you* must. Marsham has been at Dunmow in Essex this fortnight, consequently I have heard nothing of Younge.

With every tender of affectionate esteem,

I am most sincerely yours,

W. Jones.

Dr. Younge to J. E. Smith, Geneva.

My dear Sir, Paris, August 5, 1787.

Broussonèt has been extremely obliging to me since my return; he has introduced me at the Academy, and at the Society of Agriculture. I suspect the distribution of medals has ceased in the latter,

* *Entomologia Britannica,* of which the first volume, including the Coleoptera, was published in 1802; but no more ever appeared.

for I saw no such thing. The societies of this country are exactly what you describe; there is a president it is true, and members as in our societies in England; but in the French societies the members seem to have no regard to order, and the power of the president seems to extend no further than to the enforcing a momentary silence by making a more distinct noise than the members by means of a little *musical* instrument. But the great excellence of our societies in England consists in the exact limits drawn and observed between private and public business. How am I interested as a visitor about the votes of Mr. Fourcroy or Mr. Thouin, about this or that committee? yet half the time of sitting is thus spent in demanding opinions and appointing committees on private affairs. There was a communication from Gerard of Cottignac upon *Lathyrus amphicarpos*, read by Desfontaines at the Academy; but the noise and confusion was so great, that I could not understand whether it contained any thing more than the simple relation of its extraordinary fructification.

I am glad to hear of your botanical success, and of your ardour in the pursuit of natural history, which the Durazzo family still support.

I propose to leave this place in two weeks more. What I regret, after you, is that I must bid a last adieu to my very good friends, the amiable family of Mr. De Lessert at Passy. With them I am as happy as if at home. You *must* know them. I have often spoke to them concerning you. Your own name and mine will I trust be sufficient introduction to Mr. De Lessert and to Mr. Guyot. Anybody will show you their house at Passy; and on Wednesdays they dine at their town-house, No. 58 Rue Coqhéron à Paris.

Adieu! All happiness attend you!

W. Younge.

I *almost* envy you the happiness you will experience in this family; and were I not your friend I should be *downright* envious.

Rev. Dr. Goodenough to J. E. Smith, Paris.

Dear Sir, August 13, 1787.

Your silence made us very apprehensive that you were not well; the fact it appears was so. However, we were not a little gratified by reading in your letter that you had been ill but were recovered. The *we* here spoken of, means myself and Marsham, who is at present my chief communicant.—Well, when do you come home? you will find yourself much wanted. Our Natural History Society, under the auspices of our illustrious presidents Fordyce and Pitcairn, drones on in its usual course. I attended one night, and heard Dr. Fordyce read an extempore lecture upon a new arrangement of shells. Such humming, such hawing, mumbling, snuffling, such interruptions in looking for his shells to illustrate as he went on, that not a soul could tell what he would be at. After he had done, I hinted to him that I did not understand what was his aim; he began a second harangue to convince me he was right. The harangue was worse than the lecture,—I was glad of the first opportunity to turn my back. I got acquainted that night with Mr. Wilson of Snow-hill; he agreed with me the lecture upon shells could have been spared. Sticks and stones were exhibited as usual. Old Da Costa sat enjoying every thing said against Linnæus*, having before him some old yellow paper, written very close, which I afterwards understood was a lecture upon the fossil and live *Asteria*. He would not deliver it in to the society, for he meant to publish it himself. Dr. Fordyce's dilatory snorting took up the whole time, an hour (I think I could have said twice as much in ten minutes), so that Da Costa waited for another day. I have taken a capital rarity this year, *Cicada aurita*, said to be a German insect. *Silpha germanica* has also been taken. We have had the luck of taking a great variety of nondescripts, and describing many. We should have done much more, but I have been assaulted five times with blind attacks of the gout. I set out on Saturday for Hastings in Sussex, where I hope to meet with some new insects. With Marsham's assistance and my own assiduity, my

* Da Costa "appears to have taken great offence at not being chosen a member of the Upsal Academy, and conceived an antipathy to Linnæus, which the writer of this has often heard him express, but could never before account for.

"That Academy was always very select in its choice of foreign members, and subsequent events too amply justified its conduct in the present instance."—J. E. S.—See *Selections of the Correspondence of Linnæus, &c.*, vol. ii. 495.

cabinet begins to fill. Jones is at work for me; and before the end of the season hopes, he says, to get me two hundred of the Lepidoptera; so that I shall begin to be somebody. I have lately made a purchase of Allioni's *Stirpes Pedemontanæ,* at a venture; also I have set my writing-master to copy the defect in the second volume of De Geer. (How I like that work!)

Now one has you so near as France, one is apt to think that it is no distance at all. Do pray let us know when you mean to be in England; however, eager as I am to see you, study Tournefort's herbarium well, and my wish is that you would go through the whole a second time. Second thoughts are best. You need not attempt to smuggle any French wine home. It is a commodity which hangs terribly on hand,—so say the wine-merchants; and no one calls for it;—a true picture of John Bull! Tell him he must not have any thing:—Blood and thunder! but he will. Give way to him,—*fastidit et odit.* We are all here upon the tiptoe of expectation about the issue of affairs in Holland. Some positively assert that there will be war: we have no preparations for it. Mr. Pitt is said now to be for it, and a Great Personage against it. A few months ago they were of directly contrary opinions: *Tempora mutantur nos et mutamur in illis.* I hope only to hear that you are alive and well.

And am, ever yours,

S. Goodenough.

I approve entirely all that you say on sights and

saints; I envy you the former. Oh for *Orchis abortiva* and *coriophora!* I am glad you leave Pavia before your tract on the Sexes of Plants comes out: Spallanzani would assassinate you I fear.

Sir Joseph Banks to J. E. Smith, Paris.

Dear Doctor, Soho Square, August 15, 1787.

Botany flourishes here most abundantly. The Queen studies diligently under Aiton, and to much purpose. The ship which we are sending to Otaheite to bring home the bread-fruit to the West Indies, will bring many plants from thence. The garden at St. Vincent's flourishes; a new one is established at Bengal, and an intercourse prepared between the two, by the medium of England; and probably another will soon be established at Madras.

Swartz is the best botanist I have seen since Solander. I have hopes of getting him out to supply Kœnig's place in India. In short, botany may raise its head, and I think it *will.*

Believe me, dear Sir,

Most faithfully yours,

JOSEPH BANKS.

J. E. Smith to his Father.

Honoured Sir, Geneva, August 27, 1787.

Mr. Malanot was so excessively hospitable as to make me uneasy: he would let me pay for nothing, not even postage of letters, nor, what was more, coach-hire. I was obliged several times to have a coach, it being impossible (according to the execrable etiquette of Turin) to make any visit whatever but in full dress; indeed I had many great people to visit. Mr. de Sousa the Portuguese ambassador invited me to meet all the corps diplomatique, and the French ambassador was very civil to me, and invited me to dinner; but I was engaged to see La Superga and La Veneria (a hunting-seat of the King's).

I left my servant at Turin; but I have met with a Milanese lad, whom I esteem such a treasure that I shall bring him to England. He has good friends, and was recommended to me very particularly*.

Geneva swarms with English people; but I have met with no acquaintances except a Professor Zimmerman, who travels with Mr. Harbord, Lord

* The name of the Milanese lad here mentioned was Francesco Borone. Notwithstanding his humble birth and education, and the situation of domestic servant, which he retained for several years, yet his manners and subsequent acquirements elevated him above the menial condition he originally held; and he accompanied Dr. Afzelius to Sierra Leone, and also, at his master's recommendation, after his return thence, attended Dr. J. Sibthorp to Greece; in both instances as botanical assistant and companion. While thus engaged, his existence was unhappily terminated by an accidental fall at Athens, at the early age of twenty-five, in October 1794. To commemorate this faithful and attached dependant, Sir James named a beautiful New Holland genus of plants after his name; and the letters of Dr. J. Sibthorp, which appear in a subsequent chapter, sufficiently attest the estimation in which this young votary of science was held by him.

Suffield's son: they are going to Italy, and I shall give them some letters.

I cannot see this country as it deserves for want of time.

Yours, &c.,

J. E. Smith.

P. S. I have a letter from my Swedish correspondent Dr. Acrel, which informs me that I was very near losing my Linnæan treasures by a plot of Baron Alstrœmer, who wanted to have them, and who procured authority to confiscate the whole after it was sold. How his scheme failed I know not. I had heard some rumour of this in Italy.

The Marquis Ippolito Durazzo to J. E. Smith, Paris.

Monsieur, Genes, ce 10 7bre, 1787.

J'ai réçu avec le plus grand plaisir la lettre dont vous m'avez honoré; j'y trouve des détails qui me prouvent que vous êtes content de votre voyage, et je ne doute point qu'il ne vous doive arriver de même pendant le reste de votre tour de Suisse. C'est un pays de liberté et de situations romantiques, ou les beautés de la nature triomphent, et vous ne pouvez ne pas l'aimer. J'envie le plaisir que vous avez eu sur le Mont Cenis, et je vous suis bien obligé des graines que vous m'avez envoiées. A' propos, ne m'oubliez pas à present que vous êtes à Paris pour me procurer des graines de M. Thouin, et de quelques autres amateurs, et ne laissez pas la commission à M. Broussonet, mais tachez d'avoir le tout pendant votre sejour ici, et de consigner le paquet à M. le Marquis de Spinola, qui me l'enverrà : Broussonet sait ou il demeure. Je vois que, malgré vos extases pour la botanique, vous n'oubliez pas ni vos amis, ni la pauvre Villetta qui vous a mis au rang de ses plus grands benefacteurs. Madame Teresina et ma femme vous remercient infiniment de votre souvenir, et me chargent de mille compliments de leur part.

Mr. Cattaneo et le Docteur Caneferi en font de même, et souhaiteroient de vous voir bientot iç̧i de retour. Mais vous vous eloignez au contraire, et voilà le mal. Voilà sur tout mes regrets quand je me rapelle les momens, sur tout que nous avons passés ensemble à la Villetta dans une société aimable, et à la bonté que vous aviez pour moi. J'espère que vous voudrez bien me la continuer pendant le sejour que vous allez faire en Angleterre. Vous me marquerez, j'espère, tout de suite votre addresse à fin que je puisse entreprendre une correspondence regulière en bon Anglois. En attendant je verrai Mr. Zimmerman et Mr. Harbord, et je vous remercie d'avance du livre qu'ils m'apporteront.

Phaseolus Caracalla is now flourishing in an amazing glorious manner, covering the stair-wall at this Villetta. I shall in a few days pick up a fine specimen for you, and conserve it in dry paper till the time of sending to you the apples. Pray send me some commands that I can have my part in your commissions from Italy. I am with the greatest respect and consideration,

Your faithful Friend and obliged Servant,

IPPOLITO DURAZZO.

Thomas J. Woodward, Esq. to J. E. Smith, Paris.

Dear Sir, Bungay, Oct. 11, 1787.

I received your very friendly and highly entertaining letter on my return from a short tour, in the course of which I had seen your father at Norwich; and from him I heard of your arrival at Paris, and some hints of your agreeable tour to Mont Cenis, which must have been most delightful. I should have highly rejoiced at spending that time with you, and still more at visiting the glaciers. I have lately viewed Saussure's second tome of his Voyage dans les Alpes, which is extremely interesting; and I can fancy myself following your steps in the delightful valley of Chamouny, and up Montanvert to the Glacier du Bois, which I suppose was the one you visited, as your father said you made no very near approach to Mont Blanc, on the summit of which it would be my ambition to set my foot if I visited that country. Saussure must be a very entertaining companion in that tour, of which he seems to have examined every inch. He is a botanist, as appears from his mentioning the scarce plants in several parts of the Alps; but according to him, a part of Mount Jura is the best botanizing ground in the environs of Geneva. I am highly obliged by your promise of some Alpine specimens towards completing the English series, and the very friendly manner in which you mention me will highly enhance their value. We seem to think quite in consonance on this subject; for I do assure you that when I have happened to turn to one of your Scotch specimens during this journey you have been engaged in, the recollection of the donor, and the considering the plant as a *priscæ pignus amicitiæ*, has given me greater pleasure than it would have done to have recurred to the gathering it myself. I pass over Rome and Naples as ground mentioned by every traveller, but your *séjour* at Genoa with your noble friend must have been delicious; and I will indulge you when we meet in talking of Genoa as much as you can desire, provided one of those enchanting and accomplished sisters I have heard of, has not such a share in it as to divert your thoughts from your native country; for that same native country must now employ your thoughts and attention, unless you can find time for another trip.

When I go to Norwich I shall show Pitchford your letter and account of *Melampyrum sylvaticum* (which I much doubt if it has ever been found in England); but he will not now require your oath, either on the Cross or Bible, as he is at length convinced of the other being the *pratense*. The grand difficulty was to persuade him that, though Hudson had quoted Ray, it did not necessarily follow that Hudson and Ray meant the same plant; for you know he looks upon Ray's infallibility as only second to that of the Pope; and you and I, without thinking Ray so absolutely infallible as he does, may set him still higher than His Holiness in that point of view.

I have little botanical news to tell you; but you will be glad to hear that Crowe found three specimens of *Ophrys Loeselii* on St. Faith's Bogs this summer: they were far distant from the spot on which Pitchford found his, and Crowe left them untouched; they were growing on the very wettest part of the bog, and actually in the water. Mr. Sole of Bath has found several on Hinton Moor near Cambridge, where Ray mentions their growing. Roots have been sent to Curtis and to Dickson, and are grown in Curtis's and the Museum garden.

I understand the full force of your expression "*really new*" in regard to Dickson's Scotch discoveries. I do not look upon it as very clear that all are so; but as he has the game entirely in his own hands, he must play it as he likes. I understand he has said he almost wishes he had never found them, as he finds such great difficulty in making them out to his satisfaction. I wish the discoveries do not more confuse than elucidate these difficult tribes of plants. I wish much to gain a knowledge of the *Fuci*, &c. which are at present very little known, and on which few authors have written; for, except Gmelin and Lightfoot, I know none from whom much information is to be gotten. In my late tour I stayed a day at Cromer for low water, but I found little there. The *Ulvu plumosa* of

Hudson was growing on the stones, and makes a beautiful appearance in the small puddles left by the recess of the tide, representing a miniature cypress-tree, of about two or three inches high, growing in the water. It is improperly described as pinnated, as the *pinnæ* grow all on the stem like a tree ; *cypressiformis* would be a much more proper name for it. The *Ulva fistulosa* of Hudson, the *Fucus verrucosus* of Gmelin (which seems to be the *albidus* of Hudson), and some other common ones were growing. Do you know any authors who have written much on these subjects, besides those I mentioned ? I hope if you do not procure Leers before you return, that you will contrive a correspondence with the Paris bookseller, that you may receive it when he can procure it. I shall be obliged to you for the 4th Fasciculus of Retz. The most useful matter I found in him was the clearing up *Astragalus arenarius ;* though I am by no means clear that our plant is his *danicus*, as Withering gives it ; it certainly is not *arenarius*, Linn. My commissions give you a great deal of trouble, but I know your friendship will excuse it.

Believe me most sincerely yours,

T. Woodward.

J. E. Smith to Dr. Younge.

Dear Doctor, Paris, October 13, 1787.

I left Turin August 12th, with Mr. de Sousa, the Portuguese minister, Dr. Bellardi, &c. for Mont

Cenis, where I stayed six days, lodged at the hospital most comfortably;—the rest of the party in a tent with the Chevalier St. Real, intendant of Morienne, who with an officer had been six weeks in that place making philosophical observations of various kinds. The hospital stands 996 toises above the sea. One day we ascended little Mont Cenis, 558 toises higher, but I did not go to the top. Nothing could be more charming than the plain of Mont Cenis; it was all flowery with the choicest alpine plants; and I rambled about every day among the neighbouring hills and thickets, loading myself with treasures for myself and my friends, among whom you need not fear being principally remembered. The weather was delightfully cool after the suffocating heats of Turin. The fine air, good milk, and trout, pleased me extremely. The abbé, who governs the hospital like a patriarch of old, was as hospitable as possible, and my companions extremely agreeable. After our long walks we reposed on a turf of *Dryas octopetala*, and Mr. de Sousa never failed to keep us in perpetual cheerfulness with his good-humoured sallies. I stayed two days longer than the Turin party; my Milanese lad overtook me here, and I left Mont Cenis in company with the Chevalier de St. Real and his companion. We passed slowly through Savoy, visiting many of his acquaintances by the way, and were very hospitably received. At one place we met with a magnetizer, at whose whimsies, when not too contemptible, we laughed among ourselves. At length we arrived at St. Jean de Morienne, the little capital of the country, where I stayed two days at the house of the Chevalier, and then left him with regret; he is a very intelligent friendly man.

The manners of this country reminded me of old English hospitality, which we know only by report; but cleanliness is not here considered as a first-rate virtue. At Geneva I saw Messrs. Bonnet, De Saussure, Senebier, De Luc, &c. &c. I scarcely ever met with so many scientific people together. Bonnet is a most interesting man, all complaisance and enthusiasm, and very communicative; but, alas, almost deaf and blind! He is not like the vulgar traducers of Linnæus, but allows him all his merit, and has his portrait in his own chamber, with a few other first-rate naturalists. De Saussure is a rough man, and at sword points with De Luc. Dr. Butini senior is one of the most pleasing and apparently ingenious physicians I ever saw; and his son, to whom Batt gave me a letter, is a very clever man, in whose company I was very happy during the week I spent at Geneva.

I visited the glaciers about Mont Blanc, had very fine weather, and another rich botanical harvest. Passed by Martigny, Bex, &c. over the lake to Lausanne, which I need not describe, as it is the road you took. Tissot received me very civilly, and gave me his little publication on the Vapours. I was much pleased with Berne, and the Bears, to whom I hope you paid your compliments. I visited the unmarked grave of Haller, and got acquainted with his son; but nobody at Berne pleased me so much as Wyttenbach the clergyman, whose acquaintance

I shall earnestly cultivate. He is a good naturalist, has a very pretty museum, and is a most pleasing companion. He gave me a letter to Mr. la Chenal at Basle, with whom I was pleased, no less for his knowledge and manners, than for his politeness in presenting me with my grand desideratum, the *Phytobasanos,* for which I ought indeed to thank you, as I suspect he had put it by for me in consequence of what you said; at least he took an early opportunity to show me his own copy, and when I said I had long sought that book in vain, he immediately produced the duplicate. He promised to help me all in his power with the *Syst. Veget.*, and will inform Schreber and Jacquin that I mean to publish it.

I saw all the venerable relics of Holbein and Erasmus which you mentioned to me, and was much pleased with the three days I spent at Basle. At Strasburg I passed two days entirely with Professor Herman, one of the best and most zealous naturalists I have met with.

At Nancy I saw Mr. Willement, a good botanist, from whom, as well as from most of the other people I have mentioned, I expect assistance for the Linnæan Society. In the greenhouse at Nancy is the bust of Stanislaus the banished king of Poland, to whom Louis XV. gave this province for a refuge. Under it are the two lines which I have heard you quote as written for Louis XIV., and I wish to know whether they were so, or really original at Nancy.

> "Inter - - - - - - - succosque salubres,
> Quam benè stat populi vita salusque sua."

Pray send me them complete at your leisure, as I have no copy of them.

I got to Paris, September 19th, and am employing myself at the Cabinet du Roi, where my work is nearly finished. I cannot too highly thank you for procuring me the acquaintance of the De Lessert family, of which however I have hitherto only seen the female part, the gentlemen being all at Lyons at present, but they are expected home soon. Parr, who knew the young men at Edinburgh, took me one morning in his whiskey to Passy; Mrs. De Lessert invited us to dinner a few days after, and I never spent a more agreeable day. I looked over the herbarium of Rousseau with great pleasure. I find his letters on botany were addressed to Mrs. De Lessert.

I am very much pleased with my servant in every respect; he is a remarkably clever lad, and at present very good. He is beginning to learn drawing, and succeeds in everything he attempts. Broussonet is gone into the country this morning on some agricultural business.

Here are great reforms going on. L'Ecole Militaire is to be abolished, and the house made an hospital. We have been at Versailles and saw the King and Queen, but the waters have ceased playing for the winter,—a great disappointment.

Adieu, my good friend; believe me I shall ever think with pleasure on the time we have spent together. I could scarcely have hoped that so long a connection of the same kind with any one would have left me so many pleasing ideas.

J. E. Smith.

Dr. Younge to J. E. Smith, Chelsea.

My dear Doctor, Liverpool, Nov. 7, 1787.

Your good fortune in meeting with the most intelligent men in the different towns of France through which you passed, I rejoice at. I did not see any gentleman at Nancy, nor the greenhouse which you mention, though I spent nearly a whole morning in visiting the different parts of that charming town. I am glad to be particularly informed respecting an inscription, which I remember only as related by my father, and, as it was always my idea, applied by an English gentleman, then on a tour in France, to Louis XIV., on an inscription being publicly requested for his statue.

This supposed statue I never saw, and therefore suspect the inscription might be originally intended for the place where you saw it. It is thus:

"Inter vitales herbas, succosque salubres,
Quam benè stat populi vita salusque sui."

The wit of this couplet consists, it may be truly said, in a verbal pun; but there is a smartness and elegance in the conceit, which after ten years acquaintance with I still continue to admire. I am happy to be able to furnish you with what may be a correct copy of this couplet.

Your affectionate friend,

W. Younge.

In November 1787, Sir James returned to his native country; and in 1793 he published his "Sketch of a Tour on the Continent," in three volumes, octavo*.

His enthusiastic admiration of whatever he saw beautiful in nature and excellent in art, for which he appears to have possessed an intuitive discrimination, renders his "Sketch of a Tour on the Continent" one of the most engaging narratives of the kind which has come from the press. It attracted the approbation of every one who took it up; yet by a fatality often attending what is excellent, it appears to be less known than it deserves. There is something very romantic in the description of his approach to Genoa.

"Traversing," he says, "majestic cliffs among groves of olive- and carob-trees, and thickets of oleander and myrtle,

> "'I felt as free as nature first made man,
> When wild in woods the noble savage ran.'"

His entrance by the gate of St. Thomas, walking alone into the Strada Balbi, to the palace of the father of his friend the Marquis Ippolito Durazzo, —his kind, yet magnificent reception, the state in

* The author's father left the following observation upon this work, in his common-place book.

"A Sketch of a Tour on the Continent in the Years 1786 and 1787, by James Edward Smith, M.D., F.R.S., Member of the Royal Academies of Turin, Upsal, Stockholm, Lisbon, &c. &c. President of the Linnæan Society." In 3 volumes.

"I must leave the world to judge of the merits of this work of my most dearly beloved son; it is for me to contemplate them with a tender father's partiality in silence."

which they lived, give an impression which reminds the reader of some characters in the novels of Richardson.

Under the protection of this family Sir James had the opportunity of seeing the great council and the final election of a Doge, the only occasion on which strangers can be admitted at all to that august assembly.

While wandering among the classic scenes of Italy, his admiration for the works of art, whether of painting or sculpture, was frequently called forth, and here his enthusiasm breaks out in language which inspires a similar enthusiasm in the reader. The strength of a first impression will be shown in his sensations when in the chapel of the Albergo : " My attention," he observes, " was entirely withdrawn from a fine piece of sculpture I had been looking at, by an accidental glance to the left, where another sculpture over a small altar riveted my eyes and every faculty of my mind, in a transport of admiration and tender compassion as fervent as ever Mrs. Siddons herself excited. This was no other than the bas-relief by Michael Angelo Buonarotti. The subject consists of two heads, about the natural size, of a dead Christ, and his mother bending over him. Words cannot do justice to the expression of grief in the Virgin. It is not merely natural in the highest degree, 'tis the grief of a character refined and softened above humanity. The contemplation of it recalls every affecting scene, every pathetic incident of one's whole life. Those who have watched all the agonizing turns of countenance of the great actress above mentioned, in the parts of Isabella and Belvidere, can alone form a conception of the wonderful effect of this marble. In contemplating it, every exquisite variety of that expression seems to pass in turn over its breathing features. The reader must pardon my enthusiasm. This was the first truly fine piece of sculpture I ever saw. I had not before any conception of the powers of art. I shall have few occasions of re-lapsing into such rapture."

The following, though but a translation, yet only a mind inspired with the impression could have done justice to it. "Conscious," the writer tells us, "of my own inability to furnish adequate conceptions upon the subject, I beg leave to offer Winkelman's *Hymn*, as De la Lande has well denominated it, in honour of the famed Apollo. It is a curiosity which has not appeared in English; and while it exemplifies the genius of its author, will afford a conclusive proof that the study of antiquities is not always a dry one."

Winkelman's Hymn to the Apollo Belvidere.

"Of all the productions of art which the ravages of time have spared us, the statue of Apollo is indisputably the most sublime. The artist has conceived this performance from ideal being, and has made so much use of matter only as was necessary to execute and give a body to his thought. As far as the description of Apollo in Homer surpasses those descriptions which other poets have made after him, so far does this figure exceed all other figures of the same divinity. His stature is more than human, and his attitude expresses majesty. An eternal spring, like that of Elysium, diffuses the beauty of youth over the manly perfection of his frame, and gracefully displays itself in the noble configuration of his limbs. We must endeavour to penetrate into the empire of incorporeate beauty; seek to become creators of a celestial nature, in order to elevate the soul to the contemplation of such supernatural perfection; for here is nothing mortal, nothing subject to the wants of humanity. This body is neither warmed by veins, nor agitated by nerves. A celestial spirit, diffused like a gentle stream, circulates, if I may so express myself, over the contour of this figure. He has pursued Python, against whom he has bent, for the first time, his tremendous bow; in his rapid course he has overtaken him, and given the mortal stroke. In the fullness of sublime satisfaction, his august aspect, penetrating into infinity, extends far beyond his present victory. Disdain is impressed on his lips; the indignation he breathes inflates his nostrils, and mounts even to his brow. But unalterable peace is seated on his forehead, and his eye is all sweetness, as if he were now surrounded by the Muses, eager to offer him their caressing homage. Among all the representations of Jupiter that have reached us, there are none in which the father of gods seems so nearly to approach that dignity, in which he once manifested himself to the intelligence of the poet, as in this portrait of his son. The individual beauties of all other deities are assembled in this figure, as in the divine Pandora. This brow is the brow of Jupiter pregnant with the goddess of Wisdom, and its movement announces its will. These eyes, in their fine-turned orbits, are the eyes of Juno; and this mouth is the same that inspired the beautiful Branchus with voluptuousness. Like the pliant branches of a tender vine, his lovely hair waves around his divine head, as if lightly agitated by the breath of Zephyr; his locks seem perfumed with ethereal essence, and negligently attached at the summit by the Graces themselves. When I behold this prodigy of art, I forget all the universe, I assume a more dignified attitude, to be worthy to contemplate it. From admiration I pass into ecstasy. Penetrated with respect, I feel my bosom heave and dilate itself, as in those filled with the spirit of prophecy. I am transported to Delos, and the sacred groves of Lycia once honoured by the presence of the god; for the beauty before me seems to acquire motion, like that produced of old by the chisel of Pygmalion. How is it possible to describe thee, thou inimitable masterpiece, unless I had the help of ancient science itself to inspire me, and guide my pen! I lay at thy feet the sketch I have rudely attempted; as those who cannot reach the brows of the divinity they adore, offer at its footstool the garlands with which they would fain have crowned its head."

The reader who has not before seen this animated description will not consider it obtrusive here; and the writer feels she shall be treated with indulgence if she speaks with enthusiasm of the volumes which first disclosed to her knowledge the taste and character of their author: a poetic spirit breathes through these, and through all his writings, and gives a charm which is felt even in the strict language of scientific description.

"Poetry," says a fine modern writer, "has a natural alliance with our best affections. It delights in the beauty and sublimity of the outward creation, and of the soul. Its great tendency and purpose is to carry the mind beyond and above the beaten, dusty, weary walks of ordinary life; to lift it into a purer element, and to breathe into it more profound and generous emotions. It reveals to us the loveliness of nature, brings back the freshness of youthful feelings, revives the relish of simple pleasures, keeps unquenched the enthusiasm which warmed the spring-time of our being, refines youthful love, strengthens our interest in human nature by vivid delineations of its loftiest feelings, spreads our sympathies over all classes of society, knits us by new ties to universal being; and through the brightness of its prophetic visions, helps faith to lay hold on the future life. It is objected to poetry, that it gives wrong views, and excites false expectations of life, peoples the mind with shadows and illusions, and builds up imagination on the ruins of wisdom. That there is a wisdom against which poetry wars,—the wisdom of the senses, which makes physical comfort and gratification the supreme good, and wealth the chief interest of life,—

cannot be denied; nor is it the least service which poetry renders to mankind, that it redeems them from the thraldom of this earthborn prudence. It is good to feel that life is not wholly usurped by cares for subsistence and physical gratifications, but admits, in measures which may be indefinitely enlarged, sentiments and delights worthy of a higher being." This refinement pervaded his whole character, gave a charm to his domestic habits and social pleasures, which stood in place of the luxuries of fortune, and surpassed them.

"There is one subject," Sir James observes in the preface to his Tour, "which commonly makes a conspicuous figure in all travels to Italy,—the absurdities and abuses of the Catholic religion. On this head many a Protestant writer seems to think himself privileged to let loose every species of sarcasm, censure, and calumny, without any qualification or distinction. He censures a pretended infallible church, as if he himself and his own mode or fashion of belief alone were really infallible: he condemns a persecuting religion, while he himself persecutes it more uncharitably and unrelentingly with his pen or his tongue, than any churchman ever did heretic with fire and faggot; and he execrates those who keep no faith with unbelievers, while he betrays the confidence of friendship and hospitality, and perverts the kindness of human nature, (which gets the better even of religious antipathies,) into a tool of ridicule against those who have exercised it in favour of himself. These errors, by far more disgraceful and blameable than errors of faith, he has earnestly wished to avoid."

After such a declaration, which is justified by the forbearance, or rather disinclination, to make any offensive remarks on the religion of a country among whose members he had met with acknowledged hospitality and great kindness, he nevertheless was so unlucky as to incur very serious reprehension, from an unknown hand, upon this very subject and some others; all contained in a long letter sent him two or three years after the publication of his Tour.

This letter, however amusing it occasionally proved to the person to whom it was addressed, might be considered too long to insert, and as Sir James never knew who was the writer of it, it cannot be said that he lost a friend; but he was so unfortunate as to wound the feelings, very undesignedly, of our late Queen Charlotte, by an expression in speaking of the Queen of France, which seems never to have been obliterated from Her Majesty's remembrance, and occasioned the withdrawing of her favour, although the offending epithet was expunged in the second edition of his Tour.

It happened some time in the year 1791, that Sir James's friend, Dr. Goodenough bishop of Carlisle, being about to write a botanical paper on the British species of *Carex*, had occasion to consult the herbarium of Mr. Lightfoot. This had been bought by His Majesty George III. on the death of its original possessor, and presented to the Queen. Dr. Goodenough obtained permission to examine it:

the Queen, being present when he went to Frogmore, conversed with Dr. G. on the subject of his studies. He found the herbarium very much damaged, and recommended Her Majesty to have it looked over by some intelligent person, mentioning Mr. Dryander and Sir James Smith, as either of them capable of advising some method of preserving what remained. The name of the latter was not unknown to the Queen; he had some time before presented her, through the kindness of the Hon. Mrs. Barrington, with a copy of his Coloured Figures of Rare Plants, which both Their Majesties were pleased enough with to desire to become purchasers of three copies besides; and in one of Her Majesty's familiar visits to the late Viscountess Cremorne, the Queen carried her a copy as a present.

Lady Cremorne, perceiving with agreeable surprise that the author of this new work was a person she had honoured with her friendship so far as to have fixed on him as an acquaintance and companion for an only and beloved son, the consequence of the rencontre was, that when Dr. Goodenough gave the above-mentioned hint to Her Majesty, she instantly fixed upon him, for the purpose not merely of arranging Lightfoot's herbarium, but of conversing with herself and the Princesses on the elements of botany and zoology. These visits were at all times remembered by him with considerable pleasure; and he never spoke without veneration of the character of the Queen, her care to afford her daughters every opportunity of acquiring information, and to furnish them with the means of filling their time with worthy objects of pursuit.

Sir James sincerely esteemed his royal patroness, and was flattered to find his services acceptable to her. In every conversation the benignity and cultivated understanding of the principal personages removed every sensation of awe, and royalty itself lost none of its dignity in the polite and accomplished companions.

With these impressions he took leave of his illustrious pupils, suspecting no ill, and designing nothing disloyal, nor was it likely that any passages in his Tour should have given offence at any other time. But the events of the French revolution had filled the minds of many with suspicion and alarm: what he had said was represented to him "as injurious in these times to crowned heads;" and a passage wherein he is accused of eulogizing Rousseau, was regarded as "hostile to religion, virtue, and loyalty."

There can be no doubt that Her Majesty's mind was prejudiced against him by one who had been a mutual friend, but whose personal contests with Rousseau had warped his judgement.

That Sir James regretted this alienation cannot be matter of surprise to those who have felt what it is to be misunderstood, and who recollect that his sentiments of regard for those who had shown him kindness were no less warm than sincere. He had no other reason to regret the circumstance, for it was a disinterested connection entirely.

To the friend above alluded to, he replied, "If you consider calmly what I have said of Rousseau, you will find it rather an *apology* than an *eulogium*, and cannot be understood to palliate any of his faults or mistakes. What I have said of the unhappy Queen of France in vol. iii. p. 217 and 218, is the most favourable apology that can be made consistent with truth and the sacred interests of virtue. The other expression* I regret, and will correct it."

Those who have not seen these passages may have a curiosity to peruse what at the present day would scarcely be considered hostile to religion and virtue, likely to produce alarm, or to be construed into disaffection.

Extract from the Tour, vol. iii. *p.* 217.

"Of her political faults during her prosperity, I presume not to form an idea; for who could dive into the intricacies of one of the most intriguing of all courts? Her subsequent conduct, her plots as they are called, her *treason* against her oppressors, none that can put themselves into her situation will wonder at or blame. Her private faults I will not palliate. They were but too well known, when she was in a situation that might be supposed out of the reach of all justice, except the divine; but they will not fail now to be blackened, no doubt, where that can be done. Let it however be remembered, that the state prisons revealed no secrets to the dishonour of this unfortunate Queen, no victims of her jealousy or resentment, though they were often filled with those of the worthless mistresses of for-

* The epithet of Messalina applied to the Queen of France.

mer kings. The canting Madame Maintenon spared no pains to entrap and to confine for life a Dutch bookseller, who had exposed her character: but Marie Antoinette took not the least vengeance of the most abusive things, written and published by persons within her own power."

Vol. i. *p.* 110. "With respect to the character of Rousseau, about which the opinion of the world is so much divided, I have found it improve on a near examination. Every one who knew him speaks of him with the most affectionate esteem, as the most friendly, unaffected and modest of men, and the most unassuming in conversation. Enthusiastically fond of the study of Nature, and of Linnæus as the best interpreter of her works, he was always warmly attached to those who agreed with him in this taste. The amiable and accomplished lady* to whom his Letters on Botany were addressed, concurs in this account, and holds his memory in the highest veneration. I have ventured to ask her opinion upon some unaccountable actions in his life, and especially about those misanthropic horrors and suspicions which embittered his latter days. She seemed to think the last not entirely groundless; but still, for the most part to be attributed to a something not quite right in his mind, for which he was to be pitied, not censured. Her charming daughter showed me a collection of dried plants, made and presented to her by Rousseau, neatly pasted on small writing-paper, and accompanied with their Linnæan names and other particulars.

* Madame de Lessert.

"Botany seems to have been his most favourite amusement in the latter part of life; and his feelings with respect to this pursuit are expressed with that energy and grace so peculiarly his own, in his letter to Linnæus*, the original of which I preserve as an inestimable relic. I need offer no apology to the candid and well-informed reader for this minuteness of anecdote concerning so celebrated a character. Those who have only partial notions of Rousseau, may perhaps wonder to hear that his memory is cherished by any well-disposed minds. To such I beg leave to observe, that I hold in a very subordinate light that beauty of style and language, those golden passages, which will immortalize his writings; and a faint resemblance of which is the only merit of some of his enemies. I respect him as a writer eminently favourable on the whole to the interests of humanity, reason, and religion. Wherever he goes counter to any of these, I as freely dissent from him; but do not on that account throw all his works into the fire. As the best and most religious persons of my acquaintance are among his warmest admirers, I may perhaps be biassed in my judgement; but it is certainly more amiable to be misled by the fair parts of a character, than to make its imperfections a pretence for not admiring or profiting by its beauties. Nor can any defects or inconsistencies in the private character of Rousseau depreciate the refined moral and religious principles with which his works abound.

* Published in the "Selection of the Correspondence of Linnæus."

Truth is truth wherever it comes from. No imperfections of humanity can discredit a noble cause; and it would be madness to reject Christianity, for instance, either because Peter denied Christ, or Judas betrayed him.

"It will be hard to meet with a more edifying or more consolatory lecture on religion than the deathbed of Julia. Her character is evidently intended as a model in this respect. By that then we should judge of its author, and not by fretful doubts and petulant expressions, the sad fruits of unjust persecution, and of good intentions misconstrued.

"Nor would it be difficult to produce, from the works of Rousseau, a vast majority of passages directly in support of Christianity itself, compared with what are supposed hostile to it. It is notorious that he incurred the ridicule of Voltaire, for exalting the character and death of Jesus above that of Socrates. 'But he was insidious, and he disbelieved miracles,' say his opponents. If he believed Christianity without the assistance of miracles to support his faith, is it a proof of his infidelity? If he was insidious, that is his own concern. I have nothing to do with hidden meanings or mystical explanations of any book, certainly not of the writings of so ingenuous and perspicuous an author as Rousseau. Unfortunately for him, the whole tenour of those writings has been too hostile to the prevailing opinions, or at least to the darling interests of those in authority among whom he lived; for Scribes and Pharisees are never wanting to depress every attempt at improving or instructing the world, and the greatest heresy and most unpardonable offence is always that of being in the right. For this cause, having had the honour of feeling the vengeance of all ranks of tyrants and bigots, from a king or bishop of France, to a paltry magistrate of Berne or a Swiss pastor, he was obliged to take refuge in England. Here he was received with open arms, being justly considered as the martyr of that spirit of investigation and liberty which is the basis of our constitution, and on which alone our reformed religion depends. He was caressed and entertained by the best and most accomplished people, and experienced in a particular manner the bounty of our present amiable sovereign.

"One cannot but lament, that one of the most eminent, and I believe virtuous, public characters of that day, should of late have vainly enough attempted to compliment the same sovereign, by telling him he came to the crown in contempt of his people, should have held up a Messalina for public veneration, and become the calumniator of Rousseau!

"It is, indeed, true that a certain morbid degree of sensibility and delicacy, added to the inequalities of a temper broken down by persecution and ill health, made Rousseau often receive apparently well-meant attentions with a very bad grace. Yet, from most of the complaints of this kind, which I have heard from the parties immediately concerned, I very much suspect he was not unfrequently in the right. But supposing him to have been to blame in all these instances, they occurred posterior to his most celebrated publications. Was it not very unjust, therefore, for those who had patronized and extolled him for those publications, to vent their animosity against *them* for any thing in *his* conduct afterwards?

"Far be it from me, however, to attempt a full justification of his writings. I only contend for the generally good intention of their author. The works themselves must be judged by impartial posterity. I merely offer my own sentiments; but I offer them freely, scorning to disguise my opinion, either because infidels have pressed Rousseau into their service, or because the uncandid and the dishonest have traduced him falsely, not daring to declare the real cause of their aversion,—his virtuous sincerity."

When compassion for the sufferings of another mingles with similitude of feelings, especially on some favourite point, it never fails to produce a strong partiality. The same love of Nature, the same passion for her works, the same rapture in beholding the sublime scenery of the Alps, characterized both these men; and in reading, as Sir James sometimes did, the 7th Promenade of the celebrated author of "*Julie*," he saw, as it were in a mirror, his own sensations reflected.

"Toutes mes courses de botanique, les diverses impressions du local, des objets qui m'ont frappé, les idées qu'il m'a fait naître; les incidens qui sont mêlés, tout cela m'a laissé des impressions qui se renouvellent par l'aspect des plantes herborisées dans ces mêmes lieux. Je ne reverrai plus ces beaux paysages, ces forêts, ces lacs, ces bosquets, ces rochers, ces montagnes dont l'aspect a toujours touché mon cœur ; mais maintenant que je ne peux plus courir ces heureuses contrées, je n'ai qu'a ouvrir mon herbier, et bientôt il m'y transporte. *Les fragmens des plantes que j'y ai cueillies suffisent pour me rappeller tout ce magnifique spectacle.* Cet herbier est pour moi un journal d'herborizations, qui me les fait recommencer avec un nouveau charme, et produit l'effect d'un optique qui les peindroit derechef à mes yeux."

> "Clarens! by heavenly feet thy paths are trod,
> Undying Love's, who here ascends a throne
> To which the steps are mountains; where the God
> Is a pervading life and light—so shown
> Not on those summits solely, nor alone
> In the still cave and forest: *o'er the Flowers*
> *His eyes are sparkling, and his breath hath blown.*"

There were, however, others who saw the sentiments put forth in these volumes in their genuine light, in the sense intended by their author: Dr. Pulteney will not be accused of disloyalty, nor of any latitude of opinion beyond what has been the distinguishing mark of a British subject, true to every part of the institutions of his country; though he could not exceed the writer of these travels in his patriotic spirit.

Dr. Pulteney to J. E. Smith.

Dear Sir, Blandford, January 30, 1794.

Mr. White sent me down, about a month ago, your Tour on the Continent, and I take an opportunity, by means of a frank to my friend Dr. Garthshore, to send you these few lines, to thank you for the entertainment I have received in the perusal of it. This work will, I hope, exalt your fame, and contribute to your emolument. Indeed I cannot doubt it, as you have so judiciously proportioned your information to such a variety of readers. You will not wonder that *I* should have wished to have seen more of natural history; but I am well aware how proper it was not to introduce too much of any particular branch of knowledge, since the general taste was to be consulted, to make a work popular.

Your mode, however, of introducing what you have done on our favourite subject, is so judiciously managed, that I hope it will much tend to recommend it. This is not all I wish to applaud. Be assured, my dear Sir, that I am delighted with that spirit of freedom, and zeal for the cause of liberty, which so eminently appears throughout these volumes: your sentiments are so congenial to my own, that I honour you for the expression of them, and for the spirit you have shown on this occasion, as well as for the *prudent discrimination* you have made *every where* betweeen *Liberty* and *Licentiousness*, at a time when our neighbours are discrediting their principles, and injuring the best of causes, by their savage proceedings. I ever thought they had neither virtue nor religion enough to deserve, at present, or to attain, the noble object they aimed at.

I am, dear Sir, with true esteem,
Your faithful humble Servant,
R. PULTENEY.

Thomas Pennant Esq. to J. E. Smith.

Dear Sir, Downing, March 9, 1794.

Your travels were continued to Holyhead, and are but just returned, otherwise I should have sooner made my best acknowledgements.

From the part I have read I have little doubt of the satisfaction I shall have in the perusal to the last page. I opened accidentally on your account of the French, which suits my sentiments most exactly. I detest the savages; but think every virtuous man must be struck with horror at the abandoned manners of Louis XV. and his court. Could reformers tell where the proud waves of correction should stop, it would be happy: but the cruel inundation has made havoc indiscriminate. But Providence for wise ends often suffers its judgements to pass over the just and unjust. When we weep when we hear of the axe falling on the merciful Louis XVI., how do we exult when we see under it Brissot and a number of the other regicides.

I thank you for the fine drawings. Adieu.

Yours most faithfully,
THOMAS PENNANT.

Among the friends whom the publication of his Tour procured him must be mentioned Colonel Johnes, of Hafod. To his secluded residence in the Alpine wilds of Cardiganshire, Sir James had the happiness of making several visits, and enjoying the society of the master of that elegant mansion, whose historical researches, whose residence in many of the European courts, whose acquaintance with general literature, and whose knowledge of the world, made his companionship delightful. Here too Sir James directed the taste and encouraged the pursuit of natural history among the other acquirements of his amiable daughter, a child in years, but mature in mind at a very early period. Miss Johnes's genius for music was extraordinary at the age of ten years: but admirer as he was of Handel and of Italian airs, in both of which she excelled in no common degree, the greatest charm was to hear her fine voice among those mountain paths, singing the native songs of the country, as they botanized together on the hills; and of these the solemn music of *Morfa Rhyddlan* was the favourite of both:

> "Wild on old Havren's banks the modest violet blooms,
> And wide the scented air its fragrant breath perfumes;
> Bright shines the glorious sun amid the heaven,
> When from its cheering orb the clouds are driven;
> A form more beauteous still adorn'd the flood,
> Gwendolen's fatal form, Llewellyn's blood!"

This charming villa, containing within itself all that could gratify the taste,—inaccessible to intrusion, yet replenished with sufficient society to dispel any feelings like solitude or desolation,—Mr.

Johnes, by way of expressing his own satisfaction with the spot where all his domestic comfort centered,—had called, after the ideal abode of Johnson's Rasselas, The Happy Valley. And surely never was a more appropriate name bestowed; "All Nature and all Art" conspired to make it enchanting; and like a scene of enchantment, it has vanished away! The accomplished owner and his only child have long been numbered with the dead; and of those who formed the social party, how many are gone since the days alluded to! But his name will live as long as the works of Froissart and Monstrelet and Joinville and Brocquière, which all issued from the Hafod press, continue to occupy the shelves of our libraries.

The following letter to Sir James, from Mr. Symmons, a mutual friend, will give the reader some idea of the place and its inhabitants.

Dear Sir, Paddington, Sept. 30, 1794.

On my return home yesterday from a fortnight's excursion into Kent and Surrey, I found your very obliging letter on my table, for which, as well as every other mark of your kind and friendly attention, my best acknowledgements are due. When you come finally from Norfolk, and I learn you are settled in Marlborough-street, I shall certainly take the earliest opportunity of thanking you in person, and of further cultivating, so far as your numerous avocations will permit without inconvenience to you, an acquaintance from which I promise myself so much advantage of every kind. It must be a subject of great disappointment to Mr. Johnes, that your plans did not admit of a call upon him at Hafod; where, besides a very handsome place, situated amidst wilds of a character the most romantic and picturesque imaginable, and some curiosities in the botanic line, which might perhaps not have fallen within your observation elsewhere, you might have assured yourself of a most hearty and hospitable reception from people naturally kind, benevolent, and generous in their dispositions, and who would have felt a very particular satisfaction in contributing to yours by every means in their power. I have seen Sir George Staunton, but not so much of him as I could have wished, enough however to know that he has not visited the capital of China without making such observations as cannot but prove highly interesting in the detail, and from which I flatter myself with the hope of much delight and information at some future time, when both of us are less occupied than at present. His little boy comes with his tutor to my garden every day, and goes over the collection of plants in a regular course, with a Linnæus and a Hortus Kewensis in his hand. His memory is great, and his apprehension quick and lively, so that there can be little doubt of his progress in that, or any other study to which he applies his mind. But I have fears for his health, which seems but ill established, and cannot, in my judgement, be benefited by those continued attentions to all that diversity of languages and sciences which the baronet is perpetually pouring into him. The vessel is certainly of fine but of delicate materials, and may be prematurely broken by too frequent use. I have insensibly scribbled to the end of my paper, with a bare reservation for what may be necessary for assuring you, that I am, dear Sir, with much kindly regard,

Your Friend and Servant,

JOHN SYMMONS.

Sir James's first visit to Mr. Johnes was made in August 1795; and in a letter to Mr. Woodward he decribes his journey and reception at Hafod in these words:

Dear Sir,

I left Norwich last Monday se'nnight; and on the following Saturday went to Worcester,—continued my journey to Ludlow (32 miles) in the mail, by a very hilly and rough road. Observed *Inula Helenium* in moist places by the road side; the country rich, and beautifully varied with woods and fields. Ludlow is on a hill, very pleasantly situated; the castle, where Comus was first acted, is in ruins. On Monday morning at five I set out in a day coach for Aberystwith (80 miles), too great an undertaking by far in such a country for one day. As we proceeded, the country grew more hilly, but not craggy. 'Tis like Westmoreland or the north of Yorkshire, but without lakes. We breakfasted at Bishop's Castle, a shabby town, built of stone as in the North of England. A little way past it we ascended a hill called the Bishop's moat, from whence is one of the most extensive views I ever saw, it only wants water. The hills are lofty and swelling, not abrupt; green to the top, and cultivated half way up; not abundant in wood. We soon after entered Wales, and had another more beautiful view to the right, over an immense extent of hills and valleys to Montgomery, and far beyond, commanding a fine reach of the Severn between woody rocks like Matlock, looking as if one might take them up like a toy. We dined at Newtown, and were to have met the Aberystwith coach, but it was overturned (a common case) six miles from Newtown, and we were delayed in waiting for it all that evening, but spent our time pleasantly enough in walking about the beautiful country around, which much resembles the environs of Matlock, (not the valley itself,) and next morning I proceeded towards Aberystwith (45 miles), but the road was so very hilly we did not get there till six in the evening. The latter part of the way was very beautiful, commanding the river, and a rich vale through which it runs, with openings to the sea. The town is ugly, but abounds with company, and I met Sir George and Lady Caley there. Mr. Johnes's man and horses were waiting for me, but I did not set off till next morning, when I had a most charming ride to Hafod over the hills sixteen miles;

> "And here I rest, as, after much turmoil,
> A blessed soul doth in Elysium."

Here is a most sumptuous modern house, in a vale, among rocks, woods, and cascades of the wildest kind, and on the most magnificent scale. I know nothing like it, except some parts of the north of

Yorkshire. The house looks small on the outside, and yet the gallery is 200 feet long. The drawing-room is one of the most elegant anywhere to be seen, with Gobelin tapestry made on purpose, and chairs of the same. French glasses throughout the house. Every thing in the highest style of decoration. The library an octagon of 30 feet.

Miss Johnes, though not above ten years of age, has taken a wonderful turn for botany and entomology. She has made out almost every plant within her reach that is in Flora Londinensis, or English Botany, and has the latter almost by heart. She longs to botanize in a chalk country. She is almost equally fond of insects, and her whole delight is to walk with me about the woods, searching for mosses and insects, patiently attending to every thing I say, and telling me all her observations, doubts, queries, &c. This is the more extraordinary, as she has had no companion till now. Her mother indeed is fond of a garden and greenhouse, and her father encourages her by all the means in his power; but it is a remarkable instance of early ardour. Miss Johnes is also a capital musician.

I have been rather disappointed hitherto as to botany in Wales. What I have observed are the common plants of hilly, not alpine countries, and not a great variety of them. Mosses and Lichens are plentiful, but they are chiefly of the tree kind, yet no *filamentosi*. *L. læte-virens*, *glomuliferus*, *scrobiculatus*, *sylvaticus*, *resupinatus*, *plumbeus*, are common. I have found one specimen of *perlatus* in fructification. *Hypericum dubium* of Leers, lately observed near Worcester, is one of the most common things here, and I dare say has been overlooked for *H. perforatum*,—pray see to it. The leaves are not punctated, except a line of purple dots on the margin.

I have been this morning a walk of five miles in the grounds, among rocks, cascades, woods, every thing that is romantic, and quite natural. This family are the guardian angels of the country. Mr. Johnes has lately imported a ship-load of wheat from Bristol, to sell at cent per cent loss.

The harvest here is fine, and so it seemed all the way I came. The hop-grounds of Shropshire are beautiful.

I send you a bit of *Hypericum dubium*, and also a *Gnaphalium*, which I took at first for the true *sylvaticum*, but it must surely be only *rectum* of Engl. Bot.; pray give me your opinion. My fair pupil has promised me a tune, and I must go down.

August 31.

We are to go tomorrow to the Devil's Bridge, where I hope to find some Mosses at least. I must leave this charming spot in seven or eight days, and mean to go to Llandillo, where a friend of mine, Dr. Parr, lives, and thence to Bath to see my aunt Kindersley and her grandchildren.

There is no limestone, chalk, or gravel here, and consequently the Flora is poor. *Solidago virgaurea*, *Serratula tinctoria*, *Hypericum dubium* are prodigiously abundant; but scarcely any thing else that is at all uncommon; few Ferns; many Mosses and Algæ. I have also found few insects.

Mr. Caldwell is disappointed of his Guernsey expedition, and regrets disappointing you of Fuci. I wanted him to accompany me to Wales, but he could not.

I have just been looking over a book of twenty-six drawings of shells, &c., done for M. Paris, at ten louis each. Mr. Johnes bought it for about 130*l*. His copies and manuscripts of Froissart are extremely valuable; he is at work on an elaborate translation of that work, with notes.

Your ever affectionate friend,

J. E. Smith.

In the summer of 1796, the writer accompanied her husband to this romantic spot. In addition to the native beauty of the scene, a flock of peacocks from the neighbouring woods came frequently at early dawn, and placing themselves

> "Right against the eastern gate,
> Where the great sun begins his state,"

displayed their gorgeous plumage between the battlements of the mansion, and with their singular cry of Pavo! Pavo! enlivened and animated the otherwise silent solitude of the "morning spread upon the mountains;" nor can she ever hear the note unaccompanied with that peculiar sensation

> "Which out of things familiar, undesign'd,
> When least we deem of such, calls up to view
> The spectres whom no exorcism can bind;

The cold, the changed,—perchance the dead, anew,
The mourn'd, the loved, the lost—too many! yet how few!

How agreeable Mr. Johnes was in conversation, his letters give some idea of, for they greatly resemble his colloquial manner. Abounding in information, he was rapid, allusive, and obscure to those who had more limited knowledge than he possessed.

The letters which remain of this lively and amiable correspondent are very numerous, beginning in 1794, and ending in February 1816, only two months before his death.

The following, written after the loss of the beloved object of his affection, is remarkable for the altered style, subdued by grief, yet full of resignation to the stroke which severed the last branch from the parent tree.

My dear Friend, Hafod, August 17, 1811.

We are very much obliged by all your attentions, and wish you could have sent us better accounts of your own health. We came here this day se'nnight, and hope we may be able to endure it, although deprived of its brightest charm. It is here we can do most good, and that must now be our only endeavour, to show some gratitude for mercies already received, and they have been manifold, and to indulge a hope that we may all meet again hereafter. By placing our dependence on the sole Source of happiness and comfort, we have been wonderfully strengthened, not only to bear our loss with resignation, but with thankfulness.

My wife is still very low, though calm. She

cannot see any company; but if you could favour me with a visit, I think it would do us both good.

I never saw this place in such beauty, and I trust that we shall enjoy it as we ought, within a short time, for we daily regain strength.

Tomorrow will, I fear, unhinge us again, for we must go to church.

Our thoughts now will be attached to all that loved her, and to every thing that may remind us of her excellencies.

I do not think I shall ever again bear music!

Adieu, my good Friend, I am always yours,

T. Johnes.

A very flattering tribute of approbation to Sir James's Tour comes from the pen of the amiable and accomplished Mrs. Watt, the only child of the celebrated Ellis, whose discoveries concerning the formation and nature of Corallines give him a high rank among those whose genius enriched the science of natural history in the eighteenth century.

My dear Sir, Northaw Place, June 23, 1793.

Amidst the various avocations of a large family, a mother anxious to discharge her duty, cannot be supposed to have much leisure for the pursuit of those amusements which, in the earlier period of her life, constituted her greatest pleasures; unless from the affluence of her fortune she can transfer all the cares, and retain only the enjoyments of it. This not falling to my lot, I have been necessitated to postpone thanking you for your most obliging present of a "Sketch of a Tour on the Continent," which I received some months since. What little acquaintance I had before the pleasure of having with you, I must confess interested me very sincerely in your success through life; as you appeared to merit in the highest degree the encouragement of every one who was anxious to promote that noblest of all studies, the study of natural history. But since the perusal of your work, methinks I discern in your character some traits which, for the sake of my much-loved country, I most fervently wish were the leading ones in the character of every Englishman. You unite religious and moral with natural philosophy; you permit it to conduct you, as it was ever intended it should, to the great Origin of all; and while it enlightens your mind to contemplate the most sublime subjects, it heightens your enjoyment of that infinite variety of pleasures which the society of those of a similar turn for improvement must afford. While you have contracted no mean prejudice against those who may differ from you in opinion, your benevolence inclines you to think every individual of every persuasion entitled to your charity and good-will. Allow me therefore to assure you that, for the benefit of the world in general, as well as those particularly of your own circle of friends, both Mr. Watt and myself most unfeignedly wish you health and long life! since we are both fully persuaded, if by an All-gracious Being you are blessed with the one, your ardent and indefatigable pursuit after useful knowledge will render the other a blessing to mankind.

I return you the Linnæan and Philosophical Transactions, with my very best acknowledgements; and have, I assure you, much pleasure in subscribing myself, my dear Sir,

Your much obliged Friend,

MARTHA WATT.

To this lady Sir James was indebted for the whole correspondence of her admirable father, with various other papers, for the purpose of turning them to any public use he might think proper. He thus became possessed of the letters on both sides between Ellis and Linnæus, which were published in 1821, in "A Selection of the Correspondence of Linnæus and other Naturalists."

This excellent woman was the second wife of Alexander Watt, Esq., of Northaw, Herts. In the memoir of Ellis, in Rees's Cyclopædia, Sir James tells us, "She inherited her father's taste and character, more especially his piety and sensibility of mind, with a considerable likeness to his person. She died in childbed at Northaw, in the spring of 1795. Her will, written entirely in her own hand, and a letter to her husband, found after her decease, are worthy of the pen of a Richardson, and the character of a Clarissa."

From Mr. Davall, of Orbe, in Switzerland, an enthusiastic botanist, whose correspondence will appear hereafter, Sir James received the following notice of his travels.

Orbe, Canton de Berne, July 22, 1794.

My dearest Friend,

I have at last received your inestimable present, *your Tour*, which I have hitherto devoured, not yet having been able to read soberly; it has given me such delight as I cannot express. No doubt from my attachment I cannot be perfectly impartial; yet I can solemnly declare, that were I ignorant of its author, I should say he is the favoured child of *Nature*, and displays more solid sense than almost any other I have ever read. I have been delighted with subjects on which I understand little or nothing, such as painting, from the nature of the remarks, *ex. gr.* vol. iii. p. 35. I had, just before I read this, *seen* the *very* attitude between my boy and his nurse. You know me well enough, my dear friend, to be assured that I am incapable of flattering any one, from any motive whatever: believe me then when I say, that I have no words to express the delight your Tour has given, and continues to give me; you give throughout the whole, in my opinion, the best proofs of sound and solid sense, freedom from prejudice, and all the qualities that become a being worthy of the name of man. You desire particularly to have my opinion of what you say on Rousseau. I must frankly tell you the truth; I have never yet read any one of his works! Having rather too early in life had reason to be disgusted with *general* society, not having within my reach here any one congenial soul, I have long lived in a very recluse manner, visiting *Nature!* as well as very indifferent health and delicate eyes could permit. "ANIMÆ *causa ad lenienda vitæ fastidia.*" Scopoli Entom. Carniol. Præf. I cannot refrain in one single instance, my dear friend, lamenting a very great *error of the press* I hope,—viz. vol. iii. p. 170, in styling Van Bercherm an ingenious botanist! For *botanist* let us read, if you will, *zoologist.* I knew him well,—a good young man; the only solid knowledge he had then in natural history was relative to *Quadrupeds,* not *Mammalia.* But as for botany! I should,—as I love you, and as I wish you ever from my heart as free from error and injustice as possible,—I should, I say, have hardly been less grieved had you spoke of your cordial friend Davall as an *ingenious astronomer!*

Another testimony must still be added, though last, not least welcome to the author of the Tour; and it is memorable as being contained in almost the first letter he ever received from the historian of the *Medici.*

"I will not trouble you with enumerating the particulars which your parcel contained, all of which I highly value, but none more than the three volumes of your Tour, which display so much of your own observations and opinions. To say that these are almost invariably in unison with my own is perhaps but a very doubtful kind of commendation; and I will therefore add, that these volumes exhibit that well-tempered zeal for rational liberty, that love of science and predilection for works of art, which will always render them highly interesting to all those who can feel for the true dignity and happiness of mankind.

"Your affectionate friend,

"W. Roscoe."

Mr. Woodward in some observations upon the Tour, selects the description of his journey up Mount Cenis, as the part he preferred. "This charming chapter," he tells his friend, "I can never sufficiently admire. For a botanist the scene has, as you observe, peculiar charms: in reading this the second time, I frequently laid down the book for a few minutes to prolong the pleasure, dreading to get to the end of the chapter. The journey to Chamouny has nearly equal charms, but the subject was not so novel as Mount Cenis."

After this excess of approbation, which the mind is not always in tune to accord with, it is a bold resolve to give the reader the passage, which his partial friend so much approved.

"*Aug.* 11. About eleven at night I set out from Turin along with Mr. de Sousa, Dr. Bellardi, Dr. Buonvicino a mineralogist, and the Abbé Vasco a natural philosopher.

"*Aug.* 12. Early in the morning we found ourselves among the narrow passes about the foot of the Alps, with majestic scenery intermixed with cultivation, and here and there a not very flourishing village. Passed through Suze, the key of Piedmont, which of course is very strongly fortified; its bastions are cut out of the live rock. The country grew more hilly and romantic at every step. At the miserable village of la Novalaise we were obliged to quit our carriages for mules; and after a tedious ascent by a zigzag stony road, no way dangerous however, we reached the top, that is, the plain of Mount Cenis, towards noon. Within about a mile of the summit I found *Juncus filiformis* in a wet place on the left of the road, and *Lichen polyrhizos* on a rock near it. Not far from hence, on a small plain before we arrived at the great one, grew *Bartsia alpina*, in seed, *Trifolium agrarium* of Linn. (Dickson's Dried Plants, No. 80.), widely different from that of English writers, and many other rare plants. On our right a magnificent cascade fell close to the road. All along a great part of the way I had observed various alpine species of *Anemone* and *Pedicularis*, mostly in seed, with a novelty of appearance in the herbage highly encouraging, and a luxuriance at which (having no idea of alpine pastures) I was surprised. I lamented only the advanced state of these plants, and feared we were too late for the season; but when I found the plain of Mount Cenis all flowery with the rarest alpine productions, such as we delight to see even dragging on a miserable existence in our gardens, and the greatest part of which, disdainful of our care and favour, scorn to breathe any other air than that of their native rocks, none but an admirer of nature can enter into my feelings. Even the most common grass here was *Phleum alpinum*, and the heathy plain glowed with *Rhododendrum ferrugineum* and *Arnica montana*. Well might Clusius so beautifully say—'*Non carent altissimi montes præruptique scopuli suis etiam deliciis*' *; nor need one have the science of a Clusius to feel pleasure in such scenes. Scarcely any traveller passes the Alps in summer without either lamenting the 'neglect of his botanical studies,' or more honestly regretting that he had never attended to this source of pleasure at all. I have long ago perhaps tired the reader with my admiration of the works of art. If he had had indulgence enough for me to get thus far, he must now lay in a fresh stock of patience while I expatiate on the productions of nature; unless he should chance to be a botanist, and then all I can say will not satisfy his curiosity. Dr. Bellardi and myself were accommodated at the hospital, built for the reception of travellers in bad weather, which is now under the direction of an abbé, named Tua; the good father Nicholas, so much celebrated by Lady Miller, being dead long since. This is a good rustic sort of inn; so far from being intolerable, that the English ambassador, Mr. Trevor, and his lady, had lately resided some weeks here on a party of pleasure. Mr. de Sousa and the rest of our company took refuge in a tent not far distant, belonging to the chevalier de St. Real, then intendant of the province of Maurienne, and now of the valley d'Aost, who with an ingenious young officer of Chambery, Mr. Martinel, had spent several weeks here, and as many in the preceding summer, in order to investigate thoroughly the geography and

* " 'The most lofty mountains and most rugged precipices are not without their own peculiar charms.' " (Clus. Panon. 316.)

natural history of the environs. In society like this, no less pleasant than instructive, and amid such scenery, the time passed but too quickly; and I could not but regret the impossibility of my spending the three following summers here with the worthy intendant, who had allotted five years in all for the accomplishment of his undertaking. His observations must be inestimable: but it was thought the whole would hardly be given to the public, as the Sardinian court very justly objected to making their French neighbours too well acquainted with all the secrets of their natural bulwark.

"This hospital stands by the high-road side half way over the plain, which is at least two miles in length; and about the middle, which is its broadest part, it may be a mile across. It is entered at each extremity by a narrow pass, and surrounded on both sides with very lofty mountains capped with eternal snow. The plain itself is full of inequalities. Towards the northern extremity are two or three beautiful lakes, with an island in the principal one, clothed with shrubs and rich pasturage. This lake empties itself to the south by a small river, whose rocky channel often forms considerable cascades of great beauty, and is overhung with luxuriant herbage, and shrubberies of *Rosa alpina*, *Mespilus* (or rather *Cratægus*) *Chamæmespilus*, &c. &c. This part of Mount Cenis is seldom visited by travellers; but, being within a moderate walk from the post-house or the hospital, richly deserves attention. On the other side of the rivulet, about the bottom of the hills, are some alders, which, being sheltered by the craggy rocks, attain a considerable height; otherwise no tree in general, not even the fir, grows to any size so high on the Alps. A little further up are most delicious pastures, intersected with alder thickets, and bordered with *Cacalia alpina*, *Aquilegia alpina*, *Ranunculus aconitifolius*, *Sisymbrium tanacetifolium*, *Pyrola minor*, *Juncus spicatus*, and other rarities. This beautiful *Aquilegia*, which far exceeds our garden kind, was very sparingly in flower, and I am obliged for its detection to my faithful attendant Francesco Borone, who here imbibed that taste for botany which afterwards led him to Sierra Leone; and by whose acuteness and activity I have often profited. Some little hillocks on the left of the front of the hospital are covered with *Rhododendrum ferrugineum*, among which grew *Pyrola rotundifolia*, and in the clefts of rocks the very rare *Saponaria lutea*. (*Spicileg. Bot.* t. 5.) Here I first found *Lichen cucullatus*, Trans. of Linn. Soc. vol. i. 84, t. 4, f. 7, which I am astonished any body can confound with *L. nivalis*: the latter too grows here, as does *L. ochroleucus*, Dickson Fasc. Crypt. iii. 19. Descending towards the river I came to a most delightful little valley, like the vale of Tempe in miniature, with a meandering rivulet, scarcely three or four feet broad, running through it, and bordered with abrupt precipices not much more in height, in which were several fairy caves and grottos, their entrances clothed with a tapestry of mantling bushes of *Salix reticulata* and *retusa*. These dwarf willows grow close pressed to the rocks, whether horizontal or perpendicular, almost like ivy, and may be stripped off in large woody portions. By the rivulet, which issued in several streams from these caves, was a profusion of *Anthericum calyculatum** and *Leontodon aureum*, with many other things equally uncommon, and in full bloom.

"*Aug.* 14. We all sallied forth on foot about five in the morning to ascend little Mount Cenis, one of the most considerable hills that front the hospital on the other side of the lake Pursuing a winding path through the thickets, we came to a few cottages, in surely one of the most retired habitable spots in Europe, and which probably are seldom four months in the year uncovered with snow. Yet at this season who would not have envied their situation? No lowland scenes can give an idea of the rich entangled foliage, the truly enamelled turf of the Alps. Here we were charmed with the purple glow of *Scutellaria alpina;* there the grass was studded with the vivid blue of innumerable Gentians, mixed with glowing Crowfoots, and the less ostentatious *Astrantia major* and *Saxifraga rotundifolia*, whose blossoms require a microscope to discover all their beauties; while the alpine rose, *Rosa alpina*, bloomed on the bushes, and, as a choice gratification for the more curious botanist, under its shadow, by the pebbly margin of the lake, *Carex capillaris* presented itself. The riches of nature, both as to colour and form, which expand so luxuriantly in tropical climates, seem here not diminished but condensed. The further we ascended,

* *Tofiedia pallustris* Fl. Brit., and Engl. Bot. vol. viii. t. 536.

the more every production lessened. By the sandy bed of a torrent, which runs from the glaciers above, the very elegant *Saxifraga cæsia* seemed to emulate the glistening of the hoar-frost around it.

"At length, about eleven o'clock, we reached a small plain full two-thirds of the way to the top. Here we divided. Some of our party were adventurous enough to climb the very summit; but being already got to the utmost limits of vegetation, and near those of perpetual snow, I had no business higher. Indeed this plain appeared to be clothed with a short barren turf that promised little; nor was it till I examined it on my hands and knees, that I discovered this turf to be a rich assemblage of *Cherleria sedoides, Alchemilla pentaphyllea, Chrysanthemum atratum, Gentiana nivalis*, and other diminutive inhabitants of the highest Alps, among which one of the most beautiful is a dwarf variety of the common Eye-bright, *Euphrasia officinalis*, with large purple flowers.

"This plain was here and there sunk, on the margin of the declivity, into little hollows, watered by very small trickling rills, and in such parts vegetation appeared extremely luxuriant. *Bartsia alpina* was here but in flower, along with *Satyrium nigrum;* the latter smelling like Vanilla. I observed a pair of *Papilio Apollos* in this exalted region, fluttering about and celebrating their innocent nuptials. After enjoying from hence the view of the plain of Mount Cenis, with the lake and woods about it, we descended on the side fronting the hospital, and arrived there by six o'clock, not a little fatigued, having been all day on our legs, without any refreshment except what a servant had carried with us; but I believe our satisfaction much exceeded our fatigue.

"*Aug.* 15. This day Dr. Bellardi and myself ascended the hill called *Ronche,* immediately behind the hospital, where Professor Allioni first discovered *Viola Cenisia* and *Campanula Cenisia*. Dr. Bellardi found them this day, though I was not so fortunate; nor did I meet with any thing very desirable except *Juncus Jacquini*; and in the boggy sides of a little rivulet, in the very highest part of the mountain, a little *Carex* of great rarity, the *juncifolia* of Allioni's Flora Pedemontana. This is surely the same species as Lightfoot's *C. incurva,* though on the Alps its stem is seldom curved. I have it also from Iceland. *Juncus triglumis* grew along with it, and in other parts of the hill *Carex fœtida* of Allioni, and *C. atrata,* with *Antirrhinum multicaule.*

"Before the post-house are some remarkable white limestone rocks, on which grow *Dianthus virgineus*, and the real *Festuca spadicea* (see Trans. of Linn. Soc. vol. i. p. 111). Below these rocks, by the lake, I gathered the most beautiful *Gentiana asclepiadea,* and in the surrounding pastures *Agrostema Flos Jovis, Senecio Doronicum, Aster alpinus, Centaurea uniflora, Arnica montana,* and the *Rumex arifolius* of Linnæus's Supplement, which last is, I presume, more certainly a native of the Alps than of Abyssinia. Immediately before the hospital is great plenty of *Rumex alpinus,* and a little further on I joyfully waded up to my knees in a swamp to gather *Swertia perennis*. All the plain abounds with the beautiful *Dianthus alpinus*, the leaves of which differ so much in narrowness and sharpness from the Austrian one, that I have sometimes suspected them to be distinct species. Nothing however is more common on Mount Cenis than *Dryas octopetala*, forming thick tufts many feet in breadth, covered with its elegant flowers and feathery heads of seeds. On this elastic alpine couch we reposed when tired with walking, and the delicious temperature of the air made any shelter perfectly indifferent.

"Such are a part of the botanical riches of this interesting mountain, not to mention numerous species of *Arenaria*, *Silene*, *Achillæa*, *Astragalus*, *Juncus*, and grasses of various kinds. Of all these treasures I laid in as large a stock as I could well bring away, multiplying my own enjoyments in the anticipation of the pleasure I should have in supplying my friends at home. The selfish dealer in mysteries and secrets, the hoarder of unique specimens, knows nothing of the best pleasures of science."

In an introductory lecture at the Royal Institution Albemarle-street, in 1808, the author of the Tour observes, that "Of all studies, perhaps natural history is the most practical. Its very charm consists in the interest it gives to objects always before our eyes, but which it furnishes us with a new sense to admire, to enjoy, and to understand.

"From the earliest period of my recollection,

when I can just remember tugging ineffectually with all my infant strength at the tough stalks of the Wild Succory on the chalky hillocks about Norwich, I have found the study of Nature an increasing source of unalloyed pleasure, and a consolation and refuge under every pain. Long destined to other pursuits and directed to other studies thought more advantageous or necessary, I could often snatch but a few moments for this favourite object. Unassisted by advice, unacquainted with books, I wandered long in the dark; till some of the principal elementary works, the publications of Lee, Rose, Stillingfleet, and a few others, came in my way, and were devoured over and over again. This kind of botanical education has the advantages of the necessary drudgery of a grammar-school; it trains the mind to labour, it fixes principles and facts and terms and names, never to be forgotten. At length, however, I found I wanted something more, to apply to practice what had thus been acquired. I was then furnished with systematic books, and introduced to Mr. Rose, whose writings had long been my guide. I was shown the works of Linnæus; nor shall I ever forget the feelings of wonder excited by finding his whole system of animals, vegetables and minerals, comprised in three octavo volumes. I had seen a fine quarto volume of Buffon, on the Horse alone. I expected to find the systematical works of Linnæus constituting a whole library; but they proved almost capable of being put, like the Iliad, into a nutshell. Hence a new world was opened to me.

I found myself moreover in the centre of a school of botanists. Ever since the Spanish tyranny and folly had driven commerce and ingenuity from Flanders, to take refuge in Britain, a taste for flowers had subsisted in my native county along with them. Our weavers, like those of Spitalfields, have from time immemorial been florists, and many of them most excellent cultivators; their necessary occupations and these amusements were peculiarly compatible. And it is well worthy of remark that those elegant and virtuous dispositions which can relish the beauties of nature, are no less strictly in unison with that purity of moral and religious taste which drove the founders of our Worsted manufactory from foul and debasing tyranny to the abode of light and peace and liberty.

"Our circle of naturalists at Norwich, far from being confined to florists, had long contained some systematic botanists. They were students of Gerard, and Parkinson, and even of the learned works of Ray. In my young time this circle was peculiarly enriched by the possession of Mr. Rose, Mr. Bryant, and Mr. Pitchford, three names well known to all who are conversant with the botany of Britain. They were often favoured with the society of the learned and amiable Stillingfleet, and the correspondence of Hudson; and they may all together be considered as the founders of Linnæan Botany in England, to the promulgation of which, the publications of Rose, Stillingfleet, and Hudson, have contributed more than any others whatever; while the indefatigable practical labours of Mr.

Bryant and Mr. Pitchford were daily enriching the science with new discoveries. Thus the botany of Norfolk has become celebrated, and its Flora has proved richer, I believe, than that of any other county, because it has been more closely investigated.

"Fired by these examples, and ambitious of contributing to the common stock of knowledge, I took advantage of a journey to Yorkshire, Derbyshire, and Lancashire, to inquire what those counties afforded. I had visited them several times before, but now I wondered how any of my former journeys had afforded me pleasure. I felt like a man born blind, who first walked abroad to look about him. The wild moors, the mossy rocks, the mountainous woods, to me were 'opening Paradise'.

"Some time afterwards the country about Edinburgh, and the mountains of Scotland, afforded me a fresh harvest; and at length the classical scenes of Italy derived for me a new charm from the occasional pursuits of botany. But above all, the alpine scenery and treasures of Savoy and Switzerland have left the most pleasing impression. In those countries all the riches of Flora burst upon us at once, during their short spring and hasty summer. From the first melting of the snow, when the Jura and its brother Alps are absolutely covered with Crocuses, as with a purple robe, to the bright days of autumn, when the Raspberry and Bilberry glow on their dewy bushes, such a profusion of foliage and flowers and fruits, so much more abundant and varied than we have any experience of, is crowded into a few weeks, that the whole face of the country is like a flower-garden, as rich and brilliant as the finest collection of Cape plants about London.

"I know not how it is that alpine scenery is so enchanting to the lovers of nature, under which denomination I mean to comprehend the poet, the painter, and the naturalist, as well as those who alone have minds to taste and appreciate their performances. The purity of the air, the grandeur of the scenery, the beauty of the face of nature altogether, affect the spirits and senses in a manner that is scarcely experienced elsewhere. If we quit the hot-bed of the metropolis and its neighbourhood, to welcome the rising spring among the Derbyshire hills, we taste something of this enchantment; but infinitely more when we quit the scorching plains of Italy for the Alps, which rise above them like an immense purple rampart of clouds."

It may not be unentertaining to give at one view a general enumeration of the men of science Sir James met with in his travels, and the ideas their different characters impressed upon him. These are extracted from his Tour. The names of several have appeared already, and others will occur hereafter.

Leyden.

VAN ROYEN,—"I had an introduction to the celebrated Mr. David Van Royen, whose politeness and attention could not be exceeded. He had been professor here thirty-two years, and has lately resigned, having still permission to use the garden for his amusement. His fondness for botany continues as strong as ever."

ALLAMAND.—"The chair of natural philosophy was at this time filled by Mr. Allamand, well known by his edition of the Natural History of Buffon. A fine old man of the most agreeable manners, and with that happy mixture of politeness and cheerfulness almost peculiar to Frenchmen in the decline of life."

SANDIFORT.—"Dr. Sandifort's private library is one of the finest things in Leyden."

PESTEL.—"One of the chief ornaments of this university is Mr. Pestel, professor of jurisprudence. His *Fundamenta Jurisprudentiæ Naturalis* is a book every inhabitant of a free state ought to study. His pure system of elevated piety, his union of christianity with morality, and of manly principles of liberty with virtuous order, are not at all in the style of philosophers who insinuate atheism, or of fools who avow it."

Lyons.

VILLARS.—" A very able entomologist, whose cabinet was said to contain five thousand European species of insects. We found Mr. Villars modest, communicative, and unassuming, like most people of real knowledge and genuine taste for science."

Versailles.

LE MONNIER.—" First physician to Louis XV.,

after whom *Monniera* was named. I found him in his garden with Messrs. Thouin, Dombey, and other botanists. His herbarium is said to be uncommonly valuable."

St. Germain-en-laye.

MARECHAL DE NOAILLES.—"From Versailles we passed by Marli to St. Germain-en-laye, and slept at the house of the venerable Marechal de Noailles, the old friend and correspondent of Linnæus. Since my being there, the Marechal has decorated it with a monument to Linnæus, and has celebrated a jubilee in his honour."

Paris.

D'AUBENTON.—"With Mr. D'Aubenton I had often an opportunity of conversing, and always with pleasure and advantage."

"DE LA CEPEDE, who has published an able work on Reptiles, was also very frequently at the cabinet during my visits there."

"ADANSON, whose knowledge of botany would procure him great reputation, were he less a slave to paradox and pedantry. He generally accosted me with some attack on Linnæus, sometimes calling him grossly *ignorant* and *illiterate*; and then when I have ventured to quote Philosophia Botanica, as a proof of the contrary, abusing him as *scholastic*, I was contented with smiling, to think how one accusation destroyed the other."

A. L. DE JUSSIEU.—"M. Antoine Laurent de Jussieu

takes the lead among those who with respect to system may be called Anti-Linnæans. He inherits his taste for the science from his uncles Bernard and Joseph de Jussieu; the former of whom was Professor at Paris, and the latter made a fine collection of plants in Peru. Their books and collections descended to their nephew, who has not turned his attention to botany till within these few years; but with what very great success he has in that time studied natural orders, is manifested in his Genera Plantarum, published in 1789,—a work which will immortalize its author, and probably go down to posterity with the Genera Plantarum of Linnæus, to which it is an excellent companion. Those who can read and judge of this work need not be told that he is a true philosopher, profound in science, ardent in the pursuit of truth, open to conviction himself, and candid in his correction of others; nor will they be surprised to hear his manners are gentle and pleasing, his conversation easy, cheerful, and polite. Although we differed on many points, as on the laws of nomenclature, and the merits of the Linnæan system, yet as truth was our common object, repeated and free discussions increased our esteem for each other, and to me at least were productive of instruction as well as pleasure."

"De Lamarck is equally devoted to botany, but his character is less pleasing than that of Mr. de Jussieu. I freely acknowledge that I shrunk from the society of a man who always took occasion to attack with violence what he knew to be my most favourite sentiments."

L'Heritier.—"Among the Linnæan botanists, Mr. L'Héritier is eminently distinguished by his most superb and scientific publications, the plates of which are executed with a degree of accuracy rarely to be met with, nor are the descriptions less complete."

"Bulliard is well known by his *Herbier de la France.*"

"Desfontaines, Professor of Botany at the Royal Garden, was in 1786 lately returned from Barbary with a rich harvest of plants and insects, all of which he allowed me to examine and partake of."

Thouin.—"Mr. Thouin, who has the superintendence of the Botanic Garden, deserves my warmest acknowledgements for the very liberal manner in which he at all times allowed me access to that rich collection, as well as to his own private herbarium, which I looked over entirely with great advantage."

Broussonet,—"with whose friendship a stranger could want nothing in Paris, and whose benevolence I had not now to seek for the first time. Few naturalists equal him for zeal and abilities."

Montpellier.

Dorthes.—"A very ingenious entomologist."

Gouan.—"The old correspondent of Linnæus, well

known by his botanical and ichthyological works."

CUSSON.—"Demonstrator of botany. His death is no loss to the science, as he kept entirely from the world his father's celebrated manuscripts, and a collection of umbelliferous plants. They are now fallen into the hands of Mr. Dorthes, who is amply qualified to digest and publish them."

Cottignac.

GERARD.—"We came to this place purely to visit Mr. Gerard, author of the Flora Gallo-Provincialis, one of the best European botanists of the golden age of Linnæus. Nothing could exceed his politeness and hospitality to us. We had much conversation together about the purchase of the Linnæan collection, a never-failing topic with all the botanists I met with in my journey. Almost all I had to tell was news to him; and I felt as if paying a visit in the Elysian fields, so little did his "tales of other times" seem connected with what is now going on in the world. He spoke very highly of Linnæus and Ray; and permitted me to copy two very interesting letters from the former to himself."

Florence.

FONTANA.—"We were fortunate enough to enjoy much of the society of the Abbé Fontana, who did us the very flattering honour of spending at our lodgings most of the evenings we were at Florence; how much to our profit and entertainment, those who know his physiological enthusiasm and penetration need not be told."

FABRONI.—"Secretary to the Agricultural Society."

TARGIONI-TOZZETTI.—"One great object, in our way, was the museum of the celebrated Micheli. His collection was bought after his death by his friend Dr. Targioni, who afterwards took the name of Tozzetti for an estate. His son, who now possesses these relics, is a man of the most engaging simplicity, modesty, and benevolence of manners, and worthy to be the heir of the amiable Micheli."

Rome.

CORRÊA.—"I was so fortunate as to meet with the Abbé Corrêa de Serra, Secretary to the Academy of Sciences at Lisbon. With this gentleman I first became acquainted at Mr. de Jussieu's at Paris. He had resided twelve years at Rome formerly, and was attached to the place by all the enthusiasm which a man of so much fine taste and extensive literature must feel in such a residence, though he had since lived many years in Portugal, his native country."

FATHER JACQUIER.—"One of the most interesting of the literati whom it was our fortune to meet with, was the illustrious mathematician father Jacquier, general of the French minims at the

Trinità de Monti, the Roman editor of Newton, who, if I remember right, was obliged to make an apology in his preface for publishing such a dreadful heresy as that the earth moved round the sun, and to disclaim his belief of it. What more bitter sarcasm could have been offered to the very authority which required this apology!

"This good and venerable old man, who is since removed to a state where he will find truth needs no apology, was confined to his bed with a broken limb from a fall. Nevertheless he admitted visitors every evening, conversing with his natural cheerfulness and urbanity, and discussing scientific subjects with as much ardour as if he were just entering on his literary career. How delightful and how consolatory it is, among the disappointments and anxieties of life, to observe science, like virtue, retaining its relish to the last; smoothing the bed of age and infirmity, preserving the mind young and vigorous, alive to all its enjoyments, amid the wreck of its frail cottage; while, in communicating its own ardour and its own light to others, it tastes the happiness of a good father, who feels himself living over again in his children!"

Milan.

ORIANI.—"The Abbé Oriani, Astronomer Royal. He treated us as if he were paying a debt of gratitude, instead of humanity. I have ever since been proud to cultivate his friendship."

Pavia.

Scopoli.—"This is at present the most celebrated university in Italy. The loss of one bright ornament we have now indeed to regret; the accomplished Scopoli, who was at this time Professor of Botany here. We found him a man verging towards the decline of life, of a plain but animated countenance, not at all resembling his portrait in the Flora Carniolica, and entirely devoid of that stupid gravity so remarkable in that print. Breakfasting with him next morning, I took an opportunity of offering him any assistance the Linnæan herbarium could afford, by which he first understood it was in my possession, having but slightly read over our French introductory letters. He was quite overjoyed; gave me a most cordial embrace, and from that moment we scarcely separated during my abode in Pavia.

"One morning, at 7 o'clock, we attended a botanical lecture of Professor Scopoli's, in a room at a garden. It was in Italian, chiefly on *grasses*. He observed that there are really no limits between *Bromus* and *Festuca;* but that, nevertheless, Linnæus's arrangement of the genera and species in general was the best, as well as the first, ever seen."

"Gregorio Fontana, brother of the Abbé Fontana at Florence. Little could I imagine, when I enjoyed the pleasure of his conversation and admired the acuteness and versatility of his genius, that he should ever condescend, as he has since done, to become the translator and commentator of any production of mine*."

SCARPA.—"We were no less happy in the acquaintance of Professor Scarpa, the excellent teacher of Anatomy."

SPALLANZANI.—"The Abbé's countenance is austere and proud, in form inclining to the African.

"We anxiously wished to hear a lecture from this famous Professor of Natural History, not choosing to be presented to him; and it fortunately fell out that our curiosity was gratified. His delivery is exceedingly deliberate, drawling and monotonous; it could scarcely be heard with a grave face; but the composition of his lecture so admirable as to make us forget the rest. The subject was the different lengths of time which butterflies remain in the *chrysalis*, and how far their exit may be hastened by heat; with Reaumur's experiments of putting the insects in that state under a hen, and so producing the first butterflies perhaps, as the professor said, that were ever hatched by a bird."

Turin.

"ALLIONI, the father of natural history in this place, is still ardent in its pursuit."

BELLARDI.—"A most excellent botanist. He has much enriched the Flora Pedemontana of his master Allioni."

* "Discorso preliminaire.—1792."

CIGNA.—"Professor of Anatomy."

PENCHIENATI.—"Professor of Surgery."

MALACARNA.—"Whose unaffected plainness and native genius were very strongly marked in the little conversation I had with him."

DANA.—"The Cabinet of Natural History promises well under the auspices of this very able Professor."

DE SOUSA COUTINHO.—"The Mæcenas of Botany, and indeed of general science, at this period, the chevaliere de Sousa Coutinho, Portuguese ambassador. At his table was a weekly assembly of literary men, in whose conversation and pursuits he bore a very intelligent part, always making himself completely one of the company by his knowledge and enthusiasm, no less than by his winning affability."

GIORNA.—"With respect to literary and accomplished characters, I cannot but esteem myself peculiarly fortunate at Turin. The only entomologist I met there, besides Professor Allioni, was Mr. Giorna, to whose liberal communications my collection is much obliged, and who has all the candour and modesty of real merit."

Padua.

ARDUINO.—"Formerly Professor of Botany, but now of Agriculture. From him the *Arduina* was named, and he sent Linnæus those rare Brazil plants, chiefly described in the Mantissa."

Geneva.

DE LUC.—"The brother of our amiable natural

philosopher at Windsor. He has a superb cabinet of shells and extraneous fossils."

"SENEBIER, the historian of Geneva, is distinguished for his experiments relating to the physiology of vegetables."

BONNET.—"The most illustrious philosopher of Geneva is Mr. Bonnet. Though almost deprived of sight and hearing, he conversed long and most instructively on our favourite subjects, affording a fresh proof of the truth of what I observed in speaking of Father Jacquier."

DE SAUSSURE.—"I had an interview only of a few minutes with Mr. de Saussure, who had just then descended from the summit of Mont Blanc."

ZIMMERMAN.—"I fortunately met at Geneva Professor Zimmerman, the celebrated zoologist."

Basle.

"LACHENAL was a great friend of Haller. His botanical library is one of the best I ever saw."

SOCIN.—"Professor of Natural Philosophy, a very ingenious well-informed man."

BERNOUILLI.—"An excellent chymist and mineralogist."

Berne.

HALLER, Jun.—"He resembled his father in fondness for botany, versatility and even strength of genius; but not altogether in application."

WYTTENBACH.—"A most estimable character, who has cultivated natural history, especially mineralogy, with great success."

Strasburg.

HERMAN.—" Professor of Botany. I have seldom conversed with a man of a more acute or more enlarged mind."

SCHURER.—"Nor was I less obliged to Professor Schurer, teacher of Philosophy."

LAWTH.—" Professor of Anatomy, a man of ability."

Naples.

N. PACIFICO.—" A zealous botanist and able mathematician. His library possesses four volumes of the botanical plates of Cupani, an unpublished work of the most extreme scarcity."

DR. CYRILLI.—"His garden contains some rare plants, and his library and old herbarium were made by *Imperato*."

"DON FILIPPO CAVOLINI is the most ingenious and indefatigable naturalist in this part of the world; whose observations on corals and other marine animals rival those of our illustrious Ellis for extent and fidelity."

Genoa.

DURAZZO.—"Science is a plant of slow growth, nor is it yet a fashionable pursuit among the Genoese. The Durazzo family stand almost single as its encouragers. The Senator Marcellino Durazzo, owner of the great palace in Strada Balbi, has a romantic country-seat at Pino, where he very kindly received us, and gave an entertaining narrative of his going to France in the year 1747, to solicit the assistance of

Louis XV. against the Germans. All the Genoese coast being occupied by the enemy, he was obliged to return by sea to Marseilles; and that harbour being blocked up by the English fleet, he escaped in the disguise of a fisherman, with a million of French livres. For this service, and his good conduct when Doge, he has had the almost unprecedented honour of a statue in the council chamber, erected in his lifetime."

CHAPTER IV.

Foundation of the Linnæan Society.—Sir J. E. Smith elected President.—List of the Fellows in 1790.—Letters of Mr. Smith.—Letters from various Foreign Professors.—Charter.—President's Address on the Twentieth Anniversary.

FOUNDATION OF THE LINNÆAN SOCIETY.

IN looking round upon the literary institutions and learned academies of Europe, it will be seen that they have generally owed their origin and success either to large endowments, to royal favour, or to the commanding influence of persons already known by their scientific attainments, or their station. This Society is almost a solitary example of an institution deriving its origin from an individual, young and unknown to fame, without rank, without wealth, without support, whose ardour for the pursuit of science led him to risk the expectation of a moderate independence, by bringing into his native country, at the expense of his patrimony, those rich materials for which princes had contended, and upon which he was to establish a new Society, and give to it its name, its character and direction. At the commencement of a lucrative professional life, Sir James cheerfully abandoned the promises it held forth, to become the leader of a band of naturalists, who should follow in the steps of the immortal Lin næus, and among whom, by the influence of a candid and benevolent mind, he preserved the peace and harmony of the community he had framed, and the esteem and good-will of its members.

He had said long before, "My heart is formed for friendship, and cannot exist without it:" so his soul was formed for high intellectual pursuits, and could not exist without a fairer field than the opening of a young physician's course presented. In the present object both his tastes were gratified; and he found it in England, as he had done on the Continent, a passport to the best society.

Early in the month of February 1788, Sir James, in a letter to his father, acquaints him that he has engaged a house in Great Marlborough-street, belonging to Mr. Bendish, a Cambridgeshire gentleman; and to this he removed a few weeks after.

His leaving Chelsea was with an expectation of beginning his medical career in London; yet in the same letter, wherein he expresses a desire to perform his duty as a physician, he adds, "In that rank in science to which I may say *I have raised myself*, with every prospect of taking a lead in the studies to which I am peculiarly attached, with so many fortunate circumstances about me, and especially when I consider that all I enjoy is owing to this study, you can hardly expect I should give up that and all my hopes. You may depend upon it Natural History will always be the main object of my life; and I doubt not you will be thankful that I have so noble a one. I rely on this to give me real lasting honour, and to make me useful to mankind through ages when I am no more, as well as while I live to afford me relaxation from less pleasant cares. Do not think I mean to oppose it to my medical views; I have always meant them to co-operate."

Mr. Smith to his Son.

My dear Son,

The pursuit of science you say shall co-operate with practice—that is the idea I wish to entertain, to answer all your views; and I hope they will unite to raise you to a very high degree of eminence.

I am proud of the light you stand in; and every advance you make to fame lifts my heart with transport, and I want only to give you an independent fortune to make me perfectly happy: but as I cannot do that, nor any thing like it, I must repeat, my dear James, that a determination to depend upon yourself and to be your own master is so consonant to my own disposition, that it gives me great pleasure. I believe it springs from a better principle than pride in both of us, the love of dear Liberty, which is the birth-right of every individual of mankind, and has my strongest affection. I wish to see her universally enjoyed, and therefore must most earnestly desire it may be the portion of each of my dear children. Would to God I may be able to leave every one of them in a condition to possess it in a rational, virtuous degree!

J. SMITH.

It is probable his father saw more clearly than

he himself perceived, that physic would be a very secondary object: his first undoubtedly was to complete the design which he and several of his friends had long kept in view, the establishment of a new Society, and thus to render his possession of the cabinets and library of Linnæus subservient to the general use of natural science, and especially of botany.

With the assistance of Sir Joseph Banks, Dr. Goodenough, Mr. Marsham, and a few others, this object was carried into effect; and the first meeting of its members was held at his own house in Great Marlborough-street, on the 8th of April 1788, on which occasion Sir James delivered "*A Discourse on the Rise and Progress of Natural History;*"—an animated and most instructive address, auspicious of the prosperity of the new-formed Institution, and which affords a convincing proof that the study of nature is not a tasteless and insipid one. He was at the same time appointed President of the community, which was designated by the appropriate name of "The Linnæan Society."

"I consider myself," he observed, "as a trustee of the public, and hold these treasures only for the purpose of making them useful to the world and Natural History in general, and particularly to this Society, of which I glory in having contributed to lay the foundation, and to the service of which I shall joyfully consecrate my labours, so long as it continues to answer the purposes for which it is designed."

This Institution, venerable now by its duration,

approaching to half a century, has enrolled among its members from the beginning names illustrious as well by high birth, as by high claims to scientific distinction, in France, and Holland, and Germany; in Switzerland, in Italy, and Spain, as well as in England; and in its later days the catalogue is swelled with names from America and India.

An original list of the 36 Fellows and 16 Associates may not be unacceptable here. A few names yet survive, and are enrolled in the present year with full 500 additional ones. The Foreign Members, originally about 50, are still limited to that number.

With most of the Fellows and Foreign Members here enumerated, a correspondence, founded on the President's personal acquaintance on the continent, had begun, or continued from the time of the establishment of the Linnæan Society, increasing year by year, as the boundaries of science extended; and their letters bear testimony to the liberality of the possessor of the treasures of the great master of natural science.

The List of the Linnæan Society, 1790.

HONORARY MEMBERS.

Sir JOSEPH BANKS, Bart. President of the Royal Society.

HENRY, Earl of GAINSBOROUGH.

The Marechal Duc de NOAILLES.

FELLOWS.

JAMES EDWARD SMITH, M.D. PRESIDENT, F.R.S. *Acad. Reg. Sc. Taurin. Ulissip. Agron. Paris. Botanoph. Andegav. Corresp.*

Mr. George Adams, Fleet-street.

Robert Barclay, Esq., Clapham.

Mr. John Beckwith, Spitalfields.

James Crowe, Esq. Norwich.
Sir Thomas Gery Cullum, Bart. F.R.S. Bury.
Mr. William Curtis, St. George's Crescent.
Edmund Davall, Esq. Orbe, in Switzerland.
Mr. James Dickson, Covent Garden.
JONAS DRYANDER, M.A. LIBRARIAN, *Lib. to the Roy. Soc. Acad. Reg. Sc. Stockholm. Soc.* Dean-street.
Samuel Ewer, Esq. Lincoln's-Inn-Fields.
Mr. John Fairbairn, Chelsea Garden.
John Ford, M.D. Bond-street.
The Rev. SAMUEL GOODENOUGH, LL.D. TREASURER, F.R.S. Ealing.
William Hanbury, Esq. Kilmarsh, Northamptonshire.
Mr. Thomas Hoy, Sion House.
Aylmer Bourke Lambert, Esq. Salisbury.
Mr. John Latham, F.R.S. Dartford.
John Latham, M.D. Old Jewry.
The Rev. John Lyon, Dover.
THOMAS MARSHAM, Esq. SECRETARY, Upper Berkley-street, Portman-square.
The Rev. Thomas Martyn, B.D. F.R.S. *Prof. Bot. Cambridge*, Westminster.
Mr. Archibald Menzies, Broad-street, Soho.
Richard Pulteney, M.D. F.R.S. Blandford, Dorset.
The Rev. Richard Relhan, M.A. F.R.S. Cambridge.
John Rotheram, M.D.
Richard Anthony Salisbury, Esq. F.R.S. Chapel Allerton, near Leeds.
George Shaw, M.D. F.R.S. Frith-street.
John Sibthorp, M.D. F.R.S. *Prof. Bot. Oxford. Soc. Reg. Monsp. et Gotting. Correspond.* Oxford.
John Sims, M.D. Paternoster-row.
Sir George Leonard Staunton, Bart. F.R.S. Berner's-street.
Mr. John Timothy Swainson, Dover-place.
Mr. Robert Teesdale, Ranelagh.
William Thomson, M.D. F.R.S. *Lect. in Anat. Oxford*, Oxford.
Thomas Jenkinson Woodward, Esq. Bungay.
William Younge, M.D. Sheffield.
Mr. John Zier, Pimlico.

FOREIGN MEMBERS.

D. Adam Afzelius, *M.A. Botan. Demonstr. Upsal.*
D. Carolus Allioni, *M.D. Soc. Reg. Lond. Soc. Bot. Prof. Emer. et Acad. R. Scient. Taurin. Soc.*
D. Petrus Arduino, *Oecon̄. Prof. Patav.*
D. Antonius Bannal, *Hort. Reg. Monsp. Custos.*
D. Ludovicus Bellardi, *M.D. Coll. Med. Taurin. Soc.*
D. J. P. Berthout Van Berchem, *Soc. Phys. Lausann. Soc.*
D. P. M. Augustus Broussonet, *M.D. Soc. Reg. Lond. Ac. Reg. Sc. Paris. et Monsp. Socius, Soc. Reg. Oecon. Par. a Secretis.*
D. Cæsar Caneferi, *M.D.* Genuæ.
D. Antonius Josephus Cavanilles, Madriti.
D. Philippus Cavolini, *Acad. Reg. Sc. Neap. Soc.*
D. Cels, Parisiis.
D. Josephus Correa de Serra, *J.U.D. Acad. Reg. Sc. Ulyss. a Secretis.*
D. Rodrigus de Sousa Coutinho, *Legatus Regin. Lusit. ad Reg. Sardin.*
D. Dominicus Cyrilli, *Prof. Therap. in Lyceo Neap.*
D. Johannes Pet. M. Dana, *Hist. Nat. Prof. et Ac. R. Sc. Taurin. Soc.*
D. D'Antic, Parisiis.
D. Renatus Louiche Desfontaines, *M.D. Bot. Prof. Ac. Reg. Sc. Paris. Soc.*
D. Jacobus Anselmus Dorthes, *M.D. Soc. Reg. Sc. Monsp. Soc.*
D. Hippolitus Durazzo, *Patricius Genoensis.*
D. Felix Fontana, *Director Mus. Phys. Reg. Florent.*
D. Ludovicus Gerard, *M.D.*
D. Esprit Giorna, Taurinæ.
D. Antonius Gouan, *M.D. Med. Prof. Soc. Reg. Sc. Monsp. Soc.*
D. Millin de Grandmaison, Parisiis.
D. Johannes Theoph. Groschke, *M.D. Hist. Nat. Prof. Mitav.*
D. Carolus Ludovicus L'Heritier, *Dom. de Brutelle, in Aulâ Juvam. Paris. Reg. Consil.*
D. Johannes Herman, *M.D. Med. Chem. et Mat. Med. Prof. Argentor.*
D. Antonius Ludovicus de Jussieu, *Acad. Reg. Sc. Paris. Soc.*
D. Josephus Franciscus a Jacquin.

D. Nicolaus Josephus a Jacquin, *Soc. Reg. Lond. Soc. Bot. et Chem. Prof. Vindob.*
D. Wernerus de Lachenal, *Prof. Bot. Basil.*
D. Lambertus Lucas Van Meurs, *Anat. Prælect. Amstelod.*
D. Gulielmus Antonius Olivier, *M.D.* Parisiis.
D. Nicolaus Pacifico, *Acad. Reg. Sc. Neap. Soc.*
D. Johannes Baptista Pratolongo fil. *M.D.* Genuæ.
D. De Reynier, *Soc. Phys. Lausann. Soc.*
D. David Van Royen, *Soc. Reg. Lond. Soc. Bot. Prof. in Acad. Lugd. Bat.*
D. Scarpa, *Anat. Prof. Ticin.*
D. J. G. Schlanbusch, *Legatus Regis Dan. ad Reg. utriusq. Siciliæ.*
D. Christianus Frid. Schumacher, *Acad. Reg. Chirurg. Hafn. Adjunctus.*
D. Andreas Sparrman, *M.D. Acad. Reg. Sc. Stockholm. Soc.*
D. Olaus Swartz, *M.D. Acad. Reg. Sc. Stockholm. Soc.*
D. Nicolaus Sam. Swederus, *Regi Suec. a Sacris.*
D. Thibaud, *M.D.* Monsp.
D. Thouin, *Acad. Reg. Sc. Paris. Soc.*
D. Carolus Petrus Thunberg, *M.D. Equ. Ord. Wasiaci, Soc. Reg. Lond. Soc. Med. et Bot. Prof. Upsal.*
D. Tingry, *Chem. Prælect. Genev.*
D. Octavian Targioni Tozzetti, *M.D.* Florent.
D. Ericus Viborg, *Artis Veterin. Lector Hafn.*
D. Villars, *M.D.* Gratianop.
D. Philippus Werner, *Chirurg.* Algir.
D. Willemet, *Bot. Prof. Nanc.*
D. Jacob. Samuel Wyttenbach, *S.T.P.* Bern.

Associates.

Mr. William Boys, F.A.S. Sandwich.
Francis Buchannan, M.D. East-Indies.
William Coyte, M.D. Ipswich.
Edward Whitaker Gray, M.D. F.R.S. British Museum.
John Heysham, M.D. Carlisle.
Thomas Hope, M.D. *Professor of Chemistry*, Glasgow.
Mr. James Hoy, Gordon Castle, Scotland.

Mr. George Humphrey, Long-Acre.
Mr. Edward Hunter, Caen-Wood.
Mr. Thomas Lamb, Reading.
William Markwick, Esq. Catsfield, near Battle.
Mr. John Pitchford, Norwich.
Mr. William Sole, Bath.
Mr. James Sowerby, Mead-place, St. George's Fields.
Jonathan Stokes, M.D. Kidderminster.
Mr. Lilly Wigg, Great-Yarmouth.

"I understand," says Mr. Woodward in a letter dated the 20th of April, 1788, "that the Society has met at your house, and that your '*Introductory Discourse*' met with great approbation. I do not doubt of this happening: but do you remember a promise that I should see this same discourse; and do you suppose that my curiosity is quite laid asleep? I want to hear also somewhat about the Institution. Who are the resident members? What papers have you read there? If there have been more than the first meeting; and where? Do you likewise recollect that you promised me some observations, which you had collected upon *Lycoperdons?*—the want of which puts me to a stop as to my promised paper on that genus. Upon mature consideration, however, I have determined to confine my observations on the species to those of Great Britain, as being more able to speak with precision upon them; and when a person writes on a single subject, as little as possible should be left doubtful, the very design of a Monograph being to clear up all difficulties on the subject. Now it is evident that in the *Cryptogamia*, the moment you take the whole globe into your system, doubts and difficulties innumerable must arise; and in this class it is quite enough to know and understand the plants even of one's own vicinity.

"I hope you received my last, with my thanks for your invaluable parcel, which made me rich beyond my warmest hopes as a botanist. 'You hope the specimens will prove acceptable!' What a modest way of speaking of treasures above all price! I know not how sufficiently to thank you for such; the possession of which has always been beyond my hopes and expectations. I know however that your liberality, and, I am proud to add, your friendship, is such, that you truly gratify yourself when you confer a favour on your friend. The three Orchideæ are inestimable, and the rest of the specimens highly valuable. I sent the parcel to Pitchford directly, attending to your request of not mentioning what I had received myself, that I might not take off his surprise at the sight of *Ophrys Loeselii.* You have doubtless heard from him, with his acknowledgements. I got, however, rather found fault with for not informing him what I had got; and had something of a lecture for *wickedly* telling him that I lost no time in forwarding the parcel, as Lent was just approaching: and I did not know, if I delayed it till that period, whether he might not incur a penance for indulging in a mental feast at the time that he was restricted from corporeal indulgencies. I could not have helped this sally, if I had died for it! I have kept him in the dark yet, but shall send him an account directly. Mrs. Woodward thanks you for the holy chaplet*."

The exertions of the spring, or perhaps nothing more than might have happened without any such excitement, produced one of those inflammatory disorders to which Sir James had always been liable, and which in some shape or degree appeared every spring during his life. For these attacks, the climate and waters of Matlock always proved a specific remedy. This happy effect, combining with the natural beauty of the place, realized a fairy-land, where health and pleasure took place of weakness and fear: the cure was so agreeable, that instead of reflecting with aversion upon these periods of debility and alarm, he frequently recurred to such times, as parts of his life that were attended with the happiest sensations,—as a holiday from mental exertion, and during which his mind was refreshed, as well as his bodily strength renovated. Thither he went in the autumn of 1788.

The following letters from his father are addressed to him while at this place.

* In another letter, dated the 15th of November the same memorable year, Mr. Woodward concludes with telling him: "I must leave the caves of Neptune and bowers of Flora for a while, to inform you I was at Mr. Coke's magnificent fête at Holkham, on November the 5th. Descriptions of it you have seen in the newspapers, without doubt; it suffices therefore to tell you, that they were not at all exaggerated; the entertainment being magnificence itself, and the splendid mansion having quite the air of an enchanted castle, when illuminated for the reception of the company. Adieu, most truly and affectionately yours,

Thomas Woodward."

Mr. Smith to his Son, Matlock Bath.

Dear Son, Norwich, August 12, 1788.

I rejoice to hear that you bore travelling so well, and had begun to bathe, and found yourself so well after it. To these assistances to your health it is to be hoped, relaxation from application, the charms and amusements of the place you are in, and the company of the most amiable and affectionate of mothers, will make you so happy as to contribute no small share for you to lay in a good store of cheerfulness and joy. Would to God it may make you robust and strong! It is what I have seen the Bristol and Matlock waters do to many people.

You will get more good I doubt not at Matlock; but if you had not gone thither, I think you would have been both pleased and benefited at Setch* in such a season as this, with plenty of fine fruit, fine fish, a nice dairy, and beautiful fields. Perhaps the spa affords a water which would be beneficial; at least I think it would, and would try it if I should want any assistance of the kind.

It is said the benefit of the waters is most frequently found after people leave them; but I found benefit upon the spot, and every spot I trod a long time after. May it please Heaven you may both do so likewise! I hope you have or will take a walk towards Cromford, and turn to the right when you come at the stream; pursue it some way, it will afford you some very picturesque views. If you

* A farm of Mr. Smith's, near Lynn in Norfolk.

turn towards the left, towards Cromford, you will find a cold room, where the water that runs over it to the cotton-mill issues through the wall. It was there I saw the flies that alighted upon the sides so congealed with cold they could not get away, and afterwards petrified with it. I think I gave you one that I brought away with me.

Sir Richard Arkwright was clearing the ground for his house when I was there.

Dr. Johnson wanted very much to see Matlock by moonlight; you must have had that pleasure. I have been reading his Letters, published by Mrs. Piozzi. They are full of beautiful passages, and some of hers cut a very good figure.

I have not read the book you sent, and shall hear the opinion of others before I resolve to bestow my time upon it, in preference to many, many other works which I wish to read, but which I certainly shall not have time to read in this world; unless, upon the Pythagorean system of the metempsychosis, I should reappear in some other body; and then I may be thrown into a different country, where there may not be the reading I want, or no reading at all.

I admire Jebb's* works, and more the man; so

* The following note, from Mr. Smith's common-place book, illustrates his sentiments more at length on this subject:—

" *The Works, theological, medical, political, and miscellaneous, of* John Jebb, M.D. F.R.S.; *with Memoirs of the Life of the Author*, by John Disney, D.D. F.A.S., in three volumes 8vo, 1787. The excellent author of these volumes was a pattern of integrity, patriotism, and zeal for the civil and religious liberties

does your mother; very little of physic, more of politics; but most of all, the principles upon which Protestants cannot consistently subscribe to the Thirty-nine Articles, nor enforce them upon others. Next to Dr. Franklin's, I like Jebb's style; 'tis simple, plain, and persuasive; and the heart that dictated, the reader feels was an honest one.

Your brother Francis has made a most ingenious instrument or machine, to show the position of the planets, and aspect of the heavens. It is the most simple imaginable, yet the most comprehensive; it meets the admiration and approbation of the few scientific he has shown it to.

I am ever yours,

James Smith.

of mankind. He, like Mr. Lindsey and Dr. Disney, resigned preferments in the Established Church, as incompatible to hold with the principles they had espoused; and was one of the most active and able amongst the committee at the Feathers Tavern, to conduct the petition to Parliament for the abolition of subscription to the Thirty-nine Articles of the Church of England; which was unsuccessful. He was the foremost in promoting the attempts to reform the Commons House of Parliament, and to shorten its duration to one year, agreeable to the ancient usage of the Constitution. Upon these occasions he made the most respectable figure by his speeches and his publications, which are now given complete in these volumes. They contain, in regard to religion, the truest and clearest principles of Protestantism, independent of any particular sect or opinions; and in regard to the English Constitution, the most enlightened and soundest maxims upon which its liberties depend. His thoughts upon religion and parliaments are admirable; his style clear and nervous in a high degree. I think it very much resembles the great, the excellent Dr. Franklin's; and like such comprehensive and upright minds, their principles were very much the same.

Mr. Smith to his Son.

My dear Son, Norwich, Sept. 20, 1788.

I congratulate you most heartily upon your return to London in such good health. You will hear from your other correspondents how much Norwich has been amused and pleased with Mrs. Siddons, performing nine nights to audiences so crowded, so heated, that I could get in only once, and that was to the upper gallery, to see her in Belvidera, in Venice Preserved. I never was so charmed with acting, no not by Garrick; and had I not consulted my health, ease, and pocket, should have attended at every different part she performed, which is saying a great deal for me, who am not, nor ever was, fond of the theatre, nor play-reading.

We are to be very full next week at the music

Dr. Jebb was deserted by many who appeared for some time to coincide in his sentiments, and he met with very unkind treatment; but he was firm to his integrity to the last. When he left the Church, he studied physic, in which his biographer says he acquired much skill and considerable practice. He was born Feb. 16, 1736; married Miss Torkington of Little Stukeley near Huntingdon, Dec. 29, 1764; and died in Parliament-street, Westminster, March 2, 1786, in the fifty-first year of his age; eminently esteemed by the truly great and the virtuous in all the British dominions, and the United States of America; and extremely beloved by those who were so happy as to be acquainted with him; for he had the character of being a very amiable man.

"The definitions in the third volume, beginning at page 253, and ending page 258, of an Atheist, a Deist, a Jew, a Mahometan, a Christian, a Papist, a Dissenter, and a Church-of-Englandman, are admirable."

meeting,—a very novel scene to Norwich. A considerate person must marvel to see, on the one hand such extravagance and dissipation, on the other so much distress and complaint, from the decay of trade, and the increased load of taxes: but I believe it has been observed as the characteristic of a luxurious sinking empire, and is often that of a sinking individual.

Perhaps you will smile when I tell you I read Rousseau's works a second time. In reading his Heloïse, I became very much interested in his character, and pleased with his genius. I read the volume you lent me with more attention the second time; the first I hurried it over, and found I had missed a thousand beautiful passages: now, I studied him, and found in his Heloïse a store of the finest thoughts, and most profound observations of any book I almost ever read. I cannot alter my opinion of the man. I think him a heterogeneous composition of great vices, and fewer virtues, but of a sublime genius, and a penetrating faculty into the human heart, that no writer has developed with so much perspicuity and ingenuity. He carries you to the bottom of it, and will not leave you till he has made you thoroughly persuaded you understand it yourself. His descriptions are amazingly strong. Sterne had him certainly in view, but he is so minute as to leave nothing to the reader's imagination, and puts one in mind of the laboured exactness of the Flemish painters. Rousseau's pictures have infinitely more force, by not making each trait so very distinct. Moore says, " All dress was meant for fancy's aid." Rousseau's own religion is somewhat equivocal. It is Julia's pleases me. Her character is inimitable, and her last moments are the most pious, the most rational, and the most touching surely ever drawn by pen. I do not know whether I shall read all Heloïse again, but I certainly shall the last volume. I shall study it attentively, and I shall like to mark all the beautiful thoughts in it, and make them very familiar to my memory. There are many sentences infinitely surpass Rochefoucault's, or any proverbialist but the sacred. It is a pity the work is not quite fit for young persons, females especially. He has said so in the preface, and warned them against it; but there are parts I could select, which every person, young and old, might be much edified by. I was most wonderfully attached to the latter part of Heloïse, so as I never was to any book before; and in the mood in which I then was, from the solitude of my family, it made a deeper impression, I suppose, than it would at another time.

Adieu! God bless and prosper you!

JAMES SMITH.

From this period, Sir James gave lectures on botany and zoology at his house in Great Marlborough-street, where he was honoured with the attendance of the Duchess of Portland, Viscountess Cremorne, Lady Amelia Hume, and the Honourable Mrs. Barrington, besides professional men. He also delivered a course of lectures on botany at Guy's Hospital, for several successive years.

At this time a flow of letters from foreign members and professors greatly enriched his correspondence, and ceased not till his death. The few which immediately follow are from men highly distinguished for scientific attainments, and as proficients in the pursuit which engaged so much of the attention of him to whom they are addressed.

A. L. de Jussieu to J. E. Smith, P.L.S.

Monsieur, Paris, ce 19 Mars, 1788.

Je vous dois des remercimens bien sincères pour le zèle et la promptitude avec lesquels vous avez bien voulu m'obliger en faisant des recherches dans votre herbier à mon intention : les notions que vous me communiquez sur plusieurs genres me sont très utiles ; et je mets beaucoup de prix aux échantillons que vous m'avez envoyés, quoique la plus part soient très incomplets comme vous l'observez vous-même : ils m'aideront à reconnoître les mêmes plantes dans mon herbier, si je les ai, ou du moins à me faire une idée plus juste de leur caractère.

Mr. Tournier, auteur de la Collection des Portraits d'Hommes illustres vivans, compte donner le mois prochain la figure de M. Sparrmann : il est venu me demander un échantillon ou un dessin du *Sparrmannia* qu'il veut joindre à la gravure : je n'ai pu le satisfaire, n'ayant point cette plante qui n'est d'ailleurs figurée nulle part; mais je lui ai promis de vous écrire pour obtenir cette communication, et je suis porté à croire que vous voudrez bien nous obliger tous deux en ce point. Dans le cas, ou vous ne le pourriez pas, je vous prierai d'avoir recours à Mr. Banks qui se feroit probablement un plaisir de fournir l'encadrement du portrait de M. Sparrmann: je vous observe qu'on voudrait le publier à la fin d'Avril prochain.

Mon travail avance beaucoup, quoique j'ai été beaucoup détourné par diverses occupations: je viens de perdre mon beau-père après une longue maladie pendant laquelle je lui ai donné tous mes soins: c'est même en partie cette circonstance qui a retardé ma reponse à votre lettre. Vous savez que chez moi les sentimens du cœur ne sont point etouffés par l'ambition de l'esprit: on prétend que les sciences dessèchent l'âme; je crois que c'est selon la manière dont elles sont traitées. Ceux qui font de la science un métier, négligent pour elle tout le reste: ils la regardent simplement comme un moyen de parvenir, et sacrifient souvent d'autres intérêts à celui-çi. De la vient que tous les savans ne sont pas aimables, que peu sont communicatifs, que la rivalité entr'eux se change souvent en inimitié. J'ai cru m'apercevoir que vous ne traitez pas ainsi la science par ce que vous dites de l'aimer pour elle-même, pour l'agrément qu'elle peut vous procurer, pour l'occasion qu'elle vous offre de vous rendre utile à vos concitoyens en leur offrant le tribut de vos connoissances. Ne vous écartez jamais de ce plan, et vous vous en trouverez bien: je vous avoue que c'est le mien: mon temps se partage entre l'étude et d'autres devoirs domestiques que je ne regarde pas comme étranger pour moi. J'aime à cultiver l'amitié; je consens volontiers à interrompre mes affaires pour m'occuper de celles des autres lorsqu'il en résulte une utilité pour eux ; et dans cette disposition les occasions se présentent souvent ; le travail particulier en souffre à la vérité ; mais il va toujours, quoique plus lentement. Depuis que je ne vous ai vu, j'ai encore fait à mon travail des additions et des changements utiles ; et j'en ferois encore plus si j'avais vu vôtre collection et celle de M. Banks ; mais il m'est impossible pour ce moment de m'absenter. Je crois, que pour satisfaire les élèves du Jardin de Paris, je publierai cette année mon *genera*, à le corriger dans une seconde édition, si elle a lieu. Croyez que lorsque je pourrai aller vous voir, je le ferai bien volontiers : je serai aussi charmé de voir M. Banks, que j'estime depuis long-tems à cause de sa manière de traiter les sciences. Rappellez-moi, je vous prie, à son souvenir, et recevez les assurances du sincère attachement et de la considération très distinguée avec lesquels j'ai l'honneur d'être,

&c., &c., &c.

A. L. De Jussieu.

D. Rodrigo de Sousa Coutinho to J. E. Smith.

Monsieur, Turin, ce 13 Août, 1788.

J'ai eu l'honneur de recevoir votre aimable et intéressante lettre, en date du 21 Juillet ; et, en vous témoignant toute ma sensibilité pour votre souvenir et pour vos gracieuses expressions, permettez que je vous assure qu'il faut toute la connoissance que j'ai de mes faibles talens pour ne point me méconnoître d'après ce que vous daignez m'écrire. Je ne sais point trouver des expressions pour vous remercier de l'amitié avec laquelle vous avez voulu agréger mon nom à celui des plus grands hommes d'Europe, où il ne peut figurer à aucun autre titre qu'à celui du vrai zèle qui m'anime pour la perfection et encouragement des sciences naturelles qui seules font le vrai bonheur de l'homme, et qui ont même dans le moral contribué beaucoup à la reconnoissance des droits dont l'ignorance et la superstition avoient partout depouillé les hommes. Permettez que j'ajoute ici l'offre ingénu et entier de mes faibles services pour votre société dont vous êtes si dignement le Président, et qui, j'espère, portera la plus grande lumière dans la science qu'elle cultive de préférence, et qui surement en a bien besoin. Je prendrai la liberté de vous adresser incessamment le genre *Arenaria*, tel qu'il est décrit dans la Flora d'Allioni, et je vous enverrai deux exemplaires de chaque espèce: je vous préviens cependant que ces espèces que M. Allioni a marquées sont encore sujettes à bien des difficultés; sa vue et ses infirmités ne lui ayant point permis d'y porter son exaction connue. Sur votre lichen que j'ai confié au Dr. Bellardi, je suis parfaitement de votre avis; et je le crois très distinct quoique très approché du *nivalis*. J'ai communiqué votre lettre au Dr. Bellardi qui a été très sensible à vos expressions sur la réponse au Dr. de Chambery, qui est un vrai charlatan, et qui par des phrases couvre ses mauvaises raisons.

Excusez si je vous prie de faire mes complimens à MM. Harbord et Zimmermann, surtout à ce dernier, dont les connoissances statistiques me le rendent très estimable.

J'ambitionne le bonheur de lire vos œuvres que vous m'annoncez; mais malheureusement je ne connois que la tardive voie de mer pour les recevoir, la même dont je me servirai pour vous adresser les plantes ci-dessus mentionnées, et que je vous expédirai ou par la voie de votre consul à Gênes ou par celle du notre. Le souvenir que vous gardez de ma bibliothèque m'oblige à vous répeter que je dois à vos tuteurs Anglais que j'ai toujours estimé de préférence le peu de lumières que j'ai dans les sciences naturelles, et dans celle de mon métier en particulier; et vous sentez bien avec quelle justice et avec quel plaisir je devois vous montrer ceux que je considère à si juste titre comme mes maîtres.

J'oubliois qu'une longue lettre va vous distraire de vos savantes occupations, qui vous assurent tous les jours d'avantage la haute estime dont vous jouissez déja, et qui m'inspire les sentimens du respect et attachement avec lesquels j'ai l'honneur d'être,

Monsieur Smith,

&c., &c., &c.

D. Rodrigo de Sousa Coutinho.

Esprit Giorna to J. E. Smith, P.L.S.

Monsieur, Turin, ce 29 Août, 1788.

J'ai reçu sur la fin de Juin passé votre obligeante lettre; et vous m'aurez déjà, à l'heure qu'il est, accusé avec raison de négligence et d'impolitesse de n'y avoir pas encore répondu. Je serais au déséspoir, Monsieur, que vous m'accusiez de manquer aux sentimens d'amitié et de respect que je vous dois: quant au reste, j'espère que vous trouverez bonnes les excuses que je vais vous faire ci-après.

Je vous dois une infinité de remercimens de l'honneur que vous avez bien voulu me faire en m'associant à votre Académie Linnéenne, honneur que je ne dois certainement pas à mes mérites, mais tout à votre bonté, car mon amour propre n'arrive pas à me flatter jusqu'à ce point.

Les collègues qu'il vous a plu me donner attendront assurément des grandes choses de moi après les assurances que votre bon cœur leur en aura données, et ils se verront trompés: en effet, comment répondrai-je à leur attente, moi, que mon état occupe aux mathématiques, et qui n'étudie l'histoire naturelle que par amusement? Dans ce cas, Monsieur, je ne vois qu'un parti à prendre, c'est de laisser à votre bonté même, qui m'a engagé dans cet embarras, le soin de faire mes excuses.

Je ne prétends pourtant pas avec cela de me refuser à donner à votre Société ce que mes faibles talens pourront découvrir de nouveau dans la vaste science qui fait son objet: au contraire, l'honneur que j'en reçois me servira d'aiguillon qui m'engagera à me donner avec plus d'interet à l'histoire naturelle, et à faire tout mon possible, *si Dii tibi otia dabunt*, pour contribuer en quelque sorte aux progrès de cette science: en attendant, pour vous marquer la sincérité de mes intentions, je tâcherai de faire peindre d'après nature quelques insectes que j'ai seul, et que je crois inconnus jusqu'à présent, comme aussi quelques oiseaux qu'on a pris en Piémont, et qui ont échappé aux ornithologistes ; et, s'il m'est possible de les avoir à tems, je vous ferai passer ces dessins avec leurs descriptions dans une caisse d'oiseaux arrangés à ma façon, que je dois envoyer à Londres vers la fin d'Octobre prochain.

Dans l'édition que vous vous proposez de faire des ouvrages du célèbre Linnée, je ne sais pas si vous contez de donner les figures de tous les insectes, ou au moins de ceux qui n'existent pas encore dans les autres auteurs ; en ce cas, je pourrais vous procurer nombre de ceux inconnus de Linnée, et nommés par Fabricius, et par Müller, qui les ont tirés du Piémont.

Nous avons ici M. le Baron Daviso, amateur et collecteur de minéralogie, qui, entendant que je devois vous écrire, m'avoit chargé de vous prier s'il était possible de faire quelques échanges avec vous, ou avec quelqu'autre de votre connaissance en minéraux. Il offre de vous envoyer le premier une collection de métaux et minéraux du Piémont, désirant en obtenir de ceux des isles Brittaniques. Il m'avoit promis une liste de tous ceux qu'il possède, et de ceux qu'il désire pour vous lui transmettre.

Il est parti pour un voyage dans les Alpes : je le croyais de retour à la fin de Juillet : je l'ai attendu ; et je l'attends encore à présent. Voilà, Monsieur, la cause du retard à vous répondre, et voilà l'excuse que j'avais à vous faire.

Faites agréer, Monsieur, mes très humbles respects aux membres de votre très louable Société, et particulièrement à M. Marsham, que vous me dites occupé à l'Entomologia Brittannica, et qui feroit pour plusieurs raisons très bien, (comme vous le remarquez) de commencer par donner les Coleoptères, et ensuite les autres classes successivement. L'auteur ainsi n'est pas surchargé d'ouvrage : il ne risque pas d'être des fois prévenu par un autre. Le public achète plus facilement un volume que deux ou trois ; et, dès qu'il a pris le premier, il est engagé de prendre les autres.

En attendant, disposez de moi, Monsieur, en tout ce dont vous me jugerez capable de vous rendre services de ces côtes ici ; et vous me trouverez toujours tel que je me proteste avec les sentimens les plus sincères.

ESPRIT GIORNA.

Professor Scarpa to J. E. Smith.

Monsieur, Pavie, 12 Septembre, 1788.

Toujours en attente de vos ouvrages, que vous m'avez fait l'honneur de m'envoyer, je n'ai pas encore rien reçu : je suis dans la plus grande inquiétude à cause qu'il m'est retardé le plaisir de les lire, et parceque je ne puis rien savoir le précis sur ce que Fabricius a écrit au sujet de l'oüie des insectes; car malheureusement nous n'avons pas encore ici les Actes de Copenhagen de l'an 1783, et quand même ils y seroient, il n'y a personne ici qui connoit la langue Danoise. Comment faire donc? Il faut nouvellement avoir recours à vôtre amitié en vous priant de me traduire en Anglais le précis du mémoire, ou de le traduire tout-a-fait, s'il n'est pas bien long, et de me l'envoyer par la poste. Surtout, s'il y a quelque planche, comme il y en aura certainement, d'en ôter les en les placant contre un vitre, ce qui se fait aisément, et de me l'envoyer aussi. Cet incident est la cause du retard de mon ouvrage, qui seroit déjà sous presse; et si vous, Monsieur, ne daignez pas d'y prendre part, il tardera peut-etre encore des années. Je ne vous fais pas des excuses sur ma franchise de vous incommoder, parceque, ayant vécu parmi les Anglais, je connois parfaitement bien leur aimable libéralité.

Je vous fais plutôt mes congratulations sur votre nouveau établissement pour les progrès de l'histoire naturelle; et je suis très sensible à l'honneur que vous me faites de vouloir bien m'agréger à cette illustre compagnie.

Pour faire connoître d'avantage les fautes de l'ouvrage de Spallanzani sur la Digestion, j'ai trouvé bon de traduire le mémoire de M. Hunter sur le même sujet. Spallanzani a fait d'abord une réponse qui repousse M. Hunter un peu vilainement. Je vois cependant, qu'il seroit necessaire que M. Hunter donne une plus grande étendue à ses idées sur la digestion, qui, dans le mémoire, sont trop serrées et concises, et qui donnent lieu à des interprétations arbitraires. Il faut démontrer evidentement que l'ouvrage de Spallanzani n'a en rien fait avancer la science sur la digestion, et ce travail ne me paroit pas difficil pour un physiologiste, sur tout pour M. Hunter.

Le professeur Fontana me charge de vous saluer très distinctement.

&c., &c., &c.,

SCARPA.

P. S. La chimie et la botanique n'est pas encore remplacé chez nous après la mort de Scopoli. On s'est flatté d'avoir ici M. Murray de Gottinge.

A. L. de Jussieu to J. E. Smith.

Monsieur, Paris, ce 20 Mars, 1789.

Je profite du départ de M. Olivier, naturaliste très recommandable et fort versé dans la connoissance des insectes, pour me rappeller à votre souvenir. Comme il part dans deux heures, je n'ai qu'un moment pour m'entretenir avec vous, et pour vous remercier de l'envoi que vous m'avez fait du *Sparmannia* l'année dernière. Votre petit échantillon me l'a fait retrouver sur-le-champ dans mon herbier, et maintenant je le connois bien : ce genre vient dans ma famille des *Tilleuls* entre le *Triumfetta* et le *Sloanea*. Je vous avois annoncé que j'allais imprimer un *Genera* disposé par familles : ce travail, commencé en Mai dernier est maintenant achevé et imprimé presque entièrement : il ne me reste plus que l'introduction élémentaire à terminer ; et j'espère pouvoir offrir l'ouvrage au public et à vous au mois de Mai prochain. Vous y verrez que j'ai profité des connoissances que vous m'avez communiquées ; et je serai également disposé dans la suite de faire le même usage de vos avis. Mes genres passeront le nombre de 1800, parceque j'y joins ceux d'Aublet, de Forskal, de Forster, de Commerson, et les nouveaux de mon herbier.

Vous m'annoncez dans votre dernière, que vous avez formé une Société Linnéene, c'est à dire, purement botanique, et que vous voulez bien m'y adopter : je serai sûrement flatté de tenir à une société qui s'occupe des progrès de la science, et je ferai mon possible pour mériter son suffrage et son estime. J'espère qu'elle voudra bien me pardonner si je ne suis pas tout-à-fait Linnéen : je m'écarte de ce grand homme dans sa partie systématique, qui me paroit éloigner la science de son vrai but ; mais je fais grand cas de sa nomenclature, de ses genres, de ses espèces ; et je crois qu'il a rendu en ce point un vrai service à la botanique, quoique souvent dans ses genres il omette les caractères essentiels tirés des insertions.

Si votre herbier de Linné fils est maintenant en ordre, je me recommanderai encore à vous pour avoir connoissance de ses genres. Je suis, comme vous savez, peu exigeant : les plus petits échantillons me suffisent : pourvu que je puisse reconnoitre les caractères, je suis content ; et j'avoue que j'ai grand besoin de cette communication pour mieux connoître les plantes du supplement, dont la description n'est pas toujours assez claire.

Je finirai ici ma lettre pour la faire tenir sur le champ à M. Olivier. Recevez, je vous prie, les assurances du sincère attachement et de la considération très distinguée, avec lesquels j'ai l'honneur d'être,

&c. &c. &c.

A. L. De Jussieu.

The Abbé Cavanilles to J. E. Smith, P.L.S.

Monsieur, Paris, ce 29 Juillet, 1789.

Agréez, je vous prie, mes remerciments pour la complaisance que vous avez montrée dans votre lettre que je viens de recevoir; elle porte l'empreinte de la franchise et de l'honnêteté qui vous caractérisent. Nous pourrons fort bien penser chacun de sa façon, et travailler cependant à l'ouvrage de la science cherie, avec la seule différence que mes ouvrages seront toujours inférieurs en mérite aux vôtres. Cette raison, Monsieur, doit vous engager à corriger mes fautes, et à donner des planches, qui mettront le public en état de s'assurer de la perfection et de la vérité. Mes *Tourreas* ont été gravées d'après les échantillons trouvés chez M. de Jussieu (Herb. Comm.), et j'ai cru rapporter à ce genre le *T. lanceolata,* malgré que ses anthères soient placées un peu plus haut que dans les autres espèces: vous aurez remarqué que je nomme *tuyaux,* tubus, ce que vous appellez nectaire d'après notre maître, Linnée. J'ai rapporté dans mes ouvrages les raisons qui m'ont obligé à répeter une expression trop vague et par conséquent obscure.

J'ai lu avec autant d'attention que de plaisir, ce que

vous m'écrivez sur les Passiflores: *unusquisque suo sensu abundet*, c'est une loi que le filosofe doit suivre.

Mais puisque vous avez la bonté de parler avec franchise, je m'en vais faire autant, et soumettre à votre tribunal mes raisons. Nous trouvons dans les ouvrages de Linnée ces principes : 1°. *Genus dabit characterem, non character genus.* 2°. *Genus est characteris causa, non character generis.* 3°. *Omnia genera sunt naturalia.* Fixons d'abord la véritable notion de ce mot *genre;* selon la bonne logique *genre* représente une idée commune à plusieurs espèces: il est par conséquent l'ouvrage de notre ame, qu'elle a fait d'après l'examen et la comparaison des caractères trouvés dans les individus: il n'existe donc dans la nature, ou tout est individuel; et tout son être est renfermé dans notre esprit; en sorte que s'il n'existoit aucun esprit, il n'existeroit non plus aucun genre. D'après cet exposé, je trouve que cette proposition, *omnia genera sunt naturalia*, est absolument fausse. Si pour la soutenir on dit que c'est la nature qui a fait les individus, et par conséquent les caractères que notre esprit choisit pour former l'idée générale, genre, on peut répondre que l'idée générale représentée par ce mot *genre*, n'embrassant tous les caractères des individus, doit être tout-à-fait différente de ce que la nature a fait.

Et en vérité pour former un genre qui embrasse toutes les Passiflores, nous serons obligés de mettre de côté comme inutile et gênant: 1°. le calice (*involucrum* Linnæi); 2°. les folioles ou divisions de celuy-ci; 3°. les franges ou couronnes tantôt simples, tantôt nulles, et quelquefois triplées; 4°. les rangs des semences, variant de trois à quatre; 5°. l'envelope partiel de chaque semence, qui est nul dans certaines espèces. Et nous serons en fin obligés de nous renfermer dans les étamines, germe, stigmats, et fruit. C'est à dire, nous serons obligés de séparer par notre imagination ce que la nature a réuni, et par conséquent nous donnerons un ouvrage different de celui de la nature.

D'après ce que je viens de dire je ne sais pas comment peut on avancer que le genre est la cause de caractères, puisque ce sont les caractères qui forcent notre esprit à former les genres; puisque ce sont les caractères qui ont une existence réelle dans la nature, et que les genres n'existent hors de nous mêmes. Vous convenez, Monsieur, que Linnée n'a pas observé souvent les lois botaniques qu'il établit dans ses ouvrages: vous dites aussi que son traité sur les *Malvacées* n'est pas la meilleure chose qu'il a fait. . Je crois qu'il ne devoit pas avancer comme il a fait dans son *Genera*, p. 306: *Nos genera distinguimus a calice, qui magni momenti est, et limites absolutos ponit:* mais je crois avoir démontré contre M. L'Héritier, que si on neglige le calice dans les Malvacées pour former les genres de cette famille, on ne fera que confondre les plantes que le port et les caractères séparent tout à fait.

Si j'osois vous dire que *genitalia et fructus sunt idem* dans une grande quantité des genres reconnus différents par tous les botanistes, les Malvacées, les Ombellifères, les Cruciformes, les Azédaracs, les Liserons, &c.; et si de plus j'avois le bonheur de vous le prouver, je crois, Monsieur, qu'alors mon idée sur les Passiflores ne seroit pas si mauvaise. J'ai dit dans ma réponse à M. Medicus, que la nature n'a pu faire aucun genre, que tous sont l'ouvrage des botanistes, et que ceux-ci les ont construit pour mettre de l'ordre dans ses recherches, et pour trouver des points d'appuy ; d'ou ils partent pour découvrir les richesses de la nature. Je vois que notre manière de penser est différente, mais je ne puis pas imaginer que ça soit un obstacle, qui nous doit aussi séparer. Mon respect, Monsieur, pour vous, et mon attachement sera sans bornes, et soyez en sur que je n'aurai jamais à me plaindre de vous, ni vous de moi.

J'ai déjà terminé ma 9me dissertation, dont le fond principal sont les *Banistères.* J'ai un doute sur la détermination du *Banisteria fulgens,* Linn., que vous pouvez dissiper, comme possesseur de son herbier. Je vous prie de me dire si la feuille que je vous envoye appartient à cette espèce.

Vous souhaitez, Monsieur, posséder la partie botanique de l'Encyclopédie. Rien de plus facile : M. Pankouk a ouvert une nouvelle souscription pour chaque dictionnaire, au prix de 12 francs le volume, si je ne me trompe pas. M. de Lamarck ajoute à la fin de chaque volume un index Latin des genres y compris ; par ce moyen vous aurez le plaisir de savoir dans un instant tout ce qu'il contient.

Permettez, Monsieur, de vous renouveller tous mes remerciments, et de vous assurer de mon profond respect, avec lequel j'ai l'honneur d'être,

Monsieur, vous tres humble, &c.

D. ANTOINE JOSEPH CAVANILLES.

Sir C. P. Thunberg to J. E. Smith.

Upsaliæ, d. 21 Dec. 1789.

Permittas velim, Vir illustris, ut hisce gratam tibi meam significem mentem, ob honorem et amicitiam mihi præstitam, ea occasione, qua membrum Societatis vestræ honorificæ et utilissimæ electus sum. Hacce quoque utor occasione, ad te mittendi quasdem meas Dissertationes academicas, inque his duo ex meis generibus Japonicis, quibus forsan non possides in herbario, scilicet, *Houtuyniam* et *Cleyeram*, quam ultimam nescio sane an sit vere diversa a *Ternstroemia*. Nihil mihi magis gratum erit, quam si ullo modo tibi et Societati, ad quam nonnullas meas quoque mitto Dissertationes, utilis esse possim; cumque credibile sit, ex Capensibus et aliis Indicis plantis, præsertim novis inque Supplemento descriptis meis plantis, multas in tuo herbario Linnæano desiderari, vellem lubenter tot tibi, Vir amicissime, suppeditare, quot mihi in duplo esse possunt. Sique ex tuis duplicatis nonnullas mihi deficientes mecum communicare velles, gratissimum hoc mihi foret munus. Rogo itaque ut deficientium catalogum mihi mittas; sique hoc non renuis, ego tibi meum catalogum deficientium reddam. Interim tuo favori et amicitiæ memet commendatum, tibi addictissimum, habeas.

The following letter, and the only one from the celebrated naturalist, Gerard, author of the *Flora Gallo-Provincialis*, "one of the best European botanists of the golden age of Linnæus," will not be considered long or wearisome by those who recollect Sir James's visit to the good old man at Cotignac, situated in a beautiful part of the South of France, where he went on purpose to see him, and felt, as he tells us, as if "paying a visit to the Elysian Fields; so little did his tales of other times seem connected with what is now going on in the world."

Monsieur,

A la satisfaction dont vous me procurates la jouissance pendant le très court séjour que vous fîtes à Cotignac, se joignît bientôt le regret causé par votre départ, qui ne me permit pas de cultiver votre amitié comme je l'aurois desiré, ni de vous procurer les renseignements que ma position particulière me permettoit de vous présenter rélativement aux plantes de cette contrée, qui ne sont point caractérisées comme elles devroient l'être dans la dernière édition du *Systema Vegetabilium.*

Mais, à la réception de votre lettre et des ouvrages dont vous avez la bonté de me faire part, que je reçois avec toute la reconnoissance dont je suis susceptible, ce regret a fait place à un souvenir qui me sera toujours infiniment précieux par le cas que je fais de votre amitié, et par le désir que j'ai de la cimenter autant qu'il dépendra de moi.

C'est pour vous en donner une preuve que je vous communiquerai les observations critiques dont le système de Murray me fournira le sujet, relativement à plusieurs plantes de Provence, et que des nouvelles observations faites depuis très peu de temps pourront rendre plus importantes. Occupé essentiellement depuis plus d'une année à perfectionner l'ouvrage que j'avois publié autrefois sur les plantes de Provence, j'ai, pour ainsi dire, entrepris un nouveau travail plus pénible que le premier qui fourmille de négligences. J'ai passé en revuë toutes les espèces : j'y en ai ajouté plus de 300, et comme il n'est presque aucune plante sur laquelle je n'aye quelque chose à dire, je pense que cet ouvrage sera au moins une fois aussi volumineux que celui qui fut imprimé en 1761. Je n'ose me flatter pourtant qu'il se publie, quoiqu'un libraire d'Aix me témoigne qu'il en auroit envie ; mais comme mon travail est assez avancé, je le poursuis avec autant d'empressement que si j'étois bien assuré qu'il ne se présentera aucune difficulté.

Je vous offrirai donc les prémices de ce travail, relativement à l'édition du *Systema Vegetabilium*, de laquelle vous vous occupez; et ce sera avec d'autant plus de plaisir qu'il ne sera pas entièrement perdu si mon projet ne s'exécute pas.

J'apprends avec plaisir que vous avez fait un long séjour en Italie ; et je ne doute point que vous n'ayiez mis à contribution les productions végétales de cette riche contrée ; je n'ai pas eu moins de satisfaction en sachant que vous vous proposiez de faire connoître par des figures les plantes de l'herbier de Linné, qui ne l'étoient nullement. C'étoit presque un engagement que vous aviez contracté en acquérant cette magnifique collection, et un hommage que vous seul êtes capable de rendre au célèbre naturaliste qui a en partie dissipé les ténêbres dont la botanique etoit environnée. Vous aviez en même temps formé le dessein de publier sa correspondance littéraire avec quelques rétranchements. Je pense et je souhaite en même temps que ce projet s'exécute, et qu'un commerce épistolaire qui doit avoir été fort étendu, et dont le sujet doit avoir roulé principalement sur des objets scientifiques, ne soit pas perdu pour le public naturaliste, surtout dans un temps où l'histoire naturelle prend toujours plus de faveur, et où la réputation du naturaliste Suédois s'est singulièrement accrue depuis surtout qu'il n'existe plus.

Vous rendez justice à mes intentions bien plus qu'à mes talents en voulant m'agréger à la Société Linnéenne, dont vous m'aviez sans doute parlé. J'ai lieu de croire que le centre de cette institution doit être à Londres. Je recevrai avec beaucoup de plaisir tout ce qui aura trait à cet établissement, parceque je suis curieux d'apprendre quelles sont les personnes qui s'adonnent à la botanique dans un pays où les sciences sont cultivées avec tant de succès.

Je ne distingue point le *Daphne* que vous m'avez envoyé du *D. alpina.* Ce petit arbuste que vous avez trouvé à Naples vient aussi en Provence sur les plus hautes montagnes, et en Languedoc à sept à huit lieux de Montpellier : c'est certainement le *Thymelea saxatilis Oleæ folio* Tourn. : il est assez bien représenté dans les ouvrages posthumes de Gesner : je dois pourtant observer que le nôtre a ses feuilles plus lisses et presque aussi blanchâtres en dessus qu'en dessous : au premier abord on le prendroit pour une autre espèce, mais, vues à la loupe, les feuilles présentent les mêmes bossettes. Le tube extérieur du calyce dans l'une comme dans l'autre espèce est velu, mais encore plus dans celle d'Italie, qui a ses feuilles plus roides, plus dures, dont les bords sont réfléchis, ce que j'attribuerois à un climat plus sec et plus chaud.

J'avois adressé, il y a plus d'un an, un manuscrit qui contenoit des observations critiques sur une traduction de Pline, au médecin du Marcgrave d' Anspach, appelé Schmidel, qui, comme vous savez, a publié quelques ouvrages de botanique : d'après la promesse qu'il m'avoit faite qu'il seroit l'éditeur de cet ouvrage dont il m'avoit accusé la reception, je me flattois qu'il me donneroit de ses nouvelles. J ai attendu inutilement jusqu'aujourd'hui, sans avoir pû être informé s'il est même encore en vie : si vous aviez appris quelque chose sur son compte, je vous serois très obligé de m'en faire part.

Conformement à vos intentions j'ai l'honneur de vous adresser quelques remarques au sujet de la nouvelle édition du *Systema Vegetabilium,* dont vous vous occupez. J'ai fait ces remarques à la hâte, afin de vous les faire parvenir plutôt. Par la même raison je ne les ai point achevées ; mais je vous promets que je ne perdrai pas de vue les objets restants. Il y aura toujours des corrections à faire dans un ouvrage de cette nature ; et pour quelles soyent faîtes avec connoissance de cause, il faut necessairement être a portée des plantes de tel et tel pays. Si Linnæus avoit pû toujours voir par lui même, et s'il ne s'en étoit pas souvent tenu à des relations infidelles, on trouveroit moins à redire dans son *Species Plantarum.* Il est certains genres, tels que ceux des *Cistus,* des *Tithymales,* qui exigent une refonte générale à l'égard même des espèces Européennes : il y auroit encore une réforme bien essentielle à faire dans ce systême sexuel ; qui seroit de supprimer la *Polygamie* (puisque la plûpart des plantes hermaphrodites qui sont surchargées de fleurs sont en partie stériles et en partie fertiles), et de réduire la *Monœcie* aux genres dont les fleurs mâles sont en châton et d'une structure différente des fleurs femelles ; car là où la disposition des fleurs avortantes est la même que celles des fleurs fertiles, je n'admettrois point de différence essentielle, parcequ'il est naturel que lorsque la nombre des fleurs excéde la force naturelle de la plante, la plûpart de ces fleurs doivent couler, soit que le germe ne noue pas, ou qu'il avorte dans son principe.

Je vous prie, Monsieur, d'excuser cette longue lettre en faveur du plaisir que j'ai eu de m'entretenir avec vous. Cette occasion étoit trop précieuse pour ne pas en profiter autant qu'il dépendoit de moi. J'espère qu'elle me procurera l'avantage de recevoir des nouvelles de votre santé à laquelle je ne saurois trop m'interesser : je me flatte aussi que vous voudrez bien employer quelques moments libres pour me faire part de tout ce qui a trait chez vous à l'histoire naturelle, et particulièrement à la botanique : quoique vous n'ayiez pas besoin d'exciter ma reconnoissance, je vous la manifesterai bien volontiers dans toutes les occasions qui pourroient me rapprocher de vous ; et je serai toujours très empressé de vous réitérer les témoignages des sentiments que vous m'avez inspiré, et avec lesquels j'ai l'honneur d'être, Monsieur,

Votre très humble et très obeissant Serviteur,

GERARD.

M. Desfontaines to J. E. Smith.

Monsieur, Paris, le 20 Juillet, 1790.

Un grand nombre d'amateurs d'histoire naturelle de notre capitale se sont réunis et cotisés pour faire exécuter à leurs frais le buste de Linné en marbre: ce buste sera vraisemblablement placé au Jardin des Plantes, soit au pied du cèdre de Liban que vous connoissez, soit dans l'école même de botanique. Vous devez bien penser que je n'ai pas été un des derniers à souscrire. Ce monument honorera plus les naturalistes Français que Linné; parceque ses ouvrages suffisent à sa gloire; mais je vois avec plaisir, que nous rendons enfin justice au père de l'histoire naturelle, et à l'un des plus rares et des plus beaux génies de notre siècle: nous espérons que cet hommage public fera enfin taire en France les ennemis de cet homme si célébre, et qu'ils n'oseront plus que murmurer sourdement. La basse jalousie qui règne en ce pays contre les hommes célébres, est réellement un fléau: je vous crois bien plus philosophes que nous à cet égard. Les Anglais savent honorer le mérite indépendamment de temps et de lieux: je ne vous dis point cela par flatterie, je le pense aussi; et vous en donnez sans cesse de preuves. Quelqu'uns de nos détracteurs de Linné voudroient se mettre de niveau ou même au dessus de lui ; mais ils n'en approcheront certainement jamais. Je voudrois bien, que vous donnassiez une édition complette de ses ouvrages sur la botanique, à laquelle vous ajouteriez des notes critiques, et des corrections que vous jugeriez convenables : je suis persuadé qu'elle auroit le plus grand succès. Vous seul pouvez remplir dignement cette tâche ; vous possédez les manuscrits et l'herbier de Linné : le public en est instruit, ce qui donneroit encore un prix infini à cette entreprise : je ne doute point qu'elle ne fit tomber absolument toutes les autres éditions que l'on donne, et que l'on se propose de donner. Ce projet est certainement utile ; et je suis persuadé qu'il seroit parfaitement rempli. Vous pourriez ajouter un supplément qui renfermeroit les genres et les espèces inconnues à Linné. Un pareil livre deviendroit le manuel des naturalistes : je crois que cela mérite attention de votre part : vous rendriez un service important à l'histoire naturelle ; et vous en retireriez beaucoup de gloire. Mais, voici, Monsieur, où je voulois en venir particulièrement : les personnes qui se sont réunies pour faire exécuter le buste de Linné m'ont chargé de leur fournir une liste exacte, s'il est possible, de tous les ouvrages du naturaliste Suédois : je n'en connois qu'un petit nombre. J'ai cru ne pouvoir mieux faire que de m'adresser à vous pour vous prier de m'aider à remplir cette tâche. Envoyez moi donc, s'il vous plait, le titre, ainsi que les diverses éditions, des œuvres de Linné que vous connoissez : je vous aurai beaucoup d'obligation. Je sais qu'il en existe en langue Suédoise, qu'il en a fait sur des sujets étrangers: je voudrois les connôitre tous: je présume que vous les possédez, ou qu'ils se trouvent dans la bibliothèque de M. Banks: j'attends donc cette grâce de vous, Monsieur; et je vous prie de vouloir me la rendre le plutôt que faire se pourra. Je me rappelle aussi que vous avez bien voulu me promettre quelques plantes de l'Inde: toutes celles que vous m'enverrez seront déposées chez moi; et personne au monde ne les publiera: je vous en donne ma parole.

Je désire bien avoir quelque jour le plaisir de vous aller voir dans votre patrie, et renouveler la connoissance que j'ai eu le bonheur de faire avec vous pendant votre séjour à Paris.

Voulez vous bien présenter mes hommages à M. Banks, et agréer les sentimens de l'estime et de la considération parfaite avec lesquels j'ai l'honneur d'être,

Monsieur, &c., &c., &c.,

DESFONTAINES.

P.S. Les sciences languissent beaucoup ici: tous les esprits sont tournés vers les affaires publiques: je désire ardemment qu'elles puissent s'arranger convenablement le plutôt possible: j'ose l'espérer. Il est bien difficile de cultiver les sciences quand on n'est pas tranquille.

Je vous prie de m'honorer d'une réponse le plutôt que vous pourrez: vous me ferez grand plaisir. J'ai vu votre dernier ouvrage chez M. l'Héritier. *Cimex Pantilius* n'est pas très content: vous lui donnez rudement sur les oreilles : j'espère que cela lui apprendra à vivre : nous en avons ri de bon cœur.

J'ai eu l'honneur de vous écrire pour vous prier de faire mes sincères remercimens à la Société Linnéene, et pour lui témoigner combien je me trouvois honoré d'être admis au nombre de ses membres. Voulez vous bien leur réitérer les assurances de mon respect, et de mon attachement inviolable.

Sir George Staunton to J. E. Smith.

Dear Sir, Buxton, Sept. 1, 1790.

Not only have I to thank you for your kind letter, but also to request you will present my acknowledgements to your friend for his very obliging invitation to his house; and though a variety of engagements will prevent me from availing myself of it this year, I shall, if I go to Scotland, as I am inclined to do next year, pay my respects then, when my little fellow will be perhaps somewhat better able than at present to distinguish the value of such a botanical collection.

Since you left us he has had the opportunity only of seeing Mr. Eyre of Hassop's collection, where he saw and ate of the fruit of the *Musa*, of that kind which is called the *Fig-Banana* in the West Indies: this was a great curiosity to him, who, though he saw the plant at Kew, never saw the fruit before. We saw at Bakewell impressions of several plants in argillaceous earth and iron-stone, which were found in working the mines of this country.

The *Arundo Bambos*, one of the exotic *Euphor-*

biæ, were distinguishable enough, and ascertained by former visitors to the collection of Mr. White Watson, a zealous naturalist; and my boy thought he could distinguish another impression to be that of the *Dodecatheon Meadia.* There are a great many other impressions of plants, not settled, and which I wish you would see: many of them appear to be of Ferns.

The same gentleman has a very curious collection of what are called petrified shells, which appear to be formed by water charged with calcareous earth, as is the water which forms stalactites, entering into the shells, and depositing its earth, which takes, as it loses its water and hardens, the form of the containing shell, in some instances covering the petrifaction. My little fellow is too great a novice in conchology to ascertain the species of the shells.

The neighbourhood of Buxton promises very little in any part of Botany; and your young friend is yet too much of a tyro to make much progress without a guide. We all, on every account, wish for your presence. When you return to London we may trouble you to settle the names of a few specimens for us.

I am, dear Sir, very much yours,

GEORGE STAUNTON.

Sir C. P. Thunberg to J. E. Smith.

Dum non æque facile scribere scio, ac quidem intelligo linguam Anglicam, spero fore, ut tu, vir illustrissime, permittas, ut hac utar lingua antiqua, atque tibi significem gratias meas maximas, tam ob litteras tuas humanissimas ad me datas, quam ob amica illa mutui commercii promissa: quantum in me est, conabor, ut cum acceperim desiderata tua botanica, nonnullæ, tam Capenses, quam Indicæ plantæ a me mittantur, quæ in vasto et pretioso tuo herbario adhuc desiderari possunt. Hacce occasione, catalogum meum mitto desideratarum plantarum; atque significo, summopere mihi gratas fore omnes illas accessiones, quas pro me sensim parare tibi, vir æstumatissime, placebit. Nec dubito, quin plures erunt, qui mecum gaudent, quod in tuis manibus venerint collectiones Linnæanæ, atque quod utiles reddantur erudito orbi, ac dudum cœpisti modo adeo honorifico illas reddere. Editione nova *Systematis Vegetabilium* infinite tibi omnes botanicos devincies, quæ ultima adeo vitiosa evasit, ut cito emendari mereatur et debeat. Opto sane, ut aliquo modo, tibi, vir illustris, in hisce terris utilis esse possim, atque vel tantillum tuam in me amicitiam atque faventem animum demereri valeam. Occupatus sum in elaboranda *Flora Capensi*, cujus quatuor priores classes absolvi, paratas pro prelo. Doleo vero, quod adeo pretiose, adeo digne hoc opus edere non valeam ac quidem Capensis Flora mereri posset, cum Upsaliæ occasiones illæ deficiant, quibus edi alibi posset, nitidissima charta et iconibus. Forsan Berolini tomus primus figuris minus bonis impressus erit. Pro Actis Societatis Linnæanæ descriptionem generis, *Dilleniæ*, cum satis nitidis figuris tribus paravi; sed ad te, vir celeberrime, mittere nolui, antequam sciam si hæcce per tabellarium mittere possim, sique acceptæ sint. Genus hoc sex constat speciebus, quarum tres omnino ignotas habeo, quarumque figuras conficere feci: una exstat in *Systemate Linnæi*, scilicet *Indica*, quam non vidi, nisi e figura, et duæ species depinguntur a Rumphio. Nuperrime accepi Ordines Naturales D[ni] Jussieu, de quibus adhuc judicium ferre nec audeo, nec debeo: videtur tamen mihi genera nimis multiplicasse. Ego semper allaboro, ut genera naturalia non separentur: ex. gr. *Mahernia* et *Hermannia, Diosma* et *Hartogia, Galium* et *Valantia,* et cætera.

Upsaliæ, d. 3 Sept. 1790.

J. P. M. Dana to J. E. Smith.

Monsieur, De Turin, ce 23 9bre, 1790.

Mes amis ne m'ont point manqué ce 21me 9bre dans l'assemblée de notre Académie Royale; et après un combat d'une heure employée en votations pour le choix d'un academicien étranger à la place qui se trouvoit vacante, vous avez été élu pour associé étranger, de préférence à M. Pallas et autres savans du premier ordre; et en cela je dus me servir de votre lettre dernière, où vous aviez offert d'envoyer un mémoire sur les fougères, qu'on a eu la bonté de considérer comme s'il eût été expédié, puisque son retard dépendit uniquement de vous avoir écrit que le volume 8me était pour sortir (comme il est à présent sorti) de sous presse. D'ailleurs vos mérites et votre célébrité auroient suffi à vous préférer (comme elle a fait) à ses anciens correspondents et à tant d'autres concurrens, qui, dans un nombre si borné des places d'académiciens, marquerent également l'empressement d'être nommés, si toute l' académie fût composée de botanistes ; mais comme les académiciens de cette assemblée étoient 19, et qu'il n'y avoit que moi et M. Allioni botaniste, de cette académie, pendant qu'il y faut au moins les quatrequints des voix pour être nommé, vous maintenant y avez été admis à la pluralité de 17 contre 2 voix. Ceci ayant été fait en bonne forme, je vous en donne avis ; et vous en sera expédié le diplome par le sécrétaire en peu de jours.

Je vous prie, Monsieur, à être assuré des sentimens de la véritable estime, et de la sincère amitié, avec laquelle je serai toujours avec le plus parfait attachement, et je suis, Monsieur,

&c., &c., &c.,

MEDECIN DANA.

Louis Bosc to J. E. Smith.

Paris, ce 15 Janv. 1791, an 3me de la Liberté.

Broussonet ou autres vous ont sans doute appris qu'après avoir élevé dans le Jardin des Plantes un buste à la mémoire de Linnæus, les anciens membres de la Société Linnéenne s'étoient de nouveau réunis, et s'étoient constitués sous le titre de *Société d'Histoire Naturelle de Paris*. Cette société a aussitôt sa naissance développé une grande ardeur pour l'avancement de la science qui fait l'objet de ses travaux; et nous sommes prêts à faire imprimer un fascicule qui contiendra beaucoup de choses intéressantes. Je ne vous parlerai pas de nos projets; vous en verrez les détails dans la préface de notre fascicule: mon but aujourd'hui est simplement de vous faire passer les diplomes qui intéressent votre personne et celles de trois de vos amis. C'est Broussonet qui vous a fait inscrire sur la liste des associés; c'est lui qui auroit dû vous envoyer ces diplomes; mais il est si occupé d'affaires politiques qu'à peine pourrons nous le voir. Nous espérons que vous voudrez bien concourir à l'amélioration de nos fascicules, en nous fournissant des mémoires, et à notre instruction, en correspondant avec nous d'une manière active.

Je profite de cette occasion pour vous envoyer quelques descriptions d'insectes nouveaux ou mal connus, pour la Société Linnéenne. Je vous prie de les corriger avant de les communiquer; car je suis fort peu assuré de ma manière d'écrire en Latin.

On vous remettra aussi un ouvrage d'entomologie nouveau, du débit de quelques exemplaires duquel j'ai été chargé par l'auteur. J'ai pensé que vous et M. Banks seroient bien aises d'en avoir chacun un exemplaire. Il coûte 30 livres chaque. Si vous n'en voulez pas, tâchez de le faire vendre à quelque personne de votre connoissance.

Ne doutez pas de mes dispositions à faire tout ce qui vous sera agréable dans ce pays.

Je vous salue. LOUIS BOSC.

Professor Zimmerman to J. E. Smith.

Dear Sir, Brunswick, July 16, 1791.

I feel myself under the greatest obligation for the undeserved honour of having been chosen a member of the Linnæan Society. As I am mainly indebted for it to your favour and friendship, I will not miss any opportunity to prove to you my acknowledgement and gratitude.

I take the liberty to direct to you a few copies of the beginning of an excellent work of my learned friend Mr. Lichtenstein of Hamburgh. In a few days the translation of Smellie's *Philosophy of Natural History*, which was undertaken under my care, will be finished. I made many additions to that anti-Linnæan work, part of which are refutations. I shall have the honour of sending a copy to the Linnæan Society, as perhaps some among her members understand our language. I cannot conceive how Mr. Smellie could undertake a work without being master of more modern languages, for a great part of the new discoveries were lost to him.

My son pays his respectful compliments to you, and acknowledges his great obligations for the kindness and care you bestowed on him last year. He is going to Göttingen, where he will study law, but properly statistics and technology.

I am, with the greatest regard, Sir,

Your humble Servant,

E. A. Zimmerman.

D. H. Stoever to J. E. Smith.

Vir doctissime, celeberrime, plurimum colende,

Inclyta, qua gaudes fama, animusque benevolens atque amicus viris qui inesse solet ingenio et doctrina excellentibus, commovent, ut equidem hisce te litteris adeam, subsidia imploraturus, tali viro saltem non indigna.

Triennium et quod excurrit jam elapsum est, ex quo vitam viri immortalis longeque de studiis et doctrina meritissimi, terræ Europæ septentrionalis quem unquam tulerint, Linnæi nimirum patris, elaborare et conscribere operam dedi. Fama quidem omnibus notus, parum erat et mancum quod de fatis ipsius memorabilibus, de indole, moribus, meritisque in universum expositum fuit. Quare studui, memoriam viri pro magnitudine rerum gestarum amplius tradere ; in quo labore ita adjutus fui a viris Sueciæ eruditis et a discipulis Linnæi, qui in Germania supersunt, ut sperem, operam me non prorsus incassum insumsisse, orbique erudito aliquo modo satisfacturum esse. Ut autem recte atque vere, quantum unquam fieri posset, omnia tradantur, opus adhuc est benevola tua ope, quam te non denegaturum esse manibus Linnæi, tam insigni studio jam a te condecoratis, a tua humanitate spero.

Tu enim, vir inclytissime, felix ille Crœsus es, cui jure emtionis contigerunt thesauri Linnæani omnes. Quum autem plurima dubia et falsa de celebri hac emtione circumferantur, quæso ut me edocere velis de sequentibus :

I. Quonam anno, quove pretio emisti ?—Cum peregrinatore Germano colloquutus, qui eodem tempore Upsaliæ fuit ac in tua societate versatus est, plura de hac re audivi, quæ autem incertiora videntur quam ut memoriæ mandare possem. Etenim nomen tuum celeberrimum, cujus sæpius in hac vita Linnæi incidit mentio, commemorando, nollem me falsi aliquid perscribere. Ceterum dicunt thesauros istos pretio 1000*pound sterl.* a te coemtos esse. Num stricte veritati respondet? Injunctum fuit a Sueco quodam divite, Mauhle, qui tunc temporis in India orientali morabatur, viro nob. Andr. Dahl, ut, pondere pecuniæ in hunc finem jam erogato, pro se coemeret thesauros Linnæi relictos, quod consilium autem successu carebat. Quomodo hoc factum? Num tu prior, vel majus pretium obtulisti? Si vis et potes, me de hac re certiorem facias, quæso.

II. Quænam omnia emisti? quibusnam ex rebus singulis composita est hæreditas illa litteraria? Bibliothecæ quantus est circiter numerus librorum? Herbarii quantus circiter numerus plantarum? collectionibus ejus naturalibus quænam est præstantissima? Quot et quænam possides manuscripta? Illud de Perlarum vel Margaritarum ortu mihi notum. Ad hæc scio, Linnæi exemplar *Materiæ Medicæ* e regno vegetabili, multis auctum et ornatum esse propriis observationibus, quum dudum constituisset eam denuo edere. Ejusmodi observationes et scholia num et in pluribus aliis ipsius libris reperiuntur? Num tu possides Diarium, quod sub titulo *Nemesis Divinæ* composuerat? Quum de hisce parum aut nihil publice sit notum, gratissimum faceres mihi orbique litterato, si certo aliquid tradere velles, non intermissurus, operam animumque tuum generosum dignis laudibus celebrare.

III. In versione *of the General View of the Writings of Linnæus, by R. Pulteney*, ante biennium a cl. M. de Grandmaison, gallice edita, relatum est, te curare novam editionem *Specierum Plantarum* et *Systematis Naturæ.* Num hoc verum? Maxime vellem hoc compertum habere, quippe vitæ adjungo catalogum scriptorum Linnæi, eorum commentariorum, versionum, &c., locupletissimum, in quo tuum addere possem. Forsan *Systema Regni Vegetabilis* intelligendum erit. Porro ex te quæram num tu præter egregios illos Fasciculos Plantarum (quorum huc usque duo publici juris facti, quosque excerptos reperis *im Magazin für die Botanik*, editum Turici a cl. Römer et Usteri) aliaque botanica, digne celebrata, alia scripta ex collectionibus Linnæanis, vel ad systema ejus facientia, nobis forsan ignota, edideris. Annotationes tuas una cum cl. Broussoneti ad Linnæi *Diss. de Sexu Plantarum* recusas esse in tom x. *Amœnitat. Acad.* ab Ill. Schrebero Erlangæ 1790 editum, te jam non fugiet. Ceterum doleo, quod cl. de Grandmaison uti non licuerit subsidiis in nostra Germania passim impressis, quibus versionem suam egregie augere potuisset.

IV. Una cum vita quum a me curatur collectio epistolarum ad viros eruditos a Linnæo datarum (congessi enim non adeo paucas) me tibi maxime devinctum redderes, si e collectione tua epistolarum ad Linnæum scriptarum unam vel alteram memorabilem, e. gr. Halleri (quum anno 1739 illi offerret munus Professoris Goettingensis) vel Marchionis clementissimæ Baadensis, vel quum vocaretur in Hispaniam, mecum communicare velles. Sed vereor, ne his quidem precibus tibi molestus sim: veniam dabis. Fortasse autem non dedignaberis mihi indicare, quot literæ Marchionis Baadensis existant.

V. Compertum habeo, Londini in honorem Linnæi institutam esse societatem, quæ te præsidem veneratur. Sed quo auctore instituta est (certe te) et quo anno? Quæso, paulo amplius me de his edoceas, quum mentio hujus instituti præclari injicienda. Ceterum addo, ad hoc vestrum exemplar præterito anno et Lipsiæ auspiciis cl. Ludwig ejusmodi societatem coalitam esse.

Sed temperandum calamo, temperandum precibus, ne te fatigem et quasi obruam. Ignosces ardori et studio veritatis, quo trahor. Per Deos, per manes Linnæi, quos colis, quos per sacrificia adeo splendida adhuc venerabiliores et honoratiores reddidisti, nunc te precor, vir humanissime, ne prorsus deesse velis votis meis. Facias me compotem eorum eo modo quo tibi placebit. Rescribas quanta et qualia tibi ex re videbuntur; omnia enim gratissima erunt, nec me publice ingratum senties. Sed responsione tua, rogo, quam primum tibi per otium licuerit, me lætitia affice; vita enim jam imprimitur, at non adeo festinanter, ut non in usum vocare possim quæ intra binos et quod excurrit menses perscribere dignaberis. Releges tunc verba tua; jucundum enim erit officium, librum tibi transmittere quam primum impressus fuerit. In responsione autem mihi annotes, quæso, *plateam* in qua habites. Ceterum rescribas vel latine, vel, si placet, anglice. *For although I am not exercised so much as to speak or write this noble language, I understand it pretty well.*

Secundum litteras, quas mense præterito ex Upsalia habui, vidua Linnæi adhuc in vivis est, vitam degens in prædio Hammarby. Filia natu minor nupsit nob. vir. Duse, litium curatori Upsaliæ. Reliquæ duæ sorores vitam agunt cœlebem. In fronte novi horti academici sumptu regio erigitur monumentum Linnæi. Amicus meus, Professor Gieseke, Hamburgensis, ejus quondam discipulus, anno proximo ejus *Prælectiones in Ordines Naturales Plantarum* edet. Plura ejusmodi nova tibi scribere possem, nisi vererer nimis leviora et minoris momenti esse.

Quod si in Germania uno vel altero modo inservire et operam tibi commodare possem, moneas, jubeas, quæso. Enixe ceterum repetens preces supra memoratas, humanitati tuæ iterum atque iterum me commendo. Vale! meque quam primum fieri possit litteris tuis lætitia affice!

Nominis tui inclytissimi cultor observantissimus,

D. H. Stoever, Doct.

Dab. Altonæ (prope Hamburgum), die Octobr. viii. 1791.

Rev. Wm. Kirby to J. E. Smith, P.L.S.*

Dear Sir, Barham, Dec. 5, 1791.

I received your letter and obliging present, which I shall keep as a precious relic, not only on account of their great original possessor, but also for the sake of the worthy donor, whose name I prophesy will not be forgotten while that of Linnæus is remembered, and the study of the works of an all-wise Creator continues to be a favourite pursuit with enlightened Europe. Pray give my compliments to Mr. Marsham, and tell him I am preparing another box of insects for him. I was much gratified lately by having an opportunity of calling upon Mr. Woodward, who very obligingly showed me his herbarium and collection of fossils. I mentioned to him a plan upon which we had some conversation when you were at Ipswich,—I mean the establishment of a provincial society for the promotion of the knowledge of the natural history of this county. He approved much of the design, and desired to be a member, in case such an institution should take place. I have a friend in an adjoining parish, who is a very ingenious naturalist, (Rev. Charles Davy†, of Caius Coll.,) with whom I have had some conversation upon the subject; and he tells me that he has had the same idea long in

* Author of a very delightful *Introduction to Entomology*, in union with Mr. Spence.

† Translator of Bourrit's *Letters upon the Glaciers*, and son of the Rev. C. Davy, of Onehouse, in Suffolk, the very learned author of *Letters on Literature*.

his mind, and that such a society, if it could be established, and the plan be generally adopted throughout the counties, would contribute more to the elucidation of the natural history of this country, than any other plan whatsoever. I should be particularly obliged to Dr. Smith for any hints he could furnish me with; and the outlines of such an institution I should esteem a great favour, if it would not engross too much time. It is the design, if a proper situation could be fixed upon, to collect specimens in every department, and to impose it as a law upon every member, to contribute a certain proportion every year, which I should hope would keep out of our society all non-effectives. I am much flattered by your obliging invitation: whenever I go to town, one of my principal pleasures will be to wait upon you.

Your sincere friend,

WILLIAM KIRBY.

Professor Scarpa to J. E. Smith.

Monsieur, Pavie, 2 Mars, 1792.

Je viens de recevoir votre obligeante lettre du 18 Août de l'année passée, avec le Discours préliminaire et les Articles qui concernent la Société Linnéenne. Parmi la liste des Associés étrangers ayant trouvé mon nom, j'en ai été très flatté; et comme c'est principalement à vous que je dois cet honneur, aussi je vous fais mille remerciments.

Dès que j'ai commencé à lire les premières lignes

de votre Discours, il ne me fut plus possible de quitter le livre jusqu'à la fin. L'excellent tableau que vous avez entrepris est dessiné avec une netteté, précision, et vérité surprenante: il y a bien long temps que je n'ai eu le plaisir de lire un morçeau si bien achevé. Les époques principales de l'histoire naturelle et les hommes illustres qui l'ont cultivé y sont marqués avec une intelligence et une critique très fine; ce qui fait voir que vous avez approfondi cette science.

Où vous parlez des établissemens, je m'attendois que vous auriez parlé du Général Comte de Marsigli de Bologne; mais peût-être ceçi n'entroit pas dans votre plan. Cependant ce grand militaire, victime de l'intrigue, etoit un assez bon naturaliste de son temps, et qui s'est dépouillé de presque tout son bien pour établir l'Académie ou Institut de Bologne, dont le bût principal est l'histoire naturelle. Son ouvrage sur le Danube, et quelques autres sur les productions naturelles de la Méditerranée ne sont pas absolument sans quelque prix. Quelque trait de votre excellent pinçeau sur les circonstances, l'état, et les talens de cet homme singulier, il me semble qu'auroit fait un très bon effet. Donnez un coup d'œuil sur l'éloge qu'en a fait M. de Fontainelle.

Je vous remercie des politesses que vous avez bien voulu pratiquer à M. Scassi: je crois qu'il sera bientôt de retour à Londres pour se rendre tout de suite ici. Je l'ai chargé de m'apporter tout ce qu'il trouvera chez vous de nouveau, en chirurgie principalement: je vous prie de l'aider dans le choix.

Je travaille actuellement à quelques planches de nécrologie fine ; mais tout chez nous va lentement ; moi encore plus que tout autre étant beaucoup distrait par un double emploi à l'université, et par la pratique.

Je désire vivement et sincèrement de vous être utile ici en quelque chose.

Je suis, avec la plus perfecte estime,

Monsieur,

Votre très humble et très obeissant Serviteur,

SCARPA.

The subjoined letter from Father Fontana will discover to the reader the impression made upon him by the person to whom it is addressed. In the third volume of his *Tour on the Continent,* Sir James acquaints his readers, that "Professor Scopoli introduced him to the celebrated Father Gregorio Fontana, Professor of Mathematics." "Little," he continues, "could I imagine, when I enjoyed the pleasure of his conversation, and admired the acuteness and versatility of his genius, that he should ever condescend, as he has since done, to become the translator and commentator of any production of mine."

Father Fontana to J. E. Smith.

Monsieur, Pavie, 5 Mai, 1792.

Je sais, Monsieur, que mon collègue et ami Scarpa vous a donné avis de la traduction Italienne que j'ai faite de votre beau *Discours Préliminaire* aux Actes de votre Société Linnéenne, qui dans ce moment vient de sortir de la presse. Le jeune homme qui vous remettra cette lettre, vous rendra aussi un exemplaire de cette version, que je vous prie d'agréer comme un hommage que je me fais gloire d'offrir à votre éminent savoir, et plus encore aux vertus aimables de votre belle âme. Ce jeune homme est M. le Dr. Valli de Toscane, qui, par son excellent caractère moral, universellement reconnu, par son amabilité, et par son savoir peu commun, mérite la considération et l'amitié de tout homme vertueux et sensible. Il est auteur de quelques petits ouvrages médicals assez estimés, et il en a publié trois à Pavie dans le peu de séjour qu'il y a fait; parmi lesquels il y a deux lettres sur l'étonnante découverte de M. le Dr. Galvani de Bologne relative à l'electricité animale, découverte qui ne fait que parôitre. M. Valli a vérifié les faits principaux, et les expériences fondamentales de M. Galvani, et est même allé un peu plus loin; car au talent d'écrire il joint le génie d'expérimenter. Agréez, Monsieur, les offres de mes services pour ce pays que j'habite.

En attendant, je suis, avec le plus entier dévouement et la plus haute estime, Monsieur,

Votre très humble et très obéissant Serviteur,

GREGOIRE FONTANA.

Professor Scarpa to J. E. Smith.

Monsieur, Pavie, ce 30 Mai, 1792.

Votre lettre du 8 de ce mois m'est parvenue quelques jours après la publication de la traduction

Italienne de votre excellent Discours : ainsi le Père Fontana n'a plus été en état de faire les additions que vous veniez de me marquer dans votre lettre : il me charge de vous renouveller ses sentimens d'estime et d'amitié. J'attends de Gènes l'occasion de vous faire passer une douzaine d'exemplaires de la traduction. Vous ne tarderez pas cependant d' en recevoir un, qui vous sera remis par M. Valli, docteur en médecine. Ce jeune homme, rempli de talent, mais tout-à-fait sans fortune, et qui depuis quelques années voyage à la manière de nos Capucins, c'est-à-dire à pied et en mendiant, s'est proposé de se rendre à Londres, comme en effet il est parti il y a un mois. Comme il marche bien, je crois qu'il ne tardera pas beaucoup à paroître devant vous, et s'acquitter de sa commission.

Il vous informera d'un nouveau fait de physique animale, que le docteur Galvani de Bologne vient de découvrir concernant l'électricité animale. Je tâcherai de vous en donner quelque idée. Si on prend la moitié d'une grenouille dont on a ôté la peau, et en prépare les nerfs cruraux de manière que l'origine de ces nerfs reste adhérente à l'épine, et en ôte toute autre substance intermédiaire entre l'épine et les jambes, et que après avoir isolé ce morceau d'animal on place l'extrémité d'un arc conducteur sur les nerfs, et l'autre partie de l'arc sur la jambe, les muscles se contractent vivement à plusieurs reprises et pendant long tems. Ce phénomène est encore plus frappant si on entoure l'origine des nerfs et une partie de la jambe avec une feuille métallique. Chez nous il y a qui tirent plusieurs conséquences de ce phénomène, jusqu'à croire que par ceci l'existence des esprits animaux soit demontrée. Quant à moi, je ne cesse de méditer sur ce fait singulier, et qui pour l'instant donne de quoi travailler aux électricistes; mais je n'y vois pas encore claire sur les conséquences physiologiques qu'on se hâte de déduire.

M. Scassi me fait savoir qu'il se trouve actuellement à Paris. J'espère qu'avant son départ de Londres il se sera informé des meilleurs ouvrages qui ont paru dernièrement chez vous en médecine pour me les procurer, comme je l'avois chargé. D'heure en avant je compte sur votre amitié pour être au fait de ce qui se passera de plus intéressant en Angleterre par rapport à la chirurgie et anatomie, qui forment ma principale occupation.

Les âmes honnêtes ont beaucoup apprécié votre passage sur la vie de Scopoli, et se réjouissent de voir qu'un juge si savant dans l'histoire naturelle, et impartial, comme vous êtes, ait pris avec franchise à dévoiler l'intrigue et la malignité d'un petit nombre d'ignorans, qui ont employé tous leurs efforts pour flétrir jusqu'à la mémoire de cet homme respectable.

Je suis, avec respect et attachement sincère,

Monsieur,

Votre très humble et très obeissant Serviteur,

SCARPA.

Dr. Darwin to J. E. Smith.

Dear Sir, Derby, Sept. 12, 1792.

I should sooner have written to thank you for the books you were so kind as to send me. I mentioned your Society to Sir Brooke Boothby of Ashbourne, and to Dr. Johnson of this town, two ingenious men, who wish to become Fellows of the Linnæan Society. Sir Brooke Boothby has a great collection of plants, which you probably saw when you were in this country. Sir Brooke is now at Ashbourne, so that I can acquaint him if it be agreeable to you to admit him to your Society. I received much pleasure from your Thesis *De Generatione*, though our theories will not agree.

With true esteem, I am, yours,

E. DARWIN.

Dr. Pulteney to J. E. Smith.

Dear Sir, Blandford, Oct. 14, 1792.

I am tempted to make a request to you, which I confess to be not very reasonable:—Can you do me the favour of the loan of your copy of the folio *Historia Muscorum* of Hedwig? I am engaged to give Mr. Nichols a compendious catalogue of the rare plants of Leicestershire, for his history of that county now in the press; and I wish to make my references to Hedwig, in the *Musci frondosi*, taking it for granted that his book will become classical in that way.

I had lately, in my attendance upon one of the family, an opportunity of spending a few hours in the botanical library of the late Earl of Bute. It is indeed very rich in books and dried specimens, as well as in volumes of paintings of plants; and it appears that the earl preserved his taste to the last, as I observed all the latest expensive works. There are (thick and thin) more than 300 folios, *strictly* botanical, and quartos and octavos in proportion; very many of the old authors, and some very scarce ones; a conservatory almost 300 feet long, full of fine plants, growing and flourishing in the soil (not in pots), like an Indian grove. To this add a garden of four acres, walled round, and full of hardy plants; and all this within 150 yards of the sea.

If you honour me with a plant, I must leave the choice of it to yourself, amidst the great variety of new subjects that are now pouring in upon you*.

I wish to ask whether the Fungus, called in Ray's Synopsis *Oak leather*, in the timber of houses, is well known in London, and its history investigated, and whether a specimen of it would be likely to be acceptable to the Linnæan Society?

I am, &c.

R. PULTENEY.

* In his work entitled Botany of New Holland, Sir James has named a plant of that country after Dr. Richard Pulteney, F.R. and L.S., well known by his Sketches of the Progress of Botany in England, and more especially by his Biography of Linnæus.

Rev. Henry Muhlenberg, D.D. to J. E. Smith.

Lancaster, State of Pennsylvania,
Dec. 1, 1792.

Honoured and dear Sir,

Pardon a stranger that intrudes upon your studies: an enthusiastical love of botany, and irresistible desire to know the plants of my native country, stimulate me to do it. Since a number of years I have endeavoured to explore the *Regnum vegetabile Americæ Septentrionalis,* in particular of Pennsylvania Media. Partly I have been successful, and have gathered pretty near all the plants of my neighbourhood, being upwards of 1200 in less than ten square miles. Of these I sent in 1786 a specimen, and in 1790 an *Index Floræ Lancastriensis,* to our Philosophical Society at Philadelphia, which will be published in the third volume of their Transactions; and am now making a full description at least of the indigenous plants, after Withering, Lightfoot, or Thunberg's method. But I met so many *adversaria, nova genera et novas species,* which I am not able to class according to the present editions of Linnæus, that I long ago and earnestly looked for a friend, who would kindly assist me to find out which plants are already described by Linnæus, and which are nondescripts. Some of my doubts have been cleared up by my worthy friend, Dr. Schreber, the editor of the eighth edition of the *Genera Plantarum;* but very many remain.

Dr. Stokes, in Withering's Arrangements, mentions the gentleman to whom I could address myself with the greatest prospect of success; "The learned, candid and ingenious possessor of the herbarium, library and manuscripts of the two Linnæi" could be my oracle, if his time and the multiplicity of his labours would permit him to assist me. I would send all the *plantas adversarias et nondescriptas* in good order, numbered, and beg of him to favour me with his judgement, which of the plants are already described, and by what name, and which are not described. Perhaps it would not be disagreeable to him to have even some of the described plants in his noble herbarium: new species certainly would be pleasing; so every doubt would be cleared up, and the *adversaria Americana* be lessened. This, Sir, is the reason of my troubling you; and may I hope to receive a favourable answer? If my offer is not disagreeable to you, pray do me the honour to let me know to what place I should send the dried specimens and my future letters; and be pleased to inform me whether you want any particular Pennsylvanian plant or seed. Any commands that way will punctually and with pleasure be obeyed. But, should your time and various labours not permit you to grant me the desired favour, may I, by your recommendation, find any candid and ingenious gentleman in England or Scotland to whom I could address myself?

Any letter to me can be addressed to my brother, Frederick Augustus Muhlenberg, Esq., Member of Congress, at Philadelphia, who will forward it safe to Lancaster, where I live.

In expectation of an agreeable answer, and recommending myself to your kindness, I have the honour to remain, with great esteem,

Honoured Sir,

Your most obedient humble Servant,

HENRY MUHLENBERG, D.D.

Minister of Trinity Church at Lancaster; Fellow of the Philos. Society at Philad., and of Academ. Imper. Nat. Cur.

J. E. Smith to the Rev. Henry Muhlenberg, D.D., Pennsylvania.

Dear Sir, March 6, 1793.

I have seldom received more pleasure from a letter than yours of Dec. 1st afforded me: never did I receive one so gratifying, I will not say to my *vanity*, but to better feelings. You cannot be more enthusiastically fond of botany than I am; and your letter promises me a fresh instance, in addition to many already experienced, that this study, charming in itself, is still more valuable as a key to the intercourse of the most amiable minds. To botany I owe friendships and connections I else could have had no chance of forming; and your letter, overflowing with the milk of human kindness, and with the amiable modesty of real merit, promises me one which it will be my most anxious care and ambition to deserve. One part only of your letter gives me pain, my dear Sir,—that, where you express a doubt (however slight) lest I should not attend to it as it deserves. Allow me to say, I am too covetous of your correspondence to turn it over to anybody else. I am extremely concerned that my ignorance of the American post-day occasioned me to miss the last; yet if you have doubted my politeness and sensibility to your worth, (of which I want no other proof than your letter,) I hope and trust such unfavourable impressions will now be effaced.

I flatter myself you will soon execute your intention, and make me as useful as you can:—you may depend on my utmost *care* and *accuracy*, and I'll honestly tell you my doubts; neither hurrying over the business, nor by any means pretending to know more than I do. I have all Kalm's original specimens; I also know by marks all Linnæus's *originally described* specimens, so there can be no fallacy. I have several similar correspondences, and with joy add yours to the number. Send your specimens numbered. How can I send you any little parcel of one or two works of my own?

Rely on my zeal and fidelity.

J. E. Smith.

Professor Martyn to J. E. Smith.

Dear Sir, Park Prospect, March 15, 1793.

My letter was just gone to the penny-post when yours arrived, containing acknowledgements far above the merit of the compliment which I was happy in paying you*, and which I do most sincerely assure you came from the heart. I ought to apologize for not having asked your consent, but I

* The dedication to the President of the Linnæan Society of his work on the Language of Botany, being a dictionary of the terms made use of in that science.

was fearful that your modesty might have declined receiving that tribute, which, in my opinion at least, you have every right to receive. I beg the favour of you, when you have occasion to look into the glossary, to note in the margin any errors or doubts that you observe, in order that a second edition may be more correct. If you wish for any copies to give away, they are at your service. You see by my letter that I have no scruple in asking your assistance. *One-ranked* does not strike me as a good term for *secundus:* at least I am afraid it is too bold to venture on. I am still for the periphrasis.

I am satisfied with what you say about *glaber* and *lævis.* But what English term shall we adopt for the latter? *Even* or *level* are the only terms I can think of; and, if I forget not, you proposed the former of these, but they express nothing of freedom from hairiness; for which reason I put *smooth* for *glaber*, and *smooth and even* for *lævis.*

Certainly most adjectives in *osus* have an unfavourable meaning. *Religiosus*, however, has very frequently a good meaning in Cicero. "Homines integri, innocentes, religiosi." *In Verrem.* "Huic ego testi gratias agam, quod et in reum misericordem se præbuit, et in testimonio religiosum." *In Cæcinam.* I do not know that Cicero or his contemporaries use *famosus* in a good sense; but Pliny and Tacitus certainly do.

Cicero generally uses *gloriosus* in a good sense: as, "Magnificum etiam illud, Romanisque hominibus gloriosum." *De Divin.* "Illustria et gloriosa facta." *De Fin.*

Milton therefore is sufficiently justified in his use of *gloriosissimus*, however superlatives may be applied; on which subject I can say nothing.

I do not find *animosus* ever used in a good sense. But we want an index of terminations to assist us in such inquiries. I suspect that in the Latin language, as in ours, and I suppose all others, many words which had originally a good sense have changed gradually into a bad one,—'a good man', 'an honest fellow', 'an innocent girl', 'a good sort of a woman', 'a good-natured man', are examples.

But a remarkable instance is the word *wench*, which originally signified a young unmarried lady. In Piers Plowman the Virgin Mary is so called.

I am, dear Sir, with great esteem,

Your very faithful Friend and Servant,

THO. MARTYN.

Dr. Roxburgh to J. E. Smith.

Dear Sir, Samulcotah, March 20, 1793.

Your polite and acceptable letter reached me on the 4th of August. The diandrous tree you intend to give my name to, you will find described and drawn amongst those sent to the court of Directors. I thank you for the honour you intend doing me in your next fasciculus. *Mucada* is one of our most valuable and finest timber trees; some of the wood I sent to Mr. Alexander Dalrymple last year, to try if it would answer for scales to mathematical instruments; for it is close-grained and less subject to crack or warp than any other wood in India. Weavers' beams are always made of it on this coast.

I have taken the liberty of sending you some of my newly discovered Fever Bark: it is from a new species of *Swietenia*, a most beautiful and most valuable tree as any in India. I have hitherto found this bark infallible in the cure of intermitting and remitting fevers, even after the Peruvian bark had failed.

My *Lythrum orixensis* is now in flower in my garden for the first time. It is a very beautiful, large, ramous shrub, of quick growth: the flowers are on small racemes of a bright vivid red, and seem to me to partake of the character of *Grislea* as well as *Lythrum*. I find my former description and drawing faulty: your observing to me that it will belong to the genus *Grislea*, first induced me to compare it with the description of that family.

I inclose you a specimen of an *Indigofera*, the leaves of which yield the most beautiful indigo I ever saw, and in a very large proportion. I have sent home to the Directors a drawing and description under the name of *Indigofera cœrulea*.

I am, dear Sir, with much regard, yours,

W. Roxburgh.

Dr. Roxburgh to J. E. Smith.

Dear Sir, Calcutta, August 20, 1794.

Since my return to Bengal I have found several plants unknown to me before: amongst them there will be, I think, three or four new genera and more new species. Amongst the former is the beautiful *Asjogam* of the *Hortus Malabaricus*, vol. v. tab. 59, which I think will make a very distinct new genus, and have called it after the late Sir William Jones, whose botanical knowledge well entitles his memory to this mark of regard*: amongst his MSS. was found a paper (70—80 pages in fol.), entitled Botanical Observations on select Indian Plants, which will be published in our next volume (the 4th) of Asiatic Researches. Amongst them I found a description of this lovely small tree: a copy of it I send you along with my own description and drawing, with the hopes of your giving one of them an early place in one of your valuable publications. I hope it will prove deserving of being so placed. The drawings are very exact, though not elegant. And may I further request of you to insert a better panegyric on so great a character? He was a most surprising man, and the complete gentleman.

You will have learned that the bark of *Corchorus olitorius* and *capsularis* is equal to the best flax, a fortunate discovery for India, and for our manufacturers at home. Do you find that it was ever used by the ancients, or any other people, as a substitute for flax? Rumphius just mentions this quality. From time immemorial the Bengalese have known

* Of this genus, which was established by Dr. Roxburgh in the Asiatic Researches, iv. p. 853, a singularly beautiful species, the *Jonesia Asoca*, has lately been introduced into our gardens, and was figured in a recent Number of the Botanical Magazine, plate 3018.

its value; yet they only employ it for cordage, twine, and other such uses, holding the bark of *Crotalaria juncea* in higher repute, it being stronger and more durable.

Your humble Servant,

W. ROXBURGH.

Professor Willdenow to J. E. Smith.

Condona, illustrissime vir, audaciam meam, quod nomine tuo celeberrimo librum a me sub titulo *Phytographiæ* editum inscripserim. Accipe hic librum tenuem grato animo; sed rigoroso haud subjicias rogo scrutinio: sunt hæc vegetabilia descripta e collectione mea herbarum, quæ non tam dives quam tua et aliorum botanicorum Angliæ. Si itaque in descriptionibus meis hinc inde erraverim, indicatio errorum a tua manu facta mihi erit grata. Flagranti amore florum vegetabilia prosequor; sed Germania nostra paucas, tam in agro, quam in hortis, alit plantas: hinc commercium litterarium et communicatio siccarum herbarum cum viro tam illustri mihi accepta erit. Vale, itaque, vir celeberrime; meque favore tuo haud indignum habeas oro rogoque.

Dabam Berolini, d. 27 Septembris, 1794.

Dr. Roxburgh to J. E. Smith.

Dear Sir, Calcutta, Dec. 27, 1794.

I have this instant got a few seeds of my *Ipomæa grandiflora*: it is *Munda-valli* of the *Hort. Mala-*

bar.; but certainly not *Bona-nox,* or Gærtner is wrong. His plant is my *I. Bona-nox.* You will find both, drawn and described, in my sixth hundred; viz. No. 567. and 568.

With a good deal of address, and the help of friends, I have been so fortunate as to procure living plants of Sir William Jones's *Jatawansi,* the real *Spikenard,* which he concluded to be a *Valerian* from the imperfect description of it he received from a friend; for he never saw the living plant. I have had them a month: they as yet show only a small lanced leaf or two. The names they were sent under by the Deb Rajah of Bootan, are *Jatawansi* and *Nampé* or *Nanpé.* Garcias ab Horto's figure of the drug is excellent. In short, every circumstance corroborates Sir W. Jones's plant to be the real *Spikenard of the ancients.* You will soon see, in the 4th volume of the Asiatic Researches, what he further said on this subject, when he saw Dr. Blane's account of an *Andropogon* which Dr. B. took for it. I am promised the living plants of this *Andropogon.* It grows far north, on the skirts of the mountains. It is a medicine of repute amongst the people of these countries*.

But let me stop my pen.

W. Roxburgh.

* Subsequently to the very excellent paper upon this plant, given by Sir William Jones in the Asiatic Researches, the plant here referred to by Dr. Roxburgh flowered; and, satisfied that it was really a *Valerian,* he published it as such with a description and figure. Sir William Jones's own paper, and that of Dr. Roxburgh, have both been reprinted in Sir W. Jones's Works, ii. p. 9. Further and very curious observations on the Spike-

Major Velley to J. E. Smith.

Dear Sir, Bath, Jan. 25, 1795.

I received yours, and am highly flattered by the very favourable, and I fear partial, opinion which you express of the work*; especially when you condescend to say that you have read it with great pleasure and advantage. Whatever difference of opinion may subsist upon the meaning of the word *seeds*, and which perhaps should be confined to those plants only in which a sexual influence prevails; yet the term seems to have been appropriated to the propagation of plants from the most remote antiquity, and long before the discovery of any sexual distinction in the vegetable world. Thus in the Scriptural account:—

"Καὶ ἐξήνεγκεν ἡ γῆ βοτάνην χόρτου, σπεῖρον σπέρμα κατὰ γένος [καὶ καθ' ὁμοιότητα]."—Lib. Gen. cap. i. ver. 12. (The word σπεῖρον, by the by, is hardly reducible to grammatical construction.)

The ancients, who formed their ideas from natural objects, probably borrowed the term from the mode of propagation observable in plants; and by analogy applied it to sexual generation in animals, without knowing at the time, that the same principle existed in the former.

Although I had my scruples with respect to the application of the term *seeds*, and for the same rea-

nard of the ancients, is given in the very learned *Hierobotanicon* of Olaus Celsius, ii. p. 1., where the plant is, as by Sir G. Blane, referred to the Grasses.

* Coloured Figures of Marine Plants of England, folio. 1795.

sons which you have assigned, yet I found myself unavoidably compelled to make use of it; as both Gmelin and Gærtner (but more particularly the last) have endeavoured to establish a material difference in the propagation of marine plants. You will readily perceive the force of my remark, from the following passage in Gærtner, p. 19:—"quod non alia in Ceramiis, quam per *solas gemmas* locum habeat propagationis ratio: dum contra in omni genuino Fuco *vera* reperiantur *semina* utero carnoso, a matris cortice ac medulla distinctissimo, conclusa."

And yet the same author maintains that these very seeds are produced by a *simple unassisted* power:—"quod in Fucis *genuinis* . . . . ipse uterus . . . . . . officia genitalium *utriusque* sexûs, præstet *solus*."—Vide p. 33.

One of the great points which Gærtner endeavours to establish, is, that the less perfect Fuci, (i. e. *complanati*), as well as the Confervæ, have a different and more simple mode of propagation than the former; and that it is effected only by *gemmæ*. He makes this distinction between the origin of *semen* and *gemma*:—"Quod *gemmæ* medulla sit pars *identica* medullæ maternæ, dum contra *seminis* medulla . . . . . a matris suæ distinctissima."

I think I have evidently proved that the grains in the Confervæ are as much entitled to the appellation of *seeds*, as those in the perfect Fuci: nay, more so, because the medullary substance of the *Conferva corallinoides* is a liquid, while the grains contained in clusters at the joints are *solid* and opaque.

Gærtner seems to have been a very superficial observer of the Confervæ, taking upon trust the opinion of Adanson. If I had adopted the term *gemma*, I should have perplexed the subject, and at the same time seemed to accede to Gærtner's theory. In short, as the expression had been used by Linnæus, and in general by subsequent writers, I thought it better to continue the same.

Whenever you may find the same objections raised against the term *seeds*, I hope, for the reasons assigned, that you will become my apologist.

I remain, dear Sir,

with sincere regard, &c.,

T. VELLEY.

Mr. Stackhouse to J. E. Smith.

Dear Sir, Pendarves, Sept. 16, 1795.

I was much flattered by a letter from Mr. Woodward, who informed me you approved of my work *.

As I described every plant from nature, and bestowed every attention in my power, I had some reason to hope it might be favourably received. The little amicable controversy between the Major † and myself has spurred me on to take a further peep into the *arcana* of marine plants; and it is with pleasure I can assure you that my pains have already been amply rewarded.

I brought down with me a compound microscope

* *Nereis Britannica*, 2 fasc. folio 1795—1797.

† Major Velley.

with six different magnifying powers, and have begun a course of observations on the fructification of every species which has offered itself in fruit. Instead of the *simple process of Gærtner*, I find nature more elaborate in her processes in proportion to the minuteness of the operation. In the larger Fuci (the *F. vesiculosus, spiralis, nodosus,* &c.) the *mucus*, as I called it, in which the seeds are lodged, appeared through high magnifiers to be a collection of pellucid tubes, in some places seemingly divided by *septa*. These tubes are curiously reticulated; and each species seems to have the form of the network different. The seeds at first adhere to the inside of the fruit in orbicular masses, but afterwards are dispersed through the tubes. Though it is asserted by Velley, in *F. vesiculosus* and *F. canaliculatus*, that the seeds issue out of holes in the tubercles of the fruit, yet I have reason to think that is by no means general to all the species. In some there are tubercles permanent on the surface: in others the surface is smooth; and though the thinnest slice of the outside has been taken off with a fine instrument, yet no aperture has been discoverable. The air-bladders likewise in *F. vesiculosus* and *F. nodosus* are by no means simple inflated vessels. Lightfoot calls them *hairy within*; but they have a curious process: it consists of pellucid strings stretched across, swelling at intervals into oval beads; these strings are ramified, and distended in various irregular directions. All this is discoverable by cutting out a slice from each side, and looking through the cavity. Whether this process is to prepare any impregnating vapour, or, more probably, to rarefy air for the purpose of inflation, I have not been able to discover. In the old bladders little traces remain of it, when the skin (as is the case in *F. nodosus*) acquires the thickness of sole-leather. I find likewise the inside of the pod of *F. siliquosus* to consist of a curious branched process, extending in all directions, and not simple threads stretched across, as I had described it, p. 9. I have likewise discovered the fructification of *F. digitatus* and *Polyschides* to be an imbedded oval bladder, and not a superficial one, as I had described it, covered with a tender pellicle. I have not yet seen it in maturity. A fragment of *F. palmatus* in fructification was thick set with conical protuberances, perforated at the summits, with the seeds imbedded in the internal substance. This mode is very similar to that of *F. vesiculosus*, &c., excepting that in the latter, nature has set apart a fruit. The powers of my glasses go as far as to discover the internal structure of *Ulva umbilicalis*, the thinnest and most membranaceous of sea plants; and I have discovered small dark perforated *granules*, which are undoubtedly the seeds. I shall take the liberty of inclosing two or three new species or varieties of sea plants to add to your collection; and having drawn up a paper on the *Herniaria glabra*, in which you will discover many errors of the early writers rectified, and a description more at large than I have met with, I shall beg the favour of you to present it to the Society.

I beg leave to subscribe myself, Yours, &c.

J. Stackhouse.

Professor Blumenbach to J. E. Smith.

Dear Sir, Goettingen, Dec. 25, 1795.

I return you my best thanks for the interesting acquaintance of Dr. Duncan you were so kind to procure me by your introductory letter. He spent some months with us, and, as I hope, to his satisfaction. I wish you would be tempted, dear Sir, to a second tour on the continent, and particularly to a trip to Goettingen, where I should think myself very happy to return you, in any way, the civilities you showed me during my abode in London.

You will receive here a letter of a friend of mine, Mr. Persoon, from the Cape, a very estimable accurate naturalist; and I request your acceptance of a pamphlet of mine I published lately.

A propos, you were kind enough to present me with a curious enigmatical sea-animal of the class of Vermes, out of your Linnæan museum, which I keep as a relic of a saint, as well as a token of your kindness to me: a figure of it is given in a rather indifferent book, Schröder's *Einleitung in die Conchylien-kenntniss*, T. 11. tab. 6. fig. 21, and described under the name of a magnified *Sabella* (?), under which genus it is likewise put in the new edition of the *S. N.* T. 1. p. vi. p. 3751.

I shall be happy to be of any use to you; and am, with much regard,

Dear Sir, &c.

J. Frederick Blumenbach.

* * * * * * *

In 1802, His Majesty King George III. granted to the Linnæan Society a Charter of Incorporation; and in 1811 the late King, then Prince Regent, honoured the Society in becoming its Patron.

It has been the custom from the time of its first institution to celebrate the birth-day of Linnæus by an annual dinner, when a large proportion of the English members assemble on the 24th of May to commemorate the event.

On the twentieth anniversary the President addressed the Society at their annual feast, in the following words, which are here transcribed from a MS. found among his papers. They have not before appeared in print.

"I cannot take this chair for the twentieth time of my being in so unanimous and flattering a manner chosen to fill it, without saying a few words, not merely to express my gratitude and attachment to the Society, which I trust would be superfluous, but to congratulate all its members, and particularly those who with me have watched over its infant growth, on the eminence it has now attained. I have another motive for choosing the present occasion for this purpose.

"This is the hundredth anniversary of the birth of Linnæus; and as we have chosen to commemorate his birth-day by our general annual meeting, the centenary revolution of it should certainly not be passed over in silence.

"I cannot but compare the obscurity of his birth, and the struggles of his youth, with those of our infant Institution; and the sanction and protection which he received from his countrymen and rulers, with what we have obtained from ours. Happy will it be, if the utility of our labours should prove to correspond with the benefits which he conferred on the scientific world.

"Our beginnings have been propitious; and I believe it will be found on comparison, that no literary institution ever produced more solid matter of information within the same limits of time; and this entirely by our own pecuniary, as well as scientific efforts.

"Celebrated academies have generally been assisted by royal stipends and ample funds; and their members have often had collateral views of worldly profit and promotion. We have been prompted by a disinterested zeal for science alone.

"It is extremely pleasing to my recollection to recall the day when the first idea of this Institution presented itself to my mind,—which I immediately communicated to my worthy friend, who was then my near neighbour at Chelsea. The result of our consultation was a conference with our much valued friend, now Dean of Rochester. We soon laid the plan of the rising Institution; and with the assistance of Sir Joseph Banks, and Mr. Dryander more especially, modelled it in such a manner that scarcely any alteration has since been found necessary.

"In my subsequent tour through Europe, I made it my business to associate with us the most learned foreigners; and after my return, in the spring of 1788, the Linnæan Society was founded.

"The honourable sanction we have received from

Government five years since, is fresh in your memory; and our publications are a better testimony of our deserts than any that I could lay before you. I trust we shall go on with equal zeal and increasing honour and success.

"In whatever way I can be most useful to you, I am at your disposal, as are the treasures of our great leader and instructor, which I hold only for the use of my fellow-labourers in science.

"I regret my personal absence from you for many reasons, but it is unavoidable; and I am annually reminded that a permanent residence in town would still less agree with my constitution than it did formerly. Nevertheless I enjoy by that means more leisure for my favourite pursuits, to which my whole time is devoted; and my mind is always with you. Allow me to conclude with my most ardent wishes for the prosperity of the Linnæan Society, and that we may all long witness and promote it."

At a period when the illustrious individual, in whose honour the Linnæan Society was founded, is assailed on all sides, it will be interesting to know, that, unmoved by the almost *general defection*, he, who may be considered as his principal representative, still continued to advocate the principles of the immortal Swedish naturalist; and this unaltered adherence Sir James expressed in his last introductory lecture at the London Institution in 1825, as well as in the concluding pages of his latest printed work, the English Flora, where the author alludes to "principles too little studied by the pursuers of superabundant discrimination, instead of philosophical combinations." "This," he asserts, "is the bane of natural science at the present day. Hence the *filum Ariadneum* is lost, or wilfully thrown away, and a bandage darkens the sight of the teacher no less than that of the student."

Yet Sir James cannot be said to stand alone and unsupported in his opinion. "The question," he remarks, "of the natural or artificial character of Jussieu's system has been ably discussed by the celebrated Mr. Roscoe in the Transactions of the Linnæan Society, vol. xi. p. 50; who, in showing that this method involves several as unnatural assemblages as the professedly artificial system of Linnæus, contends that little is to be gained by its adoption with respect to a conformity to nature." And in the fifteenth volume of the Society's Transactions, Mr. Bicheno, in a paper on Systems and Methods in Natural History, observes, "that the two great masters of botanical science (Linnæus and Jussieu) propose different ends, and ought not to be regarded as *rivals*. Division and separation are the ends of the artificial system; to establish agreements, is the end of the natural."

Following the same idea, the Rev. E. B. Ramsay, in a biographical notice of his lamented friend, printed in the Edinburgh Journal of Science, observes, that "there is no point on which young botanists are more mistaken than in their ideas of natural classification. They often imagine they have only to commence the study of natural arrangements, and become at once profound philosophical botanists. This is one of the signs of the times; a desire to grasp at general results and conclusions without a previous study in detail. The error in this case is putting the natural and artificial methods in opposition to each other; whereas it appears to be the object of the artificial system to *collect materials* to *form* a natural one. But it has been of late spoken of rather as something quite superseded, as something to give way to a new and a nobler structure, built upon a foundation entirely different."

Sir James moreover informs us, that "the illustrious Frenchman, Bernard de Jussieu, the early and confidential friend of Linnæus, conceived a philosophical idea of natural orders of plants, and conferred on the subject with the *only man* whom he found capable of appreciating or understanding them." In his letters to Linnæus* he even gives him the honour of having *first* formed a scheme of natural orders of plants; but it seems as if the science of botany were no sooner destined to emerge from obscurity and confusion by a just perception on

* Bernard de Jussieu writes to Linnæus, 15th Feb. 1742: "I learn with the sincerest pleasure of your being appointed Professor of Botany at Upsal. You may now devote yourself entirely to the service of Flora, *and lay open more completely the path you have pointed out,* so as at length to bring to perfection a natural method of classification, which is what all lovers of botany wish and expect."

"In the face of this testimony," observes Sir James Smith, "we trust it will hardly be asserted in future, that Linnæus owed his ideas of natural orders to the excellent writer of the above letter." —*Selection of the Correspondence of Linnæus,* vol. ii. p. 212.

these various subjects, than it becomes in danger of plunging again into darkness by a neglect of them. The natural principles of arrangement for a scientific knowledge of plants and a permanent discrimination of their families and species, have been no sooner distinguished (at the suggestion of Linnæus) from an artificial scheme for their convenient investigation, than those *different objects* are *confounded.*

"Linnæus read lectures on the natural orders of plants to his most assiduous and accomplished pupils *only*. To begin to teach botany by these orders would be like putting Harris's *Hermes* into the hand of an infant, instead of his horn-book."

CHAPTER V.

Enumeration of the Works of Sir J. E. Smith. 1. *Reflexions on the Study of Nature.*—2. *Dissertation on the Sexes of Plants.*—3. *Thesis de Generatione.*—4. *Reliquiæ Rudbeckianæ.*—5. *Plantarum Icones.*—6. *Icones Pictæ.*—7. *English Botany.*—8. *Spicilegium Botanicum.*—9. *Flora Lapponica.*—10. *Botany of New Holland.*—11. *Tour on the Continent.*—12. *Syllabus.*—13. *Insects of Georgia.*—14. *Tracts.*—15. *Flora Britannica.*—16. *Compendium Floræ Britannicæ.*—17. *Flora Græca.*—18. *Prodromus Floræ Græcæ.*—19. *Exotic Botany.*—20. *Introduction to Botany.*—21. *Articles in Rees's Cyclopædia.*—22. *Tour to Hafod.*—23. *Lachesis Lapponica.*—24. *Articles in Trans. of Linn. Soc.*—25. *Review of Modern State of Botany.*—26. *Grammar of Botany.*—27. *Linnæan Correspondence.*—28. *English Flora.*

1. In the year 1785, previously to his tour on the Continent, Sir James Smith made his first appearance as an author in publishing Reflexions on the Study of Nature, which he translated from the Latin. It is the Preface to the *Musæum Regis Adolphi Frederici* of Linnæus; a work containing descriptions of the various natural productions of the museum of the King of Sweden, printed in 1754 at His Majesty's expense.

In speaking of this most superb and expensive of all Linnæus's works, "one fact," he observes, "may be learnt from it,—that the study of nature does not necessarily tend to make a man irreligious, as some weak people have been led to believe. A number of illustrious examples might be produced to the contrary; none more eminent than the excellent author of this work."

2. In the following year Sir James published A Dissertation on the Sexes of Plants, translated also from the Latin of Linnæus; which he inscribed as a token of respect to Dr. John Hope, Professor of Botany in the University of Edinburgh.

Of both these works an edition was printed at Dublin in 1786.

3. *Disputatio Inauguralis quædam de Generatione complectens.*—The Thesis which entitled him to the degree of M.D. at Leyden, 1786.

It is dedicated to Robert Batty, M.D.

4. *Reliquiæ Rudbeckianæ.*—The President's next literary work was the republication of the wooden blocks of Rudbeck, in 1789.

In his introductory discourse, Sir James informs us, that "at Upsal, under the auspices of the great Rudbeck, was laid the foundation of what Mr. Stillingfleet has justly called an unrivalled school of natural history, and which was destined afterwards to give laws to the rest of the world. Rarely has such a variety of profound and extensive learning been united as in Rudbeck. In antiquities, especially those of the northern nations, and in the learned languages, his knowledge was unbounded. In botany he had erected to himself what might reasonably have been thought a '*monumentum ære perennius*', in one of the greatest undertakings of the kind, a collection of fine wooden cuts of all the plants then known. They were to have been arranged and named according to Bauhin's *Pinax*, in twelve large volumes folio. But two volumes were scarcely printed, when in 1702 a dreadful fire reduced almost all Upsal to ashes, and with it the work of Rudbeck, and many thousand wooden blocks, already cut, besides almost all the materials of a history of Lapland composed by his son."

All that remains of this work are a few copies of the second volume, and three only of the first, one of which is in the Sherardian library at Oxford. Linnæus was possessed of about 120 of the wooden blocks of this first volume, as well as eight or ten unpublished blocks belonging to some intended one, all which with his collections came into Sir James's hands.

The republication of this fragment of Rudbeck's great work was a tribute of respect to his profound and varied learning.

It is dedicated to Johannes Gustavus Acrel, Professor of Medicine at Upsal.

5. *Plantarum Icones hactenus Ineditæ, præcipue ex Herbario Linnæano.*—The first fasciculus, folio, was published in 1789. It extended but to three numbers. The first was dedicated to Sir Joseph Banks, Bart. The second to M. P. Auguste Broussonet, M.D. The third to the Marquis Ippolito Durazzo.

6. *Icones Pictæ Plantarum Rariorum; or Coloured Figures of Rare Plants.*—(This also reached but to three numbers,)—1790. It was dedicated to the Marchioness of Rockingham, which produced the following obliging letter from that lady.

The Marchioness of Rockingham to J. E. Smith.

August 28, 1790.

I must interrupt your agreeable studies for one moment, just to thank you for the letter you wrote me before you set out for Yorkshire. I ought not, even in a dream, to have had the faintest apprehension of Dr. Smith's ever writing a satire. His elegant and concise Dedication has relieved my foolish fancies; but if a *certain Lady's name*, and the *word botany*, had appeared in the *same page*, he will allow that it would have been a complete one. I dare not dispute with a Dr. Goodenough, otherwise, as it is in the form of the address of a letter, I should have supposed that *Most Honourable* was dignified enough; and that the *Noble* Duchesses might think his decision rather an infringement of their rights: if so, I hope my neighbour Duchess * will make him *stoop*, if *that* is *possible*, to ask pardon.

I ought to be ashamed that I have not before acknowledged your being so good as to add to my library the engravings of Rudbeck.

Mr. Lee brought me a plant he calls the Red Palm,—quite a new thing, and, to my taste, remarkably elegant. I shall long to hear what beautiful plants are flourishing and flowering where you are. My largest Portlandia is beginning again; four flowers in a cluster upon almost every branch, and there is also a fine fruit upon it; at the same time it is in such health, that one grieves to unload it of any

* Duchess of Portland.

of its buds, though some people reckon that right. I cannot but hope you will take a look at Wentworth as you pass by; and if so, do observe what Amaryllises there are there; and Henderson will tell you about the arrival of the *Euphorbia Wallenia,* and all about it. I also wish vastly to know whether a prodigious large Palm is there, which came from Gordon's, and was called the true Sago Palm; and pray tell Henderson that I should rejoice if he could find time to be spared to make me a visit. But I am quite ashamed to be so troublesome to you with all my little nonsenses.

I will only add my compliments; and conclude,

Dear Sir,

Your very sincere

M. Rockingham.

7. *English Botany.*—It was in 1790 that Sir James undertook to write a Flora of his native country, under the title of English Botany, to which his name did not appear as the author, till he publicly acknowledged it in a Preface to the fourth volume, in 1795; wherein he observes, "I have to answer for every word in this publication, except the letter-press to plates 16, 17 &18, which were furnished by another friend of the editor. No pains were taken to conceal the real author; nor was I aware that the truth, after a little time, was not generally known, till a criticism appeared in the Gentleman's Magazine for February 1793; in answer to which, in that for April following, it became necessary to own the work as entirely mine."

The public has from its commencement acknowledged the excellence of the figures which illustrate this work, by the pencil of the late Mr. Sowerby.

The plates amount to 2592. "And it is but just," observes its author, "to the memory of the worthy and lamented artist, to say that they are, on the whole, the most expressive and accurate of their kind. In the account of each species, besides corrected characters, synonyms, and descriptions, I have frequently introduced whatever might recommend the study of plants, diffuse a charm over the more dry and technical parts of the subject, improve our scientific language, or direct the contemplative mind to more important and exalted views of its Creator's Works."

The *English Botany*, from which the *Fungi** only are excluded, "has the singular merit of being the only national Flora which has given a figure and description of every species native to the country whose productions it professes to investigate."

It was brought to a termination in 1814, and consists of thirty-six octavo volumes.

For the motto to the first three volumes, Sir James chose the following quotation from Virgil:

"Quos ipsa volentia rura
Sponte tulere suâ."

The remainder of the work, with the exception of the last volume, has this:

"Viresque acquirit eundo."

* These having been published by Mr. Sowerby in a separate work, in three vols. folio.

The concluding volume has this line:

"Longarum hæc meta viarum."

Throughout the whole progress of this work, its author had not merely the ablest, he had the most friendly assistance; and it was one of the happy circumstances attached to a labour which was itself a pleasure, that it brought him the acquaintance of many amiable and accomplished characters whom he might otherwise never have known.

The subjoined letters will convey an idea of the zeal and delight with which persons of similar pursuits assist each other, and accord together; and as Sir James observes in the Preface to his Tour, long before he had much experience of the fact, "It brings those together who are connected by a most commendable, disinterested, and delightful tie,"—if indeed that may be termed "disinterested" which affords one of the purest, most engaging occupations in which a person possessing leisure and opportunity can fill up his vacant hours.

The amiable correspondent whose communication first appears, bequeathed to Sir James in 1805 his whole collection of dried plants. His executor Prof. Dugald Stewart, in a letter to Sir James, says that "Mr. Bruce died in the 80th year of his age, respected and beloved by all who knew him. His life was singularly blameless and tranquil. I avail myself," he adds, "with much pleasure, of this opportunity to express my high respect for your character and learning."

Mr. Arthur Bruce to J. E. Smith.

Dear Sir, Edinburgh, Dec. 2, 1792.

*Convallaria verticillata**, I beg leave to mention, is a true native from the *den Rechip*, a deep, wooded gully from the hills in the Stormont, Perthshire, about four miles north-east from Dunkeld. On the 7th of July the flowers were beginning to fade. Except the *Convallaria verticillata*, I cannot inform you of any acquisitions entirely new, notwithstanding in the course of last summer there have been discovered, not far from this city, some of our very rare Scots plants, hitherto unnoticed, particularly, *Astragalus uralensis*, North-ferry hills; *Orobus sylvaticus*, Bread-hill, near Edinburgh; *Anagallis tenella*, Hunter's bog, Arthur's seat; and about two years since, I found *Trientalis europæa* only six miles north from the ferry. All these things taken together strongly justify your expectation of the *Linnæa borealis* being hereafter found a native also in N. Britain†.

I am happy to inform you that the Natural History Society still continues to flourish. At present, and for some time past, chemical pursuits

* Narrow-leaved Solomon's Seal—*Engl. Bot.* fig. 128.

† This beautiful plant was discovered in 1795 by Professor James Beattie, jun., of Aberdeen, in an old fir wood, at Englismaldie. "Linnæus," Sir James Smith tells us in the seventh volume of Engl. Bot., "has traced a pretty fanciful analogy between his own early fate, and this 'little northern plant, long overlooked, depressed, abject, flowering early,'—and we may now add, more honoured in its name than any other."

have very much engrossed the attention of medical gentlemen at this University.

I am, with the highest respect,

Sir, your most obliged, &c.

ARTHUR BRUCE.

The Rev. H. Davies to J. E. Smith.

Aber, near Bangor, N. Wales,
March 4, 1793.

Dear Sir,

Be pleased to accept my many thanks for your kind attention to my letter and packets. I am truly sorry I cannot supply you with a specimen of *Anthericum serotinum*: I have not a perfect one myself. I shall be much gratified to see your genera of ferns: there are certainly some British species likewise of that family that want ascertaining.

I was observing to a friend very lately, that I discovered traits of a very masterly hand in Sowerby's English Botany; but I confess I did not know where to look for the author: I thank you for that information. I wish to know the plan more particularly. Is it to comprehend all English plants, or only the rarer ones, or such as are not already well figured? Is it intended to take in Cryptogamous plants? I have certainly several new ones of that class, and much at your service. Mr. Sowerby's excellence in figuring these genera, among others, is conspicuous in Dickson's *Fasciculi*.

To your query, whether plants could be sent to London fresh so as to be drawn, I answer, I suppose they might: when I get among them, the experiment will not be expensive. I wish to know what species you are most desirous of, besides that one already mentioned. The latter expedient which you mention, I should prefer; I mean, having them sketched on the spot, and sent with dried specimens; but that, I fear, is impracticable, as Mr. Pennant's draughtsman lives between fifty and sixty miles from the scene of action.

I beg, Sir, you will be assured, that I shall think no commission of yours a trouble, but shall be happy to promote so elegant and desirable a work as the English Botany, or any other that you interest yourself in.

I lately left Mr. Pennant in good health. I am much affected at your account of Mr. Hudson and Mr. Zier.

I am, dear Sir, with great esteem and regard,

Your much obliged,

HUGH DAVIES.

Marquis of Blandford to J. E. Smith.

Dear Sir, Bill Hill, Aug. 9, 1798.

I have been spending the last fortnight at Lymington, and did not lose the opportunity of searching for some of the rarer English maritime plants, as well as those that prefer a boggy soil. The principal plants I collected were, *Inula crithmoides*, salt marshes near Lymington; *Chironia pulchella* (vide Engl. Bot.), on a dry sea-beach near

Milford; *Chironia Centaurium floribus candidissimis*, on a dry pasture near the sea; *Anthemis maritima*, salt marshes, plenty; *Beta maritima, Santolina maritima, Cheiranthus sinuatus, Salsola Kali*, on Brownsea Island, belonging to Mr. Sturt, near Poole; *Pinguicula Lusitanica* (*villosa*), bogs in the New Forest. I likewise found a *Schœnus*, which I take to be *longus*.—I have just had notice of between 200 and 300 plants having left Jamaica from Dr. Dancer; a great many new genera. I shall be very happy to show them to you, as well as all my others, when I am settled at White Knights. I believe the *Pulmonaria maritima* and *Convolvulus Soldanella* grow on your coast: I should be very thankful if you could procure me some strong living plants of them, and be so good as to order them to be sent to Marlborough House.

I remain, dear Sir,

Your faithful, humble Servant,

BLANDFORD.

Rev. Thomas Butt to J. E. Smith.

My dear Sir, Upper Areley, Jan. 22, 1799.

I feel much obliged by the notice you have taken of my request concerning the *Anchusa officinalis*, and the trouble you have given yourself respecting it. In conformity to your wishes I immediately rode to the spot where I had found it, that by a second examination I might with more certainty determine its claim as an indigenous plant. I am now perfectly satisfied that it is no garden outcast. It is insulated in the sands of Hartley Links; nor have I found any specimen near the village. The kitchen-garden of Lord Delaval is at no great distance. He is no collector of plants; and this flower is rarely met with (at least by me), except in collections: there are no traces of garden rubbish about. It has been observed to me, that it might have been brought in the ballast of ships; as Hartley now and then sends vessels to Copenhagen, though nearly the whole of its trade is in coals to London. I took a person with me who was well acquainted with the port; and, when I pointed out the spot, he assured me that ships *never* emptied their ballast there, but always on the artificial mound, which has been in this manner raised close to the harbour. I examined this mound in the summer with some accuracy, and found several plants on it which I had not elsewhere observed in Northumberland,—*Pastinaca sativa*, *Anethum fœniculum*, &c., but not one plant of *Anchusa*. The spot where the latter grows lies two hundred yards north of this mound. It is not of great extent; but the plant flourishes there very well. I have not observed it in any other place; but I have not closely examined that part of the coast. I will not take upon me to decide this question; but I thought it proper to state the objection to you. My opinion might perhaps be construed into a wish of acquiring the honour of discovering an indigenous plant not before observed: I will only remark that, as Northumberland and Denmark lie under the same parallel of latitude, there is nothing extraordinary if the same plant proves indigenous in both countries.

My Lord Valentia desires me to present his best compliments. He hopes soon to meet you in London.

Believe me, my dear Sir,
Most sincerely yours,
THOMAS BUTT.

Rev. Thomas Butt to J. E. Smith.

My dear Sir, Upper Areley, Dec. 17, 1799.

I feel myself highly honoured by the mention made of me in a late number of English Botany, and am particularly pleased that my name should appear in a work, which, without any pretensions to inspiration, I may venture to predict will always retain its present celebrity. I am not, however, without some uneasiness, unsupported as I am by any botanist of established fame, lest I should erroneously have judged *Anchusa officinalis* indigenous.

With the same diffidence I shall now accurately describe to you the situation of another plant, which I met with this summer in my wanderings through Wyre Forest. I have little doubt of its being *Gnaphalium margaritaceum**, of which Withering gives no *habitat* since the time of Ray.

* This conjecture on the part of Mr. Butt proved quite correct; and the plant was figured in English Botany ten years after, from specimens sent from the spot here described.

In the heart of the forest is a spot of cleared ground, which constitutes a small farm : a cottage is erected on it. About a quarter of a mile (two fields breadth) distant from it I met with the plant, growing by the side of a running ditch, and covering several square yards of ground. The spot did not lie below the cottage ; nor could I find an appearance of any habitation having ever existed near the course of the little stream. I traced it to its source in the bogs of the adjoining forest, and downwards to its union with Dowles' Brook, but could find no more specimens of the plant. If you agree with me in believing it really wild, it will be easy for me to send you fresh plants next summer, though I cannot discover that they differ from the garden ones. If ever you have leisure in any future visit to Oxford, I much wish you would visit Elsfield Wood, and ascertain what probabilities there are for considering *Lonicera Caprifolium* as wild there: I think they rest on as firm foundation as any other plant with a single *habitat*. Dr. Randolf, the present Bishop of Oxford, showed great botanical ignorance, in affirming that it was only a variety of the common Honeysuckle, since no two plants have their distinct characters more strongly marked.

I will, as usual, subjoin a list of what rarer plants I have found this year. *Osmunda lunaria*, on a hill above Cheadle, Staffordshire. *Vaccinium Vitis Idæa*, very abundant in Cotten Woods, near Cheadle. A very small specimen of a fern, in the cavity of a rock below Renard's Cave in Dovedale, which I believe to be *Adiantum Capillus Veneris*. I was assured by a botanical gentleman of Cheadle, whose name I have forgot, that it did grow there. *Cynoglossum omphalodes* grows on a rock planted with trees on the left hand of the road from Blore to Ilam, close to Ilam grounds: I suppose it to be naturalized. The gardener, who has lived there many years, assured me that in his remembrance no flowers had ever been planted there. This visit to Cheadle I made in the month of May. Unfortunately the weather was bad, and the season backward; otherwise I might have been more successful in botany, as the country appeared promising. *Erodium maritimum* is by no means unfrequent about Areley: the Morfe in particular, at Bridgnorth, produces it abundantly. *Carex curta* grows in a pit near the High Trees, a farm-house between Areley and Enville. *Atropa Belladonna*, by a roadside near Mr. Child's of Kinlet. *Iberis nudicaulis*, frequent on the Morfe and Hartlebury Common. *Polypodium Oreopteris*, on Hartlebury Common and the N. W. side of Aberley Hill. *Carex strigosa*, in Clifton Woods on the river Team. *Inula Helenium*, about half a mile from Areley, but near a farm-house. *Serapias ensifolia*, a single plant, underneath the *Pyrus domestica*. This famous tree is in a very unhealthy state, in part owing to a large oak which overshadows it. I am almost inclined to adopt an opinion which I have heard respecting it,—that there once was a garden on the spot; for though there are no certain vestiges of it, I am much puzzled by finding *Ligustrum vulgare* growing there, which as yet I have not found in any other part of the forest, nor indeed anywhere near Areley. A *Pyrola* grows in the forest, which differs from *Pyrola minor* in having a longer bent pistillum. Qu. *P. rotundifolia? Scirpus fluitans,* in a ditch on Hartlebury Common near Stourport. *Campanula hederacea* and *Viola palustris,* on Hartlebury Common, found by J. M. Butt. *Gentiana campestris,* in Wyre Forest.

Mr. Moseley of Glasshampton, the gentleman who sent you *Serapias ensifolia,* is a skilful and most assiduous botanist. He has been fortunate in discovering rare plants in his neighbourhood, and would, I am sure, feel happy in communicating them to you. He showed me *Silene Anglica, Typha angustifolia, Stellaria Nemorum, Campanula latifolia,* and a species of *Mentha,* in a short walk which I took with him near his house. We have at Areley a common *Galium,* which differs from *Mollugo,* in having the serratures of the leaves directed towards the point, and from *erectum* in having a smooth and shining stem; but in a walk to Whitley, which lies in a limestone country, I found that the same *Galium* (to all appearance) had a hairy stem. I found also a variety of *Galium verum,* with flowers of a paler yellow and a loftier stem. When I first saw it growing with *G. verum,* its habit appeared so distinct, that I set it down for a different species; but, on examination, I could not find a specific difference.

You will, I am sure, excuse me for being thus particular about trifles, because you know that my sole intention is to be useful to you in the great work in which you are now engaged*. If any part of this detail answers that end, it will afford me the highest gratification.

Believe me, dear Sir,

Most sincerely yours,

THOMAS BUTT.

Professor James Beattie† to J. E. Smith.

Dear Sir, Aberdeen, June 12, 1800.

Accept of my warmest acknowledgements for your attention to my last communications and condescension in answering my impertinent queries. I am happy that the packet was found to contain some new plants, and that they came so seasonably. —The *Carices*, which you have pronounced new, were long thorns in my path. Often did I give them up; but when they again occurred, a new investigation was the consequence, and as fruitless as the former. My doubts and difficulties are now dispelled; and the genus *Carex*, from being the most obscure in the system, is now that with which I am best acquainted. I only hesitate a little as to the proper character of *C. Micheliana*. The circumstance of the female spikes being crowned with males is a striking one; but I question its constancy. I have many specimens gathered on the same spot, and (as I suppose) of the same species,

* *Flora Britannica.*

† Author of The Minstrel and The Hermit.

in which this characteristic is wanting. When that is the case, I am at a loss for a distinction between it and *C. recurva.* Hence I was led formerly to consider them as varieties, which will account for their being huddled together in my packet. I am carefully watching the progress of this and the other plants of which you wish for specimens, and in a very little time shall send them as directed. I lately found *Mercurialis perennis* with male and female flowers (monœcious). I have this season found the *Carex incurva* in a new situation, in the Links* of Aberdeen, in a wet spongy soil, half a mile from the sea. I have traced its progress :—first it puts forth a simple spike ; some time after, this spike unfolds itself, and the spiculæ become distinct. These terminate always in male flowers. As the capsules swell, the compound spike, from being oval, becomes conical, which is a necessary consequence of the upper florets being barren. The epithet *incurva* is in general very applicable, insomuch that the spike is often buried in the grass. Some specimens are more upright ; but the curvature is always observable immediately below the spike. I shall send you specimens in all the stages of growth. I may here subjoin the places of growth of the other *Carices*. *C. Micheliana,* low marshy ground in Mearns or Kincardineshire, also in wet gullies in the Heughs (cliffs of St. Cyrus, ditto). *C. binervis,* moorish ground, frequently in Aberdeenshire. *C. lævigata,* frequent in shady wet situ-

* A common below the town.

ations in Mearns and Aberdeenshire. *C. Davalliana*, on a moor two miles west of Aberdeen, along with *C. binervis*. *C. fulva*, one of the most common species of the genus, in upland pastures in Mearns and Aberdeenshire. I do not recollect if I took notice in my last of a mistake respecting the place of *Linnæa borealis* (Engl. Bot.); for "at Mearns," it should have been "old fir woods at Englismaldie (seat of Lord Kintore), on the borders of Mearns or Kincardineshire." Have you decided on the *Polypodium* which I had called *rhæticum*? It grows in the Den of Fullerton, Angus or Forfarshire, two miles from Montrose.

I have lately received the very agreeable information, from Colonel Brodie, that part of the *Flora Britannica* is published. Wishing it success,

I remain, Sir,

Your much obliged, &c.

JAS. BEATTIE.

8. *Spicilegium Botanicum, or Gleanings of Botany.* 2 fasc., folio. 1791. Dedicated to Lady Amelia Hume.

To Dr. W. Wade of Dublin the author was indebted for the following commendation of these early efforts:

"Sir,—Though I have not the honour and satisfaction of being personally acquainted with you, permit me to introduce myself by correspondence.

"Your different valuable publications I have taken in, and peruse with much pleasure and admiration, particularly your *Icones Pictæ* and *Spicilegium.* They are certainly without flattery the most superb botanical publications I ever saw.

"Your introductory paper, or rather admirable epitome of the history of natural science, in the Linnæan Transactions, charms me.

"I am delighted with all your various publications, 'from the hyssop on the wall to the cedar of Lebanon:' i. e. from your *Dissertation on the Sexes of Plants* to your *Icones Pictæ.*"

9. *Caroli Linnæi Flora Lapponica, Editio altera studio et curâ Jacobi Edvardi Smith. Illustrissimæ Societati Regiæ Scientiarum Upsaliensi, olim Sueciæ appellatæ, novam hanc Editionem obsequio quo par est DDD J. E. Smith.*

In a letter addressed to Mr. Woodward in 1791, Sir James informs him that "*Flora Lapponica* is at last gone to the press;" and adds "Do not think me too vain if I say it will be the most correct edition that ever appeared of any of Linnæus's works, his own Stockholm ones not excepted. The list of authors quoted will be infinitely more complete and correct; and as to synonyms, I have examined every one, judged of it as well as I was capable, and corrected the typographical errors, which are innumerable in the first edition. I add fifty-five new species to the Flora."

10. *A Specimen of the Botany of New Holland.* Dedicated to Thomas Wilson, Esq., F.L.S. 1793.

This work was published in numbers, of a quarto

size, corresponding with Dr. Shaw's Zoology of New Holland; the works were intended to accompany each other:—no more than sixteen plates of either ever appeared.

11. *Sketch of a Tour on the Continent in the Years* 1786 *and* 1787. Dedicated to his fellow-traveller, William Younge, M.D. F.L.S., 3 vols. 8vo, published in 1793.

A second edition appeared in 1807. This Tour was translated into the French, Italian, and German languages.

From his friends Dr. Goodenough and James Alderson, M.D., of Norwich, the author received the following commendations:

My dear Sir,—By your inquiry after my opinion of your Travels, I conclude my letter to you, which I got franked from Bulstrode the first week in January, has miscarried. I said a great deal upon the subject, to thank you for the great entertainment I have had in reading them: you show such multifarious reading that you quite surprise me. In one part, I thought I saw a passage, or rather a page or two, which I took to be in Dr. Kippis's style. I like to read on both sides of a question. No good cause is hurt by discussion.—Believe me your most affectionate and sincere friend,

Samuel Goodenough.

Dear Sir,—I have this moment finished your Travels, and am extremely obliged to you for the pleasure you have afforded me. At the same time

I cannot help saying that the simile of the Pope, when alarmed, going to the Castle of St. Angelo, equals, in my opinion, any thing to be found in my favourite Hudibras.—J. A.

12. *Syllabus of a Course of Lectures on Botany*. 1795.

13. *Natural History of the rarer Lepidopterous Insects of Georgia, from the Observations of John Abbot.* Two vols. folio, published in 1797, and dedicated to Miss Johnes of Hafod.

The following letter from Mr. Jones relates to this work:

Mr. W. Jones to J. E. Smith.

Dear Sir, Chelsea, 9th Sept. 1797.

Had you ere this pronounced the sentence of ingratitude upon me, the world in general would have justified you, appearances are so very greatly against me, for not giving you thanks for your magnificent present before; but the truth is I was not in possession of it till this morning. I went to the bookseller's no less than three times before I could get it; and even today I believe I should have been sent off without it, if I had not showed myself greatly dissatisfied. I took it under my arm, but soon found it sufficiently weighty. Did you ever see a caricature of a three-pounder going up Hampstead Hill to dinner?—a man oppressed with his own weight, yet surmounting all difficulties that he might enjoy the pleasures of the table. So I laboured abundantly with my load, in expectation of high gratification from the contents when I should get home,—and truly I was not disappointed; but I disclaim all merit, unless you maintain that such is due to every one that stands sponsor to a child, if such child should turn out a worthy member of society. No, Sir, the merit is yours; the demerit attaches to the engraver and colourer, for there are some faults; but upon the whole it has the three great requisites to a modern publication,—good letter, good paper, and showy plates.

I find myself very much gratified, as well for the present, as for the genteel manner you notice me therein, and for which I return you my most grateful acknowledgements.

I am in possession of a pair of *Papilio Lathonia.* Until the last time I examined your cabinet at Hammersmith I was unable to distinguish between *Phalæna, Padella* and *Euonymella:* I there saw evidently the distinctions, and concluded, from the experience I then had, that *Euonymella* was not English; but this year Mr. Haworth has taken it, and another person has taken one also. Ray described it. You are certainly wrong in naming the fly *Argiolus,* Tab. 15. I have both male and female among my drawings without a name. *Argiolus* is certainly different.

Dear Sir,

Your most obliged,

WILLIAM JONES.

14.—*Tracts relating to Natural History.*—In

1798 Sir James republished Reflexions on the Study of Nature, which was then out of print, together with the following essays, reviews, and descriptions, collected together in one octavo volume, which he dedicated to his valued friend, James Crowe, Esq., of Lakenham, under the above title.

Discourse on the Rise and Progress of Natural History.

Observations on the Irritability of Vegetables.

Review of Mr. Curtis's Botanical Magazine.

Review of Berkenhout's Synopsis.

Review of An Easy Introduction to Drawing.

Review of a Dutch edition of *Systema Naturæ.*

An Essay on the Genus of Dorsiferous Ferns.

Description of a new Genus of Plants called *Sprengelia.*

Description of a new Genus of Plants called *Westringia.*

Description of a new Genus of Plants called *Boronia.*

Mr. Pennant to J. E. Smith.

Dear Sir, Downing, Sept. 6, 1798.

I waited only for a frank, that you might receive gratuitously my thanks for your acceptable present of Tracts relating to Natural History: they give me great pleasure, especially your account of my favourite Gesner. Did you know that he died of the plague? When he was attacked, he collected his papers or works, fell to correcting, and died in the employ. I—but how dare I put my name in level with so great a man!—am doing the same, with my anasarca hanging upon me; but let me not alarm you. After being much weakened by the rough medicines necessary to expel the foe, I am much better, but left very weak,—the effects of fox-glove, a medicine not to be depreciated, for I may praise its efficacy. I have much reason to be content, at the age of 73. Why should I repine? a painless invalid in possession of my mental faculties! Every good wish attend you! May you live long a useful member of society, is the sincere wish of,

Dear Sir,

Your affectionate Friend,

THOMAS PENNANT.

Have you any friend who could give you an account of the colours used in the Hindostan paintings, if any are chemical?

15. *Flora Britannica.*—The acquisition of the Linnæan Herbarium soon discovered to the botanists of England, that many of our native plants had been mistaken, and that the nomenclature of our whole Flora stood in need of revision. Hence Sir James was led in 1794 to undertake a Latin *Flora Britannica*, which was completed in three volumes in 1804. This work the author dedicated to Sir Joseph Banks, Bart.

It was reprinted word for word by Dr. Römer at Zurich. "It is remarkable," says the writer of Sir James's obituary in the Philosophical Magazine, "like all his other labours, for accuracy in observing, accuracy in recording, and unusual accuracy in printing: being written in the Latin language, the information is condensed into a small compass, while it has the rare advantage of having had every synonym compared with the original author."

Davies Giddy, Esq. to J. E. Smith.

Dear Sir, June, 18, 1799.

I am extremely happy to learn that you are about to finish the work of my late friend Doctor Sibthorp. One need not hesitate in declaring the materials worthy of your attention. Some ancient Greek, unacquainted with the sexual system, has long since observed, *τα ανθη παντα Ερωτος εργα· τα φυτα ταυτα τουτου ποιηματα*. It gives me little less satisfaction to find that your *Flora Britannica* is in such forwardness; since, in addition to the learning and research every one expects, I understand the descriptions are to be derived from your own observations. I can't help wishing that Cornwall had been previously favoured with a visit.

Believe me, dear Sir,

Your much obliged and very humble Servant,

DAVIES GIDDY.

Sir Thomas Frankland, Bart. to J. E. Smith.*

Dear Sir, Wimpole Street, April 29, 1800.

I write a few lines to say how much I am gratified by the *Flora Britannica*, and flattered by your

* Of Thirkleby, near York; many years M.P. for Thirsk. This able botanist died early in the month of January, 1831, in the 84th year of his age.

notice of me in it. The descriptions which you have so judiciously added to every plant lessen the botanist's labour tenfold, and every little amplification of generic characters (*e.g.* in the Umbelliferous plants) is valuable. The new specific descriptions, also, are most important: *e.g.* I used to find *Gentiana amarella* near Netherby, "corolla 4-fida"; and what ground had I to doubt its being *campestris*, as it had been described in Hudson? I could observe no other difference, and from not knowing the real *campestris*, disbelieved there being such a plant as a distinct species.

THOMAS FRANKLAND.

"Hudson, Lightfoot, and Withering," says the Rev. E. B. Ramsay, "wrote *Floras* on the system of Linnæus; but there can be no doubt that Sir James Edward Smith was the most accomplished disciple of this school, and the best expounder of its principles. The *Flora Britannica* is perhaps the most perfect specimen existing; a work of which it has been said by no mean authority, that it is worth studying, as well for its logical precision, as for its botanical information."

J. E. Smith to the Editor of the Monthly Review.

Sir, Norwich, March 2, 1801.

The very favourable, and perhaps partial account of the *Flora Britannica*, given in your Review for January, is too intelligent in itself, not to deserve assistance from those capable of adding to its accuracy; and too flattering to its author, not to excite in him a wish of exculpating himself from any censures it may seem to contain against him. Yet these two objects would hardly have occasioned my troubling you with the present letter, had I not thought it justice to the public to avow some mistakes into which I have fallen, and at the same time to account for a longer delay of the remaining volumes than I once intended; for both which communications I have judged your Review the most proper vehicle.

The reason assigned in the Preface, for publishing the work in an incomplete state, is really and truly the only one that operated with me at the time: it was suggested by my publisher, and had the sanction of my most enlightened friends. I had then continued my manuscript far into the last class among the Mosses and Lichens; and had thought of nothing less than postponing that part, being determined to finish the work as completely as was then in my power. Since the publication, however, of the first two volumes, several new motives have presented themselves, which oblige me to defer the sequel longer than I intended. 1st, I wish to see Mr. Dickson's fourth Fasciculus of *Cryptogamia*, which is on the point of publication: 2ndly, The *Prodromus Lichenographiæ Suecicæ* of Dr. Acharius takes the lead so much in that department, it is absolutely essential to my purpose to compare specimens with that author; and this cannot be done either in the winter season, or in the present miserable state of Northern politics.

3rdly, A general work by Mr. Persoon, on *Fungi*, some sheets of which he has sent me, promises to be so important, that I wish to see it complete before I digest the British Fungi into order. Whatever reasons therefore might occasion the first delay, these, which are analogous to what you supposed, make me hope the public will in the end have no reason to complain of it. A still further advantage will accrue from my having the benefit of two seasons more (the spring of 1800 and 1801) to investigate the difficult genus *Salix*, which I have already written twice over, and in which the work will be more likely to merit the praise of labour and originality, than perhaps in any other part, though it will still contain only an imperfect sketch of the subject.

I proceed to notice some of your remarks. The order of *Syngenesia Monogynia* appears not to be founded in nature, nor useful in practice, because some *Gentianæ*, *Violæ* and *Lobeliæ* have the anthers perfectly united, others not at all. I have more to say on this subject than can be admitted here.

The genus of *Potamogeton* I am aware is but imperfectly treated. I have more than one new British species.

As to changing names,—*Radiola millegrana* is no "arbitrary alteration," or novelty, but the old generic name of Ray, retained as a specific one, and surely preferable to *linoides*, which I have proved to be false. My *Silene inflata* would certainly have been called *S. Behen*, had there not been *another already so called in Linnæus*. In the specific names of the genus *Glaucium,* I confess I have been tempted to follow Gærtner, in preferring precision, elegance, and truth, to barbarism, confusion, and error. The name of the common Wall-flower is not changed by me, but it is so called by Linnæus. On this subject, however, I entirely agree with you in principle ; otherwise I might have changed half the names in the book.

I differed from M. de Lamarck in his ideas of *Juncus acutus* and *maritimus,* because analogy led me to judge the panicle must be terminal in one if in the other ; but perhaps I may be mistaken, and have been led to think the two species more akin than they are, because of other authors having confounded them. I am sorry to say I fear I have added to the confusion concerning the *Dover Campion,* for Miller's pretendedly authentic specimen deceived me. Original ones in the British Museum, gathered at Dover, are a plant I do not know, and which is now said not to be found at Dover. We must wait in hopes of its being one day recovered, as was the case with *Ligusticum Cornubiense.*

I should claim no merit, even if I had corrected myself by the help of a much less able botanist than Mr. Curtis, on the subject of *Cerastium tetrandrum.* I have already found out my error in confounding two species under *Trifolium filiforme,* and shall correct that, and such further errors as may be detected, in an appendix to the last volume.

I am, Sir,

Your obliged and very humble Servant,

J. E. SMITH.

Liberté. Egalité.

Muséum National d'Histoire Naturelle.

Paris, le 12 Fiimaire, an 9 de la République, (3 Dec. 1801).

Les Professeurs administrateurs du Muséum d'Histoire Naturelle à Monsieur James Edward Smith, à Londres.

Monsieur,

Nous avons reçu par la voie de Monsieur Delaunay, que vous avez accueilli à Londres, divers ouvrages de votre composition destinés par vous a enrichir la bibliothèque de notre Muséum d'Histoire Naturelle; savoir la *Flora Britannica,* 2 vol. 8vo; le *Compendium Floræ Britannicæ,* 1 vol. 8vo; un recueil de Mémoires sur divers sujêts d'histoire naturelle, 8vo.

Ces ouvrages, dignes de la reputation de leur auteur, en contribuant aux progrès de la science, lui ont déja acquis la reconnaissance de tous les naturalistes. Recevez, Monsieur, le témoignage de la nôtre, pour un présent dont nous sentons tout le prix, et agréez l'assurance de notre sincère attachement.

Salut et estime.

JUSSIEU. FAUJAS. DESFONTAINES.
THOUIN. LAMARCK.

16. *Compendium Floræ Britannicæ,* dedicated to Aylmer Bourke Lambert, Esq. V.P.L.S., has gone through four editions, and, to borrow the observation of an able friend on the subject, is "become the general text-book of English botanists,

and is perhaps the most complete manual furnished on any subject." It was published in 1800.

Mr. Woodward, to whom the author presented this volume, says:

"I return you my most sincere thanks for your inestimable present; inestimable, *quia pignus amicitiæ*. I have been highly pleased with the Preface: the language is good, the style appropriate; but what I most admire is that unassuming display of knowledge which always informs so much more than it promises, and is the true characteristic of all my esteemed friend's writings, unlike the generality of modern authors, who promise mountains, and very often produce not even mole-hills."

17. *Flora Græca.*—A sketch of the origin of this classic Flora seems necessary in this place, to acquaint the reader with the occasion of its falling into the hands of the President of the Linnæan Society, to arrange into shape and give to the world.

The following is taken, in a very concise form, from the biographical memoir of Dr. J. Sibthorp, by Sir J. E. Smith, in Rees's Cyclopædia:—

"Dr. John Sibthorp, an eminent botanist, the younger son of Dr. Humphrey Sibthorp, Professor of Botany at Oxford, was born there in 1758, and was educated and took his degree at the same University.

"He passed a portion of the year 1784 at Göttingen, where he projected his first tour to Greece, and thence he proceeded to Vienna, where he studied with peculiar care the celebrated manuscript of Dioscorides, which has so long been preserved in the Imperial library, and procured an excellent draughtsman, Mr. Ferdinand Bauer, to be the companion of his expedition. The illustration of the writings of Dioscorides was Dr. Sibthorp's chief object. The names and reputed virtues of several plants recorded by that ancient writer, and still traditionally retained by the Athenian shepherds, he hoped would serve occasionally to elucidate or confirm the synonymy.

"In March 1786 Dr. Sibthorp and Mr. Bauer set out together from Vienna, and in May proceeded to Crete. Here, in the month of June, our botanical adventurers were welcomed by Flŏra in her gayest attire. The snowy covering of the Sphaciote mountains was withdrawing, and a tribe of lovely little blossoms were just peeping through the veil.

"The first sketch of the *Flora Græca* comprises about 850 plants. 'This,' says Dr Sibthorp, 'may be considered as containing only the plants observed by me in the environs of Athens, on the snowy heights of the Grecian Alp, Parnassus, on the steep precipices of Delphi, the empurpled mountain of Hymettus, the Pentele, the lower hills about the Piræus, the olive-grounds about Athens, and the fertile plains of Bœotia.'

"In December 1787, Dr. Sibthorp returned to England, and was enrolled among the first members of the Linnæan Society in the following spring.

"In March 1794 he set out on a second tour to Greece, attended by Francesco Borone" (the Milanese servant of Sir James before mentioned,) "as a botanical assistant. At Constantinople he was joined by his friend Mr. Hawkins. They made an excursion into Bithynia, and climbed to the summit of Olympus. Taygetus, the highest mountain of the Morea, and almost rivalling Parnassus, was ascended by these adventurous travellers. But Dr. Sibthorp's impaired health made it essential to him to return to his native country in the autumn of 1795. His few succeeding months were marked by the progress of an unconquerable disease; and he died at Bath, February the 8th 1796, in the 38th year of his age.

"The only work which Professor John Sibthorp published in his lifetime is a *Flora Oxoniensis*, in one volume octavo, printed in 1794.

"We have now to record the posthumous benefits which Dr. Sibthorp has rendered to his beloved science, and which are sufficient to rank him amongst its most illustrious patrons.

"By his will, dated January the 12th, 1796, he gives a freehold estate in Oxfordshire to the University of Oxford, for the purpose of first publishing his *Flora Græca*, in ten folio volumes, with a hundred coloured plates in each; and a *Prodromus* of the same work, in octavo, without plates.

"His executors, the Honourable Thomas Wenman, John Hawkins, and Thomas Platt, Esquires, were to appoint a sufficiently competent Editor of these works, to whom the manuscripts, drawings, and specimens were to be confided. They fixed upon the writer of the present article. The plan of the *Prodromus* was drawn out by Dr. Sibthorp;

but nothing of the *Flora*, except the figures, was prepared, nor any botanical characters or descriptions whatever. The final determination of the species, the distinctions of such as were new, and all critical remarks, have fallen to the lot of the Editor, who has also revised the references to Dioscorides, and, with Mr. Hawkins's help, corrected the modern Greek names."

Sir James lived to complete six of these volumes, and half the seventh, which has been published since his death by his distinguished friend Mr. Robert Brown.

The *Flora Græca* was a work, in the compilation of which Sir James had peculiar pleasure. Those who have seen the thousand beautiful delineations of the flowers of that country, by the hand of Mr. Ferdinand Bauer, may conceive it was no dull employment: it was a work of relaxation, to which he returned with impatience after any pause in its progress. It was agreeable to him to observe that many of the plants of that beautiful and classic land were the same as the most admired offspring of Flora in our own. The white lily was at all times his admiration; and he was charmed with discovering that its native bed and choice habitation is in the vale of Tempe. "The violet and primrose," Sir James tells us, "enamel the plains of Arcadia, and the *Narcissus Tazetta*, which Dr. Sibthorp was disposed to think the true poetic Narcissus, decorated in profusion the banks of the Alpheus. The barbarian horde under whose escort they were obliged to travel, had taste enough to collect nosegays of these flowers. The oaks of the

Arcadian mountains presented them with the true ancient misselto, *Loranthus europæus*, which still serves to make birdlime; whilst our misseltoe, *Viscum album*, in Greece grows only on the silver fir. May not this circumstance," he continues, "account for the old preference of such misseltoe as grows on the oak, among the ancient Britons, and consequently help us to trace the origin of their superstition to Greece?"

In illustration of the foregoing narrative, the following letters from Dr. Sibthorp and Mr. Hawkins will not be found uninteresting to the reader.

Professor Sibthorp to J. E. Smith.

Dear Sir, Pera, Aug. 9th, 1794.

I am glad you have safely received the *Flora Oxoniensis*: it was hastily drawn up; and, pressed for time, I had no leisure to give it what I could much have wished, for I fear it much wants a correcter form. I can only say this, that it contains only such plants as I have seen: I could have added more from the catalogue of Camden and others; but I was scrupulous, and wished to bear testimony only of those things I had seen. I recollect jocosely having proposed to you a botanical conversation, in which it should be debated in a council of botanists, what should be admitted and what rejected.

I think the latter classes of the Linnæan system are advantageously melted down amongst the others; but they have gone such lengths lately in Germany, in mixing and compounding the Linnæan system, as almost to have spoilt its shape and form. I dislike particularly the mixture by Gmelin of the *Icosandria* and *Polyandria;* the attachment of the stamina to the calyx or the receptacle furnishing a striking mark of distinction. Francesco would inform you that I arrived very ill, with a bilious fever and colic, at Constantinople :as soon as my health permitted me I visited the shores of the Bosphorus, the woods of Belgrade, and the sands of Domusderi on the Black Sea. I have noticed near eight hundred plants about Byzantium, and have got seeds of *Daphne pontica* and *Convolvulus persicus*, two good plants for our garden. I know not why *Epimedium* is termed *alpinum;* it is one of the most common plants in the woods of Belgrade, which are scarcely mountainous, much less alpine. The custom of setting fire to the woods in this country, to burn the brush-wood to give the grass room to grow, and furnish it with a manure from the ashes, is very unfriendly to the researches of the Cryptogamist. *Boletus lucidus* was almost the only Fungus I noticed; this is Forskal's *B. marginalis;* his *Salvia*, now in full flower, is a most ornamental plant. *Rosa centifolia* grows wild there; and the flowers of the vine, which mixes with the *Smilax* and twines round the chesnut, have a fragrant delicate odour beyond any flower I recollect. The shores of the Bosphorus are very poor in *Fuci*, nor are they rich in *Testacea*. I have collected about fifty sorts of fish, with their modern Greek names. The genus *Labrus* is rich in species, varying with many colours. *Julus* (*omnium formosissimus*) is often brought to market; its flavour inferior to most of the fish brought to table.

I go often upon the Bosphorus, while dolphins play around me. Gulls here are so tame that they sit upon the roofs of houses like pigeons. The *Procellaria Puffinus* is constantly flying up and down the canal: they call them here by the emphatical name of "souls of the damned." While I was reading in the palace garden the other day, a vulture, *Percnopterus*, perched in the tree hanging over my head; and I could not resist, not having the fear of the Egyptians before my eyes, to shoot it.

The summer has been very dry and hot. Fahrenheit's thermometer varied in my chamber last week from 84 to 86 degrees. There are few insects at present, except scorpions, mosquitoes, bugs, and *Conops calcitrans*, the happy accompaniments of this happy climate. The chase of the entomologist was almost over about a month since. I had fine sport. I write to you in good health and spirits; for yesterday my friend arrived, and today my baggage, having run "*per varios casus, et tot discrimina rerum.*" Tell Shaw, that Hawkins is in "high preservation," that he differs only in appearance from having the *labia barbata*—huge mustaches, which he is nursing for a Syrian and Egyptian tour. We are going together into Thessaly, Attica and the Peloponnesus, and shall winter at Zante. My health is much better. Remember me kindly to Sir J. Banks, Dryander, Lambert, &c

Yours sincerely,

J. SIBTHORP.

Borone is in good health, in action quite a Le Fleur.

Professor Sibthorp to J. E. Smith.

My dear Sir, Athens, Nov. 1, 1794.

I should have been happy to have sent you a pleasant letter from Athens; but from Athens I must this time write you a very mournful one. Poor Borone is no more! He was quite recovered from an intermittent fever that had attacked him a little before his departure from Constantinople, and on the evening of his unhappy fate was unusually gay, singing to a tune that Arakiel, Mr. Hawkins's servant, played upon the guitar. A little after midnight, we were waked out of our sleep by the cries of Francesco, who had fallen into the street, out of the window of the chamber where he slept with Arakiel. On the servants' going down to him, he languishingly groaned to Arakiel, who was the first that came up to him, "Ah! povero Francesco e morto!" James, the other servant of Mr. Hawkins, then coming up, he said, "Ah! James, James!" and expired.

As soon as Mr. Hawkins and myself heard that Francesco was hurt by his fall, we immediately got up, and went down to him. On taking him by the hand, I found the pulse gone, and no signs of life. We directly got him into the house, and attempted to bleed him, but without effect. His loins and back, on which he appeared to have fallen, were very much bruised; but there was not the least appearance of blood, nor could I find that any bones were broken. It had rained very hard on the preceding day, so that the street was dirty: the night was dark, with frequent flashes of lightning. The opening of the window out of which he fell was extremely narrow, and appears not above eighteen feet from the ground. To get out of it, he must previously have mounted on a box that stood near it, and then squeezed himself through. We have every reason to think all this was done in his sleep. On the opposite side of the room to this window was another, that opened upon a terrace, on which he was accustomed to walk. Perhaps, if awake, which I can scarcely conceive, he had forgotten which of the two windows led to the terrace.

You may imagine that after this we passed the remainder of the night dismally enough. The next day nothing remained but to perform the last offices to poor Francesco. He was buried in the evening at the Church of the Madonna, under the shade of a mulberry-tree. The obsequies were performed in a very decent manner by four Greek priests, who chanted over him the burial service. Mr. Hawkins and myself, the British Consul, and some Sclavonians who were here, with the servants, attended the corpse. The Archbishop, who a few days before had expressed the strongest obligations to the English nation, pitifully sent a Papas to demand fifty piastres (about twelve pounds) for his permission to bury him. The Consul remonstrated with him on the impropriety and exorbitancy of the demand; when he sent a second message to say he would take half that sum. This produced another remonstrance from the Consul, when he repented, and refused to take anything. He has since sent a hint that he would be glad of a present. We mean to send him a Greek Testament, that a metropolitan who has four suffragans may read a lesson of piety.

I regret with you most sincerely the cruel end of this unfortunate youth. He had escaped from the thieves of Italy, and from the inhospitable climate of Sierra Leone. He had been with me blocked up eight days by pirates at Mount Athos. Poor fellow! he was then very anxious to hide my money, that we might have something, he said, to return home with. I shall set off in two or three days for Zante, where I shall winter. In January I propose to visit, with Hawkins, the Morea; and in the spring, or early in the summer, to return to England. I have made considerable additions to my collection of Greek plants and animals, having visited the Bithynian Olympus, Troy, Lemnos, Mount Athos, and Negropont. During my stay at Athens, I have procured a pretty exact knowledge of the agriculture and natural history of Attica. Tell our friends in Soho Square, that I have all the labour, if not all the sweets, of an Attic bee.

J. SIBTHORP.

Professor Sibthorp to J. E. Smith.

Dear Sir, Oxford, Oct. 8, 1795.

I regret extremely not being in Oxford when you favoured me with a call. In my way through Town

I made my inquiries in Marlborough Street, but was informed you were at Norwich. Since my return I have felt myself very much indisposed, and in a very infirm state of health,—a nasty low fever, with a cough that alarms me, from some affection of the lungs. I have been down to Brighton, but bathed only in the tepid bath. I am now nursing myself at Oxford with asses milk, and gentle exercise on horseback;—have some thoughts of trying what Bristol will do.

I left Hawkins at Zante, from whence I had a very long and uncomfortable passage of twenty-four days to Otranto, in which I date the origin of my complaints, having in an excursion to Nicopolis, near Actium, caught a most severe cold,—the air of Prevesa is even by the Greeks deemed infamous. I travelled post from Ancona through the Tyrol, a beautiful Alpine country, very inviting to the botanist. At Erlang I paid a visit to Schreber, who received me with great politeness. He has got a Monograph ready on the genus *Carex*, and told me he would adopt Dr. Goodenough's names if he would favour him with a list of them. He is also ready with a Fasciculus of your *Gramina*. I promised to send him *Carex strigosa*, and, if I could get it, *Carex indica*.

I spent an hour with Hoffmann at Göttingen; he has much improved the Göttingen garden; he was compiling a Synopsis of the German Cryptogamia under the form of a *Taschen-buch*, or what we call Pocket-book. I should like to see it in a more dignified form.

Your situation is a very enviable one for a botanist; out of the smoke of London, and close to one of the best gardens in Europe*.

I left all my treasures behind me at Zante; but by a letter from our Consul I hope soon to receive them. I am quite a stranger to what has been going on in the botanical world for this twelvemonth past. Lambert informs me of great treasures brought by Masson, White and Menzies,—that the botanic world will grow too big for us.

You renew my grief in mentioning poor Borone. I was so affected that I could do nothing for some days after his death, not even continue my journal. If I recollect right, it was on the 20th of October that this melancholy event took place. He certainly had no watch, but as he was fond of wearing one, I lent him mine when we were at Constantinople. It was indeed a cruel fate, and we all lament his death. Hawkins, when I last heard, was gone into Thessaly to visit Ali Pashaw, in good health, but under alarms from robbers.

Yours most sincerely,

J. Sibthorp.

The following letters, it will be perceived, were written after Dr. Sibthorp's death, and when the care of his posthumous work was confided to Sir James Smith.

* Mr. James Lee's at Hammersmith.

J. Hawkins, Esq. to J. E. Smith.

Dear Sir, London, June 25, 1799.

I conceive that you must have had some trouble in packing up so large and miscellaneous a collection, but I hope you have received every part of it, particularly a large collection of seeds made in his last tour, and named by him on the spot, as far as the plants could be ascertained from whence they were taken.

I hope you will be able to decipher Dr. Sibthorp's hand-writing, which is not the best.

As to the want of names, I well recollect that he never affixed any to the specimens, but seemed to have a perfect knowledge of them, and therefore thought it perhaps superfluous.

You, of course, will find little difficulty in ascertaining them, after which you may refer to the Journals for their respective *habitats,* and all that relates to them.

You will find that Dr. Sibthorp subdivided the Flora Græca in his last tour into a number of provincial Floras, which has its use. There is a Flora Zacynthia, which was prepared for him by Consul Foresti before his arrival, with the vulgar names underwritten. Likewise a small collection of plants from Maina, procured by Consul Foresti from a Mainote.

In the collection made during the last tour you will find a few sub-alpine plants, collected by me in Crete.

I have specimens of three or four rock plants from the mountains of the Morea, and I left about two hundred behind me at Patras.

Flaxman is now executing a monument to the memory of our late friend; but the best, by far, will be that which you have in hand.

I remain, my dear Sir,
Most truly yours,
J. HAWKINS.

J. Hawkins, Esq. to J. E. Smith.

Dear Sir, Sunbury, Feb. 13, 1800.

I have before me your two letters, and hasten to give you that information which lays in my power. In our first tour a regular journal was kept by Dr. Sibthorp of our daily peregrinations, and of every object of natural history discovered. Thus for instance you will find a list of the plants which we found in the middle of March on the shores of the Hellespont, and a few days after on the coast of Caria; but when we came to Cyprus, the form of the journal was so far changed, that only the vulgar names and uses, and the remarkable plants or birds were noted therein which daily occurred; and the whole Flora and Fauna Cypria were drawn up in systematical arrangement at the end.

Every allowance must be made for the hurry and precipitation with which this Journal was written. The season for botanizing is so short in these southern latitudes, and such a profusion of new or rare plants present themselves at once, that it is hardly possible for the most practical botanist (and such was Dr. S.) to determine all with sufficient accuracy at the moment. Hence it is that many of these supposed new species prove perhaps on further investigation to be old, and some of the supposed old new, while the old too are mistaken one for the other. Add to this, that the necessity of moving quickly from one spot to another, and the time consumed on horseback or on shipboard, further contributed to limit the time which could be employed in the real scientifical examination of species of plants.

It is certainly a pity that Dr. Sibthorp did not mark all his specimens, or the drawings; but he trusted to his memory, and dreamed not of dying.

In our second tour a daily journal was kept as in the former, and the same attention was paid to the determination of the vulgar names and uses of plants and of animals, and particular species as they occurred were noted down; but now a more particular geographical arrangement was adopted, and the Flora and Fauna of every province was thrown together, as you will find in the journal-book in respect to the Flora Byzantina, Bithynica, Montis Athos, Attica and Bœotica, Zacynthia, Achaica, Eliensis, Messeniaca, Arcadia, Argolica, Corinthiaca, and Laconia.

The Dacian plants which you mention were collected on his journey to Constantinople by land, and probably include those plants which he observed on Mount Hæmus.

Those of Olympus were collected in the course of our expedition to the summit of that mountain, soon after our junction; and the Cretan plants are what I brought with me from Crete, and presented to Dr. Sibthorp at Constantinople.

I should think that all the plants collected in our last tour are still unmixed with those of the former; for the collection arrived after the final removal of Dr. S. from Oxford, and shortly before his death. They were consigned to the care of Dr. Werman, who found them in good preservation; but whether Dr. W. arranged them in any systematical order, or amalgamated them with the rest, I know not, but hope he did not; in which case you will find, besides the abovementioned parcels, a very curious and finely preserved collection of the plants of Zante, made for him by a Zantiote apothecary; a small collection of the plants of Maina, made by a Mainote; and those plants which Dr. S. collected in his voyage from Zante to Venice, besides all the insects, fishes, birds, shells, quadrupeds and seeds which he preserved in the whole course of his tour.

It appears, then, that you either have or ought to have two large and distinct collections of Greek plants, the result of the two tours.

Next summer, or the following, I will contrive to visit Norwich, and shall be happy to overlook your literary labours.

I am impatient to get your *Flora Britannica*.

I remain, my dear Sir,

Most sincerely yours,

J. Hawkins.

18. *Prodromus Floræ Græcæ.*—Of this work, the first volume was published in 1806, a thick octavo; and in 1813 a second volume of the same size.

"I have just received," says Dr. Goodenough, "your kind present of the first part of the Prodromus of Flora Græca. What a beautiful book—what good paper—what a distinct and elegant type! How much information, without a letter of ostentation or frippery. Indeed it is a sterling work."

19. *Exotic Botany.*—This work, published in numbers, was terminated in two volumes quarto, and also an octavo edition of the same was published in 1804. It is dedicated to William Roscoe, Esq.

For want of public patronage it did not go on, notwithstanding it contained figures of many new and beautiful plants, communicated from the Indian Alps, from Africa, and New Holland.

20. *The Introduction to Physiological and Systematic Botany*, published in 1807, has passed through six editions. Sir James dedicated this work to the Honourable and Right Reverend, Shute Lord Bishop of Durham. The death of Mrs. Barrington defeated the author's original design of dedicating the work to her.

Dear Sir, Auckland Castle, Sept. 26, 1807.

Every thing which reflects credit on the memory of a woman most justly dear to me, is in the highest degree gratifying to my feelings. In this view I beg to offer my earliest thanks for the very handsome manner in which you have expressed yourself in the Dedication respecting her and me. Allow me to assure you that I shall at all times be happy in seeing you when in London, or here, or at Mongewell.

I am, dear Sir, with much regard,

Your faithful Servant,

S. DUNELM.

In returning thanks to the author for a copy of this work, the Bishop of Carlisle proceeds to tell him: "I have read the Preface, Introduction, and part of the work. Really you write with great spirit and propriety, and let me add with great dignity, worthy your superior station in the botanical world. All natural history should be turned to the mental improvement of mankind, to make them think properly, and with precision and truth; arguments from this source are literally innumerable. Have you seen a publication of Mr. Vince's, four sermons against Atheism, where he makes astronomy speak in this high strain. I shall read your book through with great minuteness.

Yours ever,

S. CARLISLE.

Mr. Roscoe, writing to Sir James on the subject of the Introduction to Botany, says, "I am impatient to tell you how much I like what I have yet seen of it, particularly your Preface, with the concluding part of which I was not only delighted but really much affected."

"To those," observes the Author, "whose minds

and understandings are already formed, the study of nature may be recommended, independently of all other considerations, as a rich source of innocent pleasure. Some people are ever inquiring, What is the use of any particular plant? by which they mean, What food or physic, or what materials for the painter or dyer does it afford? They look on a beautiful flowery meadow with admiration, only in proportion as it affords nauseous drugs or salves. Others consider a botanist with respect only as he may be able to teach them some profitable improvement in tanning, or dyeing, by which they may quickly grow rich, and be then perhaps no longer of any use to mankind or themselves. These views are not blameable, but they are not the sole end of human existence. Is it not desirable to call the soul from the feverish agitation of worldly pursuits, to the contemplation of Divine Wisdom in the beautiful economy of Nature? Is it not a privilege to walk with God in the garden of creation, and hold converse with his providence? If such elevated feelings do not lead to the study of Nature, it cannot be far pursued without rewarding the student by exciting them.

"Rousseau, a great judge of the human heart and observer of human manners, has remarked, that 'when science is transplanted from the mountains and woods into cities and worldly society, it loses its genuine charms and becomes a source of envy, jealousy, and rivalship.' This is still more true, if it be cultivated as a mere source of emolument.

"But the man who loves botany for its own sake,

knows no such feelings, nor is he dependent for happiness on situations or scenes that favour their growth. He would find himself neither solitary nor desolate had he no other companion than a 'mountain daisy,' that 'modest crimson-tipped flower,' so sweetly sung by one of Nature's own poets. The humblest weed or moss will ever afford him something to examine or to illustrate, and a great deal to admire. Introduce him to the magnificence of a tropical forest, the enamelled meadows of the Alps, or the wonders of New Holland, and his thoughts will not dwell much upon riches or literary honours; things that

'Play round the head, but come not near the heart.'

"In botany all is elegance and delight. No painful, disgusting, unhealthy experiments or inquiries are to be made. Its pleasures spring up under our feet, and, as we pursue them, reward us with health and serene satisfaction. None but the most foolish or depraved could derive anything from it but what is beautiful, or pollute its lovely scenery with unamiable or unhallowed images. Those who do so, either from corrupt taste or malicious design, can be compared only to the fiend entering into the garden of Eden."—*Preface to Introduction.*

The letter which follows is from the late Professor of Natural History at Cambridge in Massachusetts (North America), a very amiable man, and who, like many of his countrymen, filled with enthusiasm for England, came to it with the feelings of those who from this island visit the classic realms of Greece and Rome.

Very dear Sir, London, Feb. 8, 1808.

Your favour of the 11th October gave me the more pleasure as it was unexpected, for you were so much occupied that I did not venture to ask such an indulgence. The letter which contains the kind expression of your good-will must be always a treasure to me.

On coming home one evening, I found your Introduction to Botany. This too was unexpected. I thank you very heartily, and may say *Vous me comblez* without an hyberbole. The Preface is excellent; the reflections are perfectly just, and place this delightful science in its proper light. Some paragraphs affected me in a manner which I cannot well describe, and which I have frequently experienced in reading Linné and Pliny. The terminating one, p. 338, is one of these. Did you ever observe in reading anything which affected the mind by its energy or sublimity, that the blood flowed more copiously to the heart, leaving a chill on the surface of the body? I cannot explain it, but I never read Linné's Introduction to the *Systema Naturæ* without this sensation.

I feel all that you say of botany as an introduction to amiable characters, for the little I know of it has brought me acquainted with some of the best and most worthy persons in every country that I have visited, and those whom I shall always love. The converse of such persons cherishes the flame of philanthropy within me. The study of Natural History, if pursued on right principles, thus serves to keep the heart warm, and to preserve in vigour its best affections; while in most other pursuits, the heart is hardened by the collision of sordid interest, and its affections smothered by emulation. The study of man has often given me pain; but since I have seen more of him, I feel a more glowing love of my species, and my heart stretched, as it were, with gratitude and affection. God keep it so! for a cold heart can never be happy.

I went to the Abbey to read the excellent epitaph of Lord A. Beauclerc, and was very well disposed to appreciate its beauties and receive its impressions, but I was disappointed. In a state of mind perfectly calm, I entered the church, and was proceeding to the spot, when I was stopped by one of those persons who get their bread by strangers and visitors, who demanded a shilling,—which indeed I intended to give him before I left the church; but the demand at that instant was revolting,—my tranquillity was thus disturbed, and almost as soon as I began to read, the organ was struck, and its solemn peals, accompanied with some charming voices of singing boys completely frustrated the object of my visit,—which was, to read in silence. So I transcribed the epitaph to read at home; but the first impression was ruined for ever. The four last lines are excellent, but "bid" in the penultimate one should be *bade*.

> "Dying he *bade* Britannia's thunder roar,
> And Spain still felt him when he breathed no more."

I have since received through the kindness of Mr. Lambert a packet of New Holland plants, and am truly sensible of your goodness in this particular, and very much gratified to find so many among them called after botanists, some of whom I know. These are precious objects, and will always bring you to my mind, as the communications of your friend Davall bring him present to you.

I often go with you to visit the Salicetum of your late friend Mr. Crowe*, see with pleasure the stream which flows by it, adorned with *Nymphæa;* visit in my way *Verbascum pulverulentum*, and collect the Scabious in passing through the churchyard. I live over again the happy days I passed at Norwich, and enjoy anew the pleasure I felt in witnessing the domestic felicity of my friends there. That it may long continue is my anxious and ardent wish.

I pray you to make my respects acceptable to your mother and your lady. I ardently wish the continuance of your health; all other wishes are superfluous, as they cannot add to your enjoyments: yet minds like yours may receive some satisfaction that even I rejoice in your prosperity. I do not forget any of my friends in Norwich, where every day of my *sejour* was *albo lapide notari dignus*.

I salute you with affectionate respect, and am,

My dear Sir,

Your truly obliged,

WM. D. PECK.

* James Crowe, Esq. of Lakenham House.

Dr. Swartz to J. E. Smith.

Dear Sir, Stockholm, Oct. 16, 1808.

Two years and a half are elapsed since our correspondence ceased. In order to remind you of your Scandinavian friend, I write a few lines in company with the Sequel of the Memoirs of the Academy of Science, until the last published.

Though I have not had any immediate information from you, I have several times inquired about your welfare, and from the news I not long since obtained, I think my wishes for you not entirely frustrated. If I am so happy as to hear it from yourself, so much the better.

We have long been exiled from foreign literature, except from that of your country, of which however we have had but sparing communications. Of your Exotic Botany I have got two volumes, and seen some fasciculi of the wonderfully well executed *Flora Græca*, with one part of its *Prodromus* from your hands. Lately I have also enjoyed a particular satisfaction from the sight of the beautiful *Historia Fucorum* of Mr. Turner.

It can hardly be expected that anything remarkable should appear from this corner, during a period the most cruel that can be imagined; so much more desolating, as in and outward causes seem to prepare completely the ruin. Under all this, for my part engaged to continue two current works, the Swedish Botany and Zoology, two volumes have appeared by my care during the two years past. They are merely calculated for home consumption, but have not escaped to be noticed abroad, as well as Westring's Economical History of the Lichens.

The interrupted communication with our neighbours has made us in a melancholy manner separated from our friends, with whom we are in behalf of literature connected. This has been the case with me in regard to the intended *Synopsis* or *Historia Muscorum*, in connection with my excellent and ever-lamented friend Mohr, at Kiel, who has eclipsed from this sublunary world during the gloomy interval, and left me alone! Much is done but more is remaining to do. Even Acharius has proved the adversity of the times on account of his work the *Lichenographia Universalis*, which was to be printed at Göttingen, under the inspection of Professor Schrader (because the enterprise was too great for any home dealer in typography). The manuscript was above two thirds past sent *in locum* last year, and some time after the printing began we were shut up, that in the course of nineteen months no account could be had. As for present appearances, there are but little or no hopes yet to get out of the cage, if not quite perished before—*Dum delirant.*

Thunberg and Afzelius are, as before, living at Upsala, and, as much as I know, both well. From the latter, some dissertations have appeared concerning the Swedish species of *Rosa*, on which he bestows rather more pains than some think deserving. But it must be something like the *Trifolium*. Pity that he has laid all his African treasures aside.

He is *à la tête* for a kind of institution called the Linnæan, but I do not know how far it has proceeded.

An excellent Eulogy on Linnæus has been printed by one of his former pupils, Dr. Hedin, first physician to the king. This was a panegyric held in the school at Weixio, which Linnæus first visited. On the same day a grand fête was given at the Academy of Upsala; a secular celebration of the birthday of the great man;—the only instance to this day, in Sweden, of public honours given to an individual of subjects. The museum and auditory were at the same time consecrated. Many thousands assisted at the solemnity, and most part of the Linnæan old disciples also; which gave the whole a most venerable appearance.

Mr. Turner has informed me that a new volume was soon to appear of the Linnæan Society's Transactions. He mentioned also that you had described a new genus of Mosses. I am most impatiently desirous to know what this may be, as well as the other contents of this first-rate work, so far above common praise.

Do not forget me, who remain with the sincerest esteem and affection,

Dear Sir,

Your faithful Servant,

OLOF SWARTZ.

Sir T. G. Cullum to Sir J. E. Smith.*

Bury St. Edmunds,
Sept. 27, 1814.

My dear Sir James,

I returned home from my journey to Wakefield, Harrowgate, &c. on Saturday last; and I cannot any longer delay returning you my thanks for your kind present of the Introduction to Physiological and Systematic Botany, the perusal of which has many years ago given me much pleasure, and will continue so to do whenever I take it up either for instruction or amusement: but I must forbear (as Dr. Thornton says) "expressing all the sentiments of respect and esteem entertained by me towards one so truly estimable."

I wished you had been with me at the Dropping Well (or rather rock) at Knaresborough, I think you would have found some Jungermanniæ, Lichens or Mosses; but I am so totally ignorant in the Class Cryptogamia, that I do not pretend to know and distinguish the common Lichen or Moss. Travelling along the road a fine plant or two of the *Inula Helenium* caught my eye, and the *Sedum Telephium* in a hedge-bank for a quarter of a mile. The churchyard of Aberford near Ferrybridge, and the hedges near the town, are full of the *Atropa Belladonna*, and the *Colchicum autumnale* in the meadows about Knaresborough, &c.

I called upon an old acquaintance of mine near Wakefield, the Rev. Dr. Zouch, a Prebendary of Durham, and a Fellow of the Linnæan Society. He printed (but I believe never published) A Memoir

* It was in July this year, that Sir James had the honour of being knighted by His late Majesty George IV.

of the Life of Dr. John Sudbury, who was Dean of Durham in 1661.

Dean Sudbury was born at Bury St. Edmunds, and left a considerable estate in Suffolk towards putting out boys as apprentices, and the residue to the Grammar-school at Bury. But an anecdote Dr. Zouch relates of Dean Sudbury's "recovering many books that had been embezzled and taken away by Mr. Isaac Gilpin, who also lent Gerard's Herbal, which cost 10*l.*, to Colonel Robert Lilburn, who is now in the Tower, and still detains that book from the church's library," I think is somewhat curious, in showing the estimation in which Gerard's Herbal was held at that time. The Doctor should have told us (if he knew) whether it was Gerard's own edition, or the amended one by Johnson.

But the best part of my letter is to acquaint you and Lady Smith the time when Bury Fair will be in all its splendour and gaiety: but this I shall leave to Lady Cullum; and shall therefore bid you farewell, and am, with great regard,

T. G. CULLUM.

Baron Humboldt to Sir J. E. Smith.

Monsieur le Président, Paris, ce 12 Juillet, 1816.

Je prends la liberté de vous adresser ces lignes par un de mes intimés amis, M. Kunth, Professeur Royal de Botanique à l'Université de Berlin. C'est le botaniste qui a publié les *Nova Genera et Species Plantarum* que vous avez reçu avec tant d'indulgence. M. Kunth étoit l'élève le plus chéri de Willdenow; et je suis sûr que vous serez content de son instruction et de sa modestie. Que n'ai-je été assez heureux d'avoir pu le suivre ; mais je suis enfoncé dans mes Cartes des Cordillères que je dessine moi-même. Vous savez par votre propre expérience que les graveurs ne sont pas des gens qu'il faut perdre de vue.

J'ai lu et même étudié avec le plus vif plaisir votre *Introduction* philosophique. C'est un petit ouvrage plein de vues ingénieuses et écrit avec une rare élégance de style. J'ai aussi fait avec bien de l'interêt la connoissance de l'aimable M. Sparshall, que nos entomologistes ont trouvé très habile. Malheureusement que mon indisposition m'a empêché de lui être aussi utile que j'aurois désiré l'être à une personne qui vous est chère. Je serai très flatté d'être un jour membre de l'illustre Société Linnéenne ; mais je ne voudrois pas que vous poussiez la bienveillance jusqu'à changer vos loix constitutionelles.

Excusez, je vous supplie, le petit mot que j'avais lâché dans ma dernière lettre sur ces grandes batailles qu'on me livre de trois en trois mois dans le *Quarterly Review*. J'ai été loué avec trop d'exagération à Edinbourg, il faut bien qu'on me blâme ailleurs : c'est le système de compensation. Je profite des critiques : je cultive les sciences parceque je les aime, et parceque ma position me permet une heureuse indépendance : je n'ai de l'amertume contre personne ; mais comme je tiens beaucoup à mes projets pour les montagnes de l'Inde, j'aimerois assez être en paix avec les trois Royaumes Brittanniques. Je n'ai d'ailleurs qu'a me louer des offres du Prince Régent, auquel le Roi de Prusse m'a recommandé personellement, et des procédés de Lord Castlereagh et d'autres membres du Gouvernement avec lesquels mon frère a eu des rapports si multipliés. Quand on a passé six ans dans les Colonies Espagnoles entre des Inquisiteurs qui restent et des Ministres qui tombent, couche par couche, comme des feuilles mortes, on ne craint pas, aver un cœur droit et un véritable amour des sciences, le séjour parmi un peuple noble, éclairé, et juste, comme le vôtre.

Je vous supplie, Monsieur le Chevalier, d'agréer l'hommage de mon admiration et de mon respectueux attachement.

ALEX. HUMBOLDT.

Dr. Boott to Sir J. E. Smith.

Derby, March 3, 1817.

You have no doubt, my dear Sir, often experienced the delightful satisfaction which results from having contributed to the knowledge or pleasures of any one to whom you have long owed great obligations; and I need not therefore say with what feelings I read your letter of the 27th ult., in which you give me to understand that my little present of plants was valuable to you.

Authors, of course, seldom look for personal acknowledgments from those who read their works; but I see no reason why such should not be made, when a reader and the author happen accidentally to be thrown in each other's way. Take, therefore, my dear Sir, my grateful thanks for the pleasure and instruction I have derived from your Introduction to Botany, the *Flora,* and *Lachesis Lapponica,* &c. &c. And when next you think of repaying any favour I can confer, remember that I owe my knowledge of botany to you, and that I shall always consider it a duty (a pleasant one) to return those obligations you first conferred on me. Let me, wherever fate wills it, if possible do something to contribute to your herbarium every year; and the consciousness of having afforded you gratification will be ample recompense for my exertions.

My sincere thanks are due to you for your sympathy in my late bereavement. Had you known my father you could better have appretiated the loss I have sustained. Many men pass away, and leave behind the reputation of a good name; but his virtues were not like those of ordinary men. I cannot tell you what he was: but when you are next in Liverpool, ask Mr. Martin, with whom I have the pleasure to be acquainted,—ask of him, of those who knew him, and you will find the contemplation of such a character will exalt your idea of human nature, to find it capable of such purity and benevolence. Excuse this passing tribute to his worth. Do you remember Horace's

> " . . . . Cui pudor, et justitiæ soror
> Incorrupta fides, nudaque veritas,
> Quando ullum invenient parem?
> Multis ille bonis flebilis occidit!"

It is so applicable!

With great respect, Yours ever,

F. Boott.

Professor Schultes to Sir J. E. Smith.

Dear and noble Sir, Landshut, 1821.

I considered it as a duty to science, as well as to mankind in general, and especially to you, to give a translation of your Introduction; because, with all my respect for the *Philosophia Botanica,* I do not know better elements of botany than yours. I am, though being a Christian, of the heathen opinion, that "*Maxima debetur puero reverentia.*"

Your idea of a grammar of botany is no less worthy of the classic solidity you fixed with the principles of science; it might perhaps also be considered as a true British sneer on those foreign botanists who, being totally ignorant of both the grammar of their science and of Greek and Latin, intrude themselves as authors, or rather as poets in botany. In our old time, grammar was studied before poetry, but now we see poets in botany who can scarce spell the terms of art; and thus we might say now of the *amabilis scientia* what Shakspeare once said of poetry,—botany "has bubbles as water has." What a patchwork did not become the glossology of some modern botanists! Happy that they are kind enough to provide a particular dictionary to every botanical pamphlet they send to the press. A complete dictionary of all the terms now used by different modern botanists would give a greater volume almost than any Polyglotten since the steeple of Babel, to which we should unavoidably recur if there is not any Hermes like you, that puts an end to that confusion. I always did, according to your view, defend the necessity of an artificial system, and especially that of Linnæus, for analytical examination; a natural one, even that of Jussieu, being of no use at all for that purpose.

I explain and contend for Linnæus's system in my Lectures. The plants in my garden are ranged according to Jussieu's.

A. SCHULTES.

21. *Cyclopædia.*—When Dr. Rees first undertook the arrangements of this great work, he requested Sir James to furnish the botanical articles, but other occupations deterred him from the undertaking: and upon his declining it, the Rev. William Wood of Leeds supplied that department of literature, till illness deprived him of the power of proceeding; when Sir James being again solicited on the subject, he no longer hesitated: and after the death of that excellent man in 1808, he continued his assistance till the close of this voluminous work.

In a letter dated 1807, Sir James tells Mr. Roscoe, that "he had undertaken to write the physiology, terminology, and biography of the botanical part of Dr. Rees's Cyclopædia, which he was requested to do when Mr. Wood of Leeds was ill. Indeed," he adds, "the whole of the botany fell upon me, but now he is well enough to resume the descriptive parts. I have just done Clusius and Peter Collinson."

In another letter of the same year, he says, "I have written much of late in Rees's Cyclopædia; it is a very pleasant employment to me, and has led me to launch out into subjects I might otherwise not have touched upon,—I mean only in botany and botanical biography."

In 1808 Mr. Wood died suddenly, in the midst of writing the article *Cyperus*. "Dr. Rees," adds Sir James, "then threw himself on me. As five thousand copies sell, it is of importance for the progress of science, and therefore I would not let it go out of my hands. I now write all the botanical part, and biography of botanists. I put an S. to the articles for which I wish to be responsible."

Thus beginning towards the conclusion of the letter C, Sir James continued his communications through the remainder of the alphabet; and, with the exception of a few articles by his friend the Rev. W. F. Drake, the botanical information contained in this work is exclusively Sir James's from the time of Mr. Wood's decease.

The communications he sent, without including the biography of botanists, amount to 3348 articles. The lives of botanists are 57, of which the following is a list:—

Michael Adanson.
Charles Clusius.
Peter Collinson.
William Curtis.
James Dalechamp.
John James Dillenius.
Rembertus Dodonæus.
Joseph Dombey.
John Ellis.
Louis Feuillée.
Leonard Fuchs.
Joseph Gærtner.
Alexander Garden.
John Gerarde.
Conrad Gesner.
John Gesner.

Nehemiah Grew.
John Frederick Gronovius.
Laurence Theodore Gronovius.
Stephen Hales.
M. du Hamel.
Frederick Hasselquist.
John Hedwig.
Laurence Heister.
William Hudson.
Engelbert Kæmpfer.
George Joseph Kamel.
John Lightfoot.
Charles Linnæus.
Charles Linnæus, jun.
Matthias de Lobel.
Christian Theophilus Ludwig.
Peter Magnol.
Marcello Malpighi.
John Martyn.
Francis Masson.
Peter Anthony Micheli.
Philip Miller.
Robert Morison.
José Celestini Mutis.
John Parkinson.
Leonard Plukenet.
Charles Plumier.
Richard Pulteney.
John Ray.
A. D. Rivinus.
Olof Rudbeck.
Olof Rudbeck, jun.
William Sherard.
John Sibthorp.
Sir. Hans Sloane.
Joseph Pitten de Tournefort.
John Tradescant.
William Turner.
Martin Vahl.
Sebastian Vaillant.
Anthony Vallisneri.

22. *A Tour to Hafod.*—A large folio, containing fifteen coloured engravings of views in that picturesque domain, was published in 1810, and dedicated to his friend Colonel Johnes.

23. *Lachesis Lapponica.*—In 1811 Sir James published *Lachesis Lapponica*, or A Tour in Lapland, in two volumes, translated from the original manuscript journal of Linnæus in Swedish.

In dedicating these travels to his valued friend Thomas Furly Forster, Esq., he observes, "Happy are those who, like you, can equally sympathize in his pious and benevolent affections, his disdain of hypocrisy and oppression, and his never-ceasing desire to turn his scientific acquisitions to practical utility."

"How have I been delighted," says the Bishop of Carlisle, "at reading *Lachesis Lapponica.* What a genius did Linnæus inherit as it were from Nature!"

"Λαχεσις.—Lachesis Lapponica means, I presume, the Fate or Lot of Lapland."

24. The titles of the following *fifty-two* papers from the hand of the President are taken from the *Transactions of the Linnæan Society.*

Vol. I. 1. Introductory Discourse on the Rise and Progress of Natural History. It was immediately translated into Italian by Father Fontana, under the title "*Sull' Origine e Progresso della Storia Naturale.*"—2. Descriptions of Ten Species of Lichen, collected in the South of Europe.—3. On the Festuca spadicea and Anthoxanthum paniculatum of Linnæus.—4. Remarks on the Genus Veronica.

Vol. II. 5. Remarks on the Abbé Wulfen's Descriptions of Lichens, published among his rare Plants of Carinthia, in Professor Jacquin's Collectanea, vol. ii. 112.—6. Additional Observations relating to Festuca spadicea, and Anthoxanthum paniculatum.—7. Remarks on Centaurea solstitialis and C. melitensis.—8. Remarks on the Genus Dianthus.—9. Additional Remarks by James Edward Smith.—10. Description of Sagina cerastoides, a new British plant, discovered in Scotland by Mr. James Dickson, F.L.S.

Vol. III. 11. The Botanical History of Mentha

exigua. It was translated into German in Schräder's Journal.—12. Botanical Characters of some Plants of the Natural Order of Myrti.—13. Characters of a new Genus of Plants named Salisburia.

Vol. IV. 14. Remarks on some Foreign Species of Orobanche.—15. The Characters of Twenty New Genera of Plants.—16. Observations on the British Species of Bromus; with Introductory Remarks on the Composition of a Flora Britannica.

Vol. V. 17. Description of Sowerbæa juncea, a Plant of New South Wales.—18. Observations on the British Species of Mentha.—19. Descriptions of five new British Species of Carex.—20. Additional Note to the Observations on the British Species of Mentha.

Vol. VI. 21. Remarks on the Genera of Pæderota, Wulfenia, and Hemimeris.—22. Remarks on some British Species of Salix.—23. Botanical Characters of four New-Holland Plants of the Natural Order of Myrti.—24. Description of the Fruit of Cycas revoluta.

Vol. VII. 25. An Illustration of the Grass called by Linnæus Cornucopiæ Alopecuroides.—26. Remarks on the Generic Characters of Mosses, and particularly of the Genus Mnium.—27. Biographical Memoirs of several Norwich Botanists, in a Letter to Alexander MacLeay, Esq., Sec. L.S.

Vol. VIII. 28. Account of the Bromus triflorus of Linnæus, in a Letter to Alexander MacLeay, Esq.—29. Characters of Three New Species of Boronia.

Vol. IX. 30. A Botanical Sketch of the Genus

Conchium.—31. An Inquiry into the Genus of the Tree called by Pona Abilicea cretica.—32. An Inquiry into the real Daucus Gingidium of Linnæus. —33. An Inquiry into the Structure of Seeds, and especially into the true Nature of that Part called by Gærtner the Vitellus.—34. Observations respecting several British Species of Hieracium.—35. Specific Characters of the Decandrous Papilionaceous Plants of New-Holland.—36. Characters of Hookeria, a new Genus of Mosses; with Descriptions of Ten Species.—37. Characters of Platylobium Bossiæa, and of a new Genus named Poiretia.

Vol. X. 38. Characters of a Liliaceous Genus called Brodiæa.—39. Remarks on the Sedum ochroleucum, or *Αειζωον το μικρον* of Dioscorides.—40. On a remarkable Variety of Pedicularis sylvatica.—41. Some Remarks on the Synonyms and Native Country of Hypericum calycinum.—42. An Account of several Plants recently discovered in Scotland by Mr. G. Don, A.L.S., not mentioned in the Flora Britannica nor English Botany.—43. An Account of a new Genus of New-Holland Plants named Brunonia.—44. A Description of Duchesnea fragiformis, constituting a new Genus of the Natural Order of Senticosæ of Linnæus, Rosaceæ of Jussieu.

Vol. XI. 45. Some Observations on Iris susiana of Linnæus, and on the Natural Order of Aquilaria.—46. Observations on the Genus Teesdalia.—47. Remarks on the Bryum marginatum and Bryum lineare of Dickson.

Vol. XII. 48. Some Information respecting the Lignum Rhodium of Pococke's Travels.—49. A Botanical History of the Genus Tofieldia.—50. Characters of two Species of Tordylium.—51. An Account of Rhizomorpha medullaris, a new British Fungus.

Vol. XIII. 52. Remarks on Hypnum recognitum, and on several Species of Roscoea.

25. *A Review of the Modern State of Botany, with a particular Reference to the Natural Systems of Linnæus and Jussieu.*—Printed in the second volume of the Supplement to the Encyclopædia Britannica.

After the publication of his Review of the Modern State of Botany, its author was gratified by the following letters.

Mr. Roscoe to Sir J. E. Smith.

My dear Friend, Liverpool, June 9, 1817.

I have read your dissertation on the present systems and state of botany with the greatest pleasure, and am quite gratified to find how fully I coincide in all your opinions; as it leads me to think that however short the progress I have made may be, I have at least had the good fortune not to wander far astray.

Your account of the botany of different countries, and of their different professors, is as novel as it is interesting; and the candid spirit of discrimination in which it is written will not only confer honour on the writer, but on the work of which it forms a part. You will not however be surprised that I am still more interested in the discussion respecting natural and artificial systems, which I conceive you have completely set at rest; and that too, without giving the advocates for either any just ground of complaint. I only regret that this excellent tract does not form a part of the Transactions of the Linnæan Society, where you would have appeared in your proper character, as defending the system of the great man whose name adorns the institution over which you have with such universal approbation so long presided, and over which I hope you will yet preside for many many years to come.

William Roscoe.

Dr. Swartz to Sir J. E. Smith.

My dear Friend, Stockholm, Nov. 19, 1817.

I return you my most hearty thanks for your kind letter of June the 19th, which I had the pleasure to receive not very long ago through the hands of a travelling friend of mine just returned home from your country. I acknowledge also with equal pleasure your very agreeable present of the article intended for the Supplement to the Edinburgh Encyclopædia, which article I have perused with the greatest satisfaction; and of the *Compendium Floræ Britannicæ*, new edition, which I had already sent for and just received. The copy offered to the Academy of Sciences I have delivered in your name, and am ordered to return you her best thanks. I am also very contented to have seen this very neat publication, as I am upon the point of publishing something of the same kind, comprehending the Scandinavian vegetables; as we want entirely a synopsis of our Flora, the contents of which is known but to exceedingly few. The cryptogamous parts having been greatly augmented of late, some summary is looked for earnestly.

The botanical part of Rees's Cyclopædia I long to see, and must have it at any price or rate, if it be had separately in the manner of the entomological part, written I suppose by Leach. The Cyclopædia itself will not so easily find its way to the North.

The delay you have proved by the continuation of your *Flora Britannica* cannot but be very disagreeable. You will, however, probably find means to carry your most excellent enterprise to an end wished for by all cultivators of our amiable science.

It gives me great pain to hear the valetudinarian state of my highly respected fautor Sir Joseph Banks. But it is our lot, poor beings! to leave the stage some way or other. He does it however not without leaving the brightest marks behind.

I hope you have received the last volume of our Academy's *Handlingar*, which I sent both for your account and for the Linnæan Society, together with those for Sir Joseph, to whom the parcel was addressed.

From our mutual friend Mr. Hooker I have received news at times that have interested me exceedingly. I can hardly say what I admire more in his works, his pencil or his pen. His talents are inimitable indeed. Mr. Robert Brown has also favoured me some months ago. It is with great impatience that I wait the sequel of his admirable Prodromus.

My friend Acharius, whom I had not seen since fifteen years past, having just these very days paid me a visit in the capital, and hearing that I intended to write to you, desired to add a line to thank you for your affectionate greetings; and you will receive it on the opposite side.

I remain, respectfully and cordially,

My dear Sir,

Your affectionate and very obedient Servant,

OLOF SWARTZ.

Amicissime,

Diu, et proh! nimis diu expectavi, nec a te litteras ullas accepi. Jam apud amicum Swartz per quasdam dies vitam jucundam degens ab eo audivi, quod ante horam ad te litteras mitteret: itaque non potui quin etiam verbis quibusdam ego te salutarem.

Ante aliquot annos cum Dno. Magist. Ekenstem Londinum proficiscente species aliquot Lichenum rariores includentem ad te misi. Exemplar *Synopseos* meæ *Lichenum* ante annum et quod excurrit, tuta ut opinabar via, ad te ablegavi. Per amicissimum Swartz, monographiam duorum Lichenum (*Glyphis* et *Chiodecton*) cum figuris, Actis Societatis Linnæanæ ut inserantur etiam tuæ curæ tradidi. An hæc omnia ad te advenerunt, nec ne, adhuc inscius sum. Fac itaque ut brevi certior fiam, et de tua perennante amicitia, et de harum rerum felici adventu.

Plura jam scribere vetat et temporis angustia et loci. Interim abs te exflagito, ut quam primum *Lichenographia Britannica* prodierit, exemplar hujus libri ad me perveniat. Philosophiam Lichenologicam cum Terminologia critica edere jam nunc meditor; sed antea opus laudabile Turneri et Borreri studere vellem atque inde erudiri.

Vale,

Die 18 Nov. 1817. E. ACHARIUS.

Sir Joseph Banks to Sir J. E. Smith.

My dear Sir James, Soho Square, Dec. 25, 1817.

My chief reason for troubling you with this is to tell you I have paid obedience to your mandate, by reading your article on botany in the Scotch Encyclopædia, which, conceiving it to be an elementary performance, I had neglected till now to peruse. I was highly gratified by the distinguished situation in which you have placed me, more so I fear than I ought to have been. We are all too fond of hearing ourselves well spoken of by persons whom we hold in high regard. But, my dear Sir James, do not you think it probable that the reader who takes the book in hand for the purpose of seeking botanical knowledge, will skip all that is said of me, as not at all tending to enlarge his ideas on the subject?

I admire your defence of Linnæus's Natural Classes: it is ingenious and entertaining, and it evinces a deep skill in the mysteries of classification, which must, I fear, continue to wear a mysterious shape till a larger portion of the vegetables of the whole earth shall have been discovered and described.

I fear you will differ from me in opinion when I fancy Jussieu's Natural Orders to be superior to those of Linnæus. I do not however mean to allege that he has even an equal degree of merit in having compiled them,—he has taken all Linnæus had done as his own; and having thus possessed himself of an elegant and substantial fabric, has done much towards increasing its beauty, but far less towards any improvement in its stability.

How immense has been the improvement of botany since I attached myself to the study, and what immense facilities are now offered to students, that had not an existence till lately! Your descriptions and Sowerby's drawings of British Plants, would have saved me years of labour, had they then existed. I well remember the publication of Hudson, which was the first effort at well-directed science, and the eagerness with which I adopted its use.

Believe me, Yours,

J. Banks.

26. *A Grammar of Botany.*—The Grammar of Botany appeared in 1821. This volume its author inscribed, as a memorial of his esteem, to Mrs.

Corrie of Woodville near Birmingham; and of this work Sir Thomas Frankland tells the writer, "I must not omit to notice your Dedication, as equal in elegance to any extant."

Mrs. Corrie to Sir J. E. Smith.

My dear Sir James, Woodville, Jan. 29, 1821.

On our return hither two days ago, we found your elegant present waiting our arrival. The book is a treasure in itself, and doubly valuable as a proof of your friendly remembrance. But what can I say in acknowledgement of the gratifying and unexpected distinction with which you have honoured me? I can conceive nothing so flattering to vanity, if vanity were not lost in the far stronger feeling of affectionate gratitude. The privilege of an introduction to you has been one of my first wishes for many years past, and I feel myself fortunate in no common degree in its accomplishment.

I am, my dear Sir James, with the greatest respect, your truly obliged,—may I add,

Your affectionate Friend,

SUSAN CORRIE.

An American edition of The Grammar of Botany (by Henry Muhlenberg, D.D.) was published in 1822 at New York; and a translation of it into German was made the same year.

The Bishop of Carlisle to Sir J. E. Smith.

Dear Sir James, Berners Street, Jan. 19, 1821.

I beg to thank you for your remembrance of me in sending me your Botanical Grammar. When I first heard of its being announced, I thought it was a mere repetition of what had been before published, which possibly was now getting out of print. But upon opening it I really perceived it to be a grammar of prime excellence. "*Disce docendus adhuc,*" said an ancient philosopher. So might you have fronted your Grammar with those humiliating words; for who is there who will not receive instruction? The endeavouring to unfold and explain the intricacies of Natural Orders requires a masterly hand such as yours. You, Correa, and Brown, are the only ones of my knowledge who could touch a subject of this kind; so that I look upon you as *facile principem*, and holding the Chair of the Linnæan Society, if I might say so, *tuo jure*. It is, in my judgement, admirably well done.

Yours,

S. Carlisle.

Professor Martyn to Sir J. E. Smith.

My dear Sir, Pertenhall, March 9, 1821.

I received your book, which gave me great pleasure. Your Grammar plainly speaks the hand of a master; concise, yet full; remarkable for clearness and neatness. Small as it is in size, it must have cost you some time and attention. I smile sometimes when I meet with the miserable, incorrect compilations and imitations of your former work (Introduction to Physiological and Systematic Botany). When your intended Flora makes its appearance, the British botanist will find everything that he wants in these three works of yours.

I am not such a bigot as to think lightly of the Natural Orders, imperfect as our present knowledge of them is. Had I been younger, that very circumstance would have incited me to pursue so delectable a subject; and I hope you will continue to do it. I am only sometimes vexed when they would fain persuade me that the natural system may supersede the artificial. My health is still good, but I grow sensibly weaker in body, and memory and recollection are decaying. With all this I am thankful that my eyes are yet efficient, that my spirits are good, and that I can still enjoy conversation with those who will condescend a little to my deafness. This is much for almost eighty-six.

With my hearty good wishes for your health and happiness,

I am, dear Sir, yours,

THOS. MARTYN.

Professor Schultes to Sir J. E. Smith.

Landshut, July 15, 1821.

My noble and very dear Sir,

It was no sooner than yesterday evening when I was happy to receive your very kind letter, dated April 23rd, together with the splendid presents you favoured me with. I shall immediately try to make my countrymen acquainted with it by a translation made by my son Jules, under my direction and by my side. The young man cannot begin his botanical career better than with the study of this work every science should be commenced with,—that of grammar; and I hope you will be content with his performance. He seems not to live but for botany; and if Heaven yields him health and strength, as he was till now graciously favoured with, he will not fail in zeal and ardour for the *amabilis scientia,* and return once from the Himallaya with a new world of plants. He might perhaps in a few years be favoured by our Government to gather plants on the source of the Ganges, as our Martius on the Rio Grandes.

The fellow was yet, when scarce ten years old, applauded by some of our best herborists for his botanical eye, when he climbed with them on our Alps, and brought them the rarest and smallest Alpine plants he had never seen alive, but only in my herbal.

I hope you had received since, my translation of your *Introduction;* I could not send it sooner than

I had got a copy of, and I did it in the very moment I received it. I am happy to find myself think as you do, and to have never mingled my elbows in the rather metaphysical than physical quarrel against the sexualists, though I was summoned. I answered almost in your own terms; even if the Linnæan theory on this subject were false, the Linnæan system would not be at all the worse.

I shall, according to your direction, look as soon as possible to complete our Rees's Cyclopædia.

How should it be possible, noble Sir, that any thing coming out of your hands should cost more than it is worth. The most trifling plant given by you becomes a jewel for any herbal in the world; and jewels never cost more than they are worth.

Sieber has collected three interesting herbals—Herb. Palæstinum, Creticum, and Ægyptiacum. I bought them, and am very content with; he has yet doubles enough if you wish. He did send now a young man to the Cape, to gather there for his collection. Another was sent by him to Domingo; and he wrote me today that a very ample load of dried plants, almost two waggons, arrived safely at Marseilles from Domingo. Command with him; he is a clever fellow.

I rejoice to learn no greater grief but what nature imposes on everybody has fallen upon your noble heart. It gives me the purest joy to hear your dear mother touched even the 88th; for we have by it the most reasonable hope to see you to us at the same term: "*Fortes creantur fortibus et bonis; inest parentum virtus!*"

Agree the homage of the sincerest and deep-felt respect, esteem, and even awe. I have the honour to continue for ever faithfully,

Noble Sir,

Your most obedient Servant and Friend,

J. A. Schultes.

Landshut in Bavaria, 1822.

My very dear and noble Sir,

Your plain English Flora will become a benefit, a blessing to your countrymen and countrywomen, and to their children in their furthest generations. You will become by it the very British botanical Socrates,who, as Tully said, "*scientiam e cœlo deduxit,*" *&c.;* and, as Socrates did in philosophy, thus banish the sophisms of *botanophili ephebi,* &c. (to use some sneer of holy father Linnæus) out of the realm of the *amabilis scientia.* Your country, as well as Germany and France, needs highly to receive the impressions of a new die, cut and hardened by the master-hand of a first-rate botanist, if botany should not relapse in the old chaos before the times of Bauhin, and cease to be of any use to mankind. All our modern botanists pretend to the possession of the key of Nature's mysterious temple; and what they open is nothing else (thus it seems to me) but the gloomy dwelling of Minotaurus.

"Multiplicique domo, cæcisque includere tectis,
Ponit opus, turbatque notas et lumina flexum
Ducit in errorem variarum ambage viarum."

You'll forward more the wealth of science and of

your country by your English Flora, by settling the due and true relations between terms and ideas, than all those who being only jealous of the glitter of novelties afford new confusion in the science to the old one.

Your *Grammar*, noble friend, is now already translated by my son, and shall I hope become the botanical common prayer-book of our Bavarian students. Pardon me for the expression of common prayer-book; 'twas only to have the honour to show you, that I am not of those fools who "have said in their heart, there is no God;" and that I know well the use of a common prayer-book. It was the very awe for religion that brought the note, page 8, in the translation of your *Introduction*. You cannot imagine to yourself in what a terrible abyss of paganism, polytheism, or rather pantheism (in the ill sense of the word, for in the good one *Deo plena sunt omnia*), of mysticism and Torquemadism my poor High-German countrymen are now fallen in, by mixing natural philosophy with theology; not in your manner, and in that of every sound philosopher, but doing it in their own way.

SCHULTES.

27. *Selection of the Correspondence of Linnæus and other Naturalists.*—Two volumes 8vo, published in 1821. These are dedicated to the Linnæan Society.

Professor Martyn to Sir J. E. Smith.

Dear Sir, Pertenhall, Nov. 16, 1821.

I have been reading your Selection of the Correspondence of Linnæus &c., and have been perhaps more interested in the work than most others; for though the chief part of the correspondence is antecedent to my botanical life, yet most of the parties were familiar to me in my early days, though I had personal acquaintance with few of them. These letters show how dead botany was in England in the middle of the last century, when Collinson and Ellis, two men not professionally scientific, but engaged in commerce, were Linnæus's principal English correspondents! His system can hardly be said to have been *publicly* known among us till about the year 1762, when Hope taught it at Edinburgh, and I at Cambridge, and Hudson published his *Flora*.

Go on and prosper, my dear friend, in your *Flora Græca*, and your other most useful works, in promoting the most delightful of all sciences; and believe me, &c.

THOS. MARTYN.

28. *The English Flora.*—Of the English Flora, dedicated to perpetuate his regard for one of his earliest as well as latest friends, Sir Thomas Geary Cullum, Bart., the first and second volumes were published in 1824, the third in 1825, and the fourth in March 1828.

The partial author of the obituary before mentioned, declares the English Flora to be "the last best work of the distinguished President of the Linnæan Society, consisting of four volumes octavo, and describing the Phænogamous plants and Ferns of Great Britain, though its title might imply a more limited range: "*Finis coronat opus*." There is no Flora of any nation so complete in flowering species, and none of any country in which more accuracy and judgement are displayed. If any person should in future contemplate a work of this kind, whatever the originality of his information, whatever the novelty of his subject, let him imitate this illustrious author in careful remark, in taking nothing upon trust, in tracing every synonym to its source, and lastly in arranging his matter in such a manner, by the aid of different types, as shall render it easy of reference, and point out at a glance the nature of it. However mechanical some of this may appear, it is absolutely essential to be attended to in natural history, where the subjects are infinite in number, and where aid must be derived from every mode of generalizing particulars."

The part of this work which it may be said the author himself considered the most original, and afforded him most satisfaction, was the natural order of Umbellatæ. "By a full investigation," he observes, "of all the organs of fructification, and by distinguishing the tumid bases of the styles from the *floral receptacle*, things hitherto confounded, I have characterized the Umbelliferous plants like the rest by the parts of the *flower and fruit* alone. In doing this, I have kept the exotic species in view, of which the Linnæan collection and those of many botanists of Switzerland, with the Greek herbarium of my lamented friend Sibthorp, have furnished me with almost all that are known. The principles I have adopted prove amply sufficient, being no other than those by which Linnæus was, on the whole, so successful, though he deserted them in the arrangement of the tribe in question. But what affords me most satisfaction is, that I am thus enabled to keep entire almost all his own genera."

In a letter addressed to Mr. D. Turner in January 1823, Sir James tells him that "the Umbelliferæ occupy me so that I have hardly time to write a letter; but I have the satisfaction of thinking I am doing some good, and that they will all be found capable of being arranged in good natural genera by their fructification alone. In reply to an objection of yours, please to observe these plants are (without exception perhaps) to be found in flower and seed at the same time.

"I have better opportunities of specimens in every state than most people who have laboured at this subject, having the herbaria of Linnæus, Davall, Reynier, Favrod, with my own. I have also Cusson's numerous letters explaining his ideas.

"I am happy that you think of our late conversation with the pleasure that I do, for indeed I have seldom been more pleased with any. We must have more such. I am often in the way of so many whom it is hopeless or worthless to converse with, that good sense and just feeling added to religious consolation are like grains of gold in the mud of a torrent."

The following flattering notice of this work the author received from a young and ardent admirer of his favourite pursuit, an acquaintance as unexpected as it was pleasing, and which he regretted having no opportunity of cultivating more closely.

My dear Sir James, Edinburgh, Sept. 6, 1824.

I have been purposing for some time past to write to you, as I should be sorry to let you forget me. The sincere pleasure, I will say delight, I experienced in cultivating your acquaintance during the short time you were with us at Cowes, I often think of, and regret that the distance now between us is likely to be a great obstacle to our meeting. My present excuse for writing to you is to thank you most cordially for the present you have lately made the British botanist. It may appear presumption in me to praise the two volumes of the English Flora lately published; but as an ardent admirer of the science, more especially as relating to these kingdoms, I consider it a great happiness and honour to address you, and to express my great admiration of the elegance, the neatness, the perspicuity of the language, the exquisite clearness of the scientific descriptions, and the consummate skill of the generic arrangements. The Umbellate plants are surely now *perfect*. I am sure, my dear Sir James, you will excuse my thus expressing my feelings; but, indeed, although you have many older and abler friends, you have none who admires and esteems you more for your scientific and also for your amiable qualifications.

A very clever man, Mr. P. Duncan of New College, Oxon (a great admirer of you), said of the *Flora Britannica*, "that it was a sort of logic,—so clear, so perspicuous and so convincing." I thought of this in the examination I have made of the Flora. Hoping soon to see the remainder, and wishing you all health and strength to go on with that and your other works, and with many years of happiness,

I am, my dear Sir James,

Your faithful Friend,

E. B. Ramsay.

P.S. As an ardent lover of British botany, permit me (though I am a mere novice) to take this opportunity of expressing my unfeigned admiration of the *Flora Britannica*. I offer rather my thanks than my approbation; for approbation insignificant as mine is can never increase *your* reputation.

Le Sécrétaire perpétuel de l'Académie à Monsieur le Chevalier J. E. Smith, Corréspondant, &c.

Institut de France, Académie Royale des Sciences,
Paris, le 10 Mai, 1824.

L'Académie, Monsieur, a reçu l'ouvrage que vous avez bien voulu lui addresser, et qui est intitulé *Flore Anglaise*, 2 vols. 8vo, 1824. J'ai l'honneur de vous remercier au nom de l'Académie de l'envoi de cet ouvrage, et de vous temoigner tout le prix qu'elle attache à cette publication. Il a été déposé honorablement dans la Bibliothèque de l'Institut, et M. Desfontaines a été chargé par l'Académie de lui en rendre un compte verbal.

Agréez, Monsieur, l'assurance de ma considération distinguée.

B. Cuvier.

P. S. Oserais-je vous prier de profiter de quelque occasion pour m'addresser le diploma de membre de la Société Linnéenne: m'honorant d'appartenir depuis long tems à cette association distinguée, je desirerais posséder le document qui en est la preuve.

Sir Thomas G. Cullum to Sir J. E. Smith.

My dear Sir James, Bury, March 19, 1824.

I received your kind present of the two volumes of English Flora, and shall wait with some impatience for the continuation of it; but should almost despair of seeing it, if you had not told us in page 21 that you would proceed immediately with the remainder.

It is true, as you say, that I have a love of natural science, and it has amused me a great many of the years of my long life; yet I am conscious how deficient I am in the real knowledge of natural history, as many amateurs, as they are called, are of the fine paintings they possess. I will not however conceal the pleasure I feel in your address to me, which tells the world far and near, that I have the esteem and affection of the author of the English Flora.

Our friend Professor Martyn told me that in the compilation of Miller's Gardener's Dictionary he wrote 20,000 sheets of paper. I know not how many sheets of paper the English Flora has occasioned you to write, neither can I calculate how many volumes of books you must have examined. Glad as I should be to live to see the remaining volumes of the Flora, yet I live in fear that you should hurt your health by the confinement and labour it must occasion you; I will allow in some cases "*Labor ipse voluptas,*" but I cannot think the confinement and labour of sending forth a finished Flora of five volumes is anything less than an arduous undertaking.

I have been looking into a few leaves of the Rev. T.M. Harris's, D.D. book on the Natural History of the Bible. He takes notice of the lilies of the field, and quotes you for their being the *Amaryllis lutea.*

Yours most gratefully and affectionately,

T. G. Cullum.

Sir J. E. Smith to Mrs. Corrie.

My dear Mrs. Corrie, Norwich, March 21, 1824.

I have, for a long time past, had so much employment for my pen, that except a few urgent letters of mere necessity or business, I have scarcely written one. I am still as busy at the sequel of my Flora, of which I hope you have received the first and second volumes, which I ordered to be sent for your kind acceptance. Nevertheless, it seems hard if I may not write a letter for mere pleasure,—and I feel but too sensibly that I am deep in your debt. I am in full as good health as when I left your hospitable and delightful abode, and hope to give a course of ten lectures at the London Institution in May and part of June. The Institution has claims upon me as an honorary member, and it is extremely well attended. I have many things to say in these lectures, as well as at the anniversary of the Linnæan Society, about the present state, progress, and corruptions of botany. I know not that I can stem the barbarous and muddy torrent, but will try.—Have you seen the *Systema* of my friend DeCandolle? It is a rich store of knowledge, the greatest assemblage of botanical information, as to species, characters and synonyms, in the world, but the nomenclature and its principles most corrupt! He has also published one volume of *Prodromus* of the whole work. His only rule for names is priority of date; as to bad or good, classical or barbarous, he has no taste. All Linnæan rules are disregarded. He condescends indeed to retain all Linnæan names; but all the classical names of Solander, Swartz, Schreber, &c. are abolished, to restore the vile barbarous appellations of Aublet and other illiterate French travellers. I know not how the world will bear this. I am no Hercules, to attempt to cleanse the stable, but I shall go on in my own way, and trust to the good sense of those who may come after me. DeCandolle is a most worthy man, and

I wish neither to hurt nor quarrel with him. Botanists are more sore on the subject of nomenclature than any other, especially when they are in the wrong. Linnæus was aware of this when he wrote his *Critica Botanica.* I shall be happy to hear your opinion, with any advice or corrections, on these two volumes of the English Flora. I think there must be five in all, if I have health and eyes to finish them. I am now in *Didynamia,* and hope to get through it before I go to Town. DeCandolle is most injudicious in *Tetradynamia,* with an immensity of genera and subdivisions.

We had an agreeable visit at Holkham in October last, with Mr. and Miss Roscoe, the Duke of Sussex, &c. I introduced there Mr. Hunter, a native American, brought up among the Indians, and now going back with the noble design of improving them on the wisest and best principles. If you and Mr. Corrie have not read his Memoirs of his Captivity, pray do. The Duke of Sussex and Mr. Coke were delighted with him. He came to Europe with every possible recommendation from people whom I know.

I am, dear Madam,

Your ever obliged,

J. E. Smith.

Nathaniel John Winch, Esq. to Sir J. E. Smith.

Newcastle-upon-Tyne, April 26, 1824.

My dear Sir,

Permit me to return you my best thanks for the

notice you have been pleased to take of my Northumberland and Durham Catalogue and pamphlet in your excellent English Flora, which has afforded me infinite information, especially on the difficult subject of the synonymy of the older authors,—a topic far beyond the reach of ordinary compilers. Had I been aware that no localities of rare species would have been admitted into your work without you possessed specimens, I would have sent you the last I have of *Pyrola secunda*, from the only well authenticated English habitat at present known:—I mean Ashness Gill, above Barrow Force, near the Derwent-water Lake in Cumberland, where it was gathered in 1807 by your friends Turner, Hooker, and myself. As for *Pyrola rotundifolia*, it is abundant in the romantic Dene at Castle Eden in Durham. In vol. ii. p. 369, Gibside Woods, in the same county, is mentioned as a locality of *Spiræa salicifolia*, but it is merely naturalized there, in the same way as in the Duke of Athol's woods at Dunkeld, and by Roadley Lake on the wild moors of Northumberland beyond Cambo (the birth-place of Capability Brown); but there, very old and stunted lilac-trees pointed out the exotics. While on the subject of naturalized plants, it may not be amiss to mention that I once met with a considerable quantity of *Saxifraga umbrosa* in the woods of Blair Athol, associated with *Pyrola secunda, Arbutus Uva-Ursi* and *Habenaria bifolia*, and close to *Pellidea venosa*. Here I thought the London-Pride might be considered truly wild, especially as the general habit of the plant was much altered; but, to my disappointment, the gardener informed me he had introduced the plant in various parts of the wood several years before. In a small lake, high on the Wallingford moors, *Nuphar minima* is found in plenty; and about twelve years ago, Mr. Trevelyan had plants removed to the ponds in his gardens, where it flourishes with *Butomus umbellatus*, and *Menyanthes nymphæoides* of the South of England, and *Nymphæa alba*, and *Nymphæa lutea* of the North, but still retains its diminutive size.—This, allow me to remark, is all the difference I can find between it and *N. lutea;* for the teeth of the stigma (I mean their protrusion) depends upon the age of the seed-vessel, and the approximation or separation of the lobes of the leaves also, is owing to age. *Erythræa littoralis*, I think, must be considered a good species; it is scattered over several hundred acres in Holy Island; and though I met with many gigantic specimens, still the peculiar shape of the leaf and mode of growth were retained, while *Erythræa Centaurium* grows on the sandy sea-beach near Tynemouth, without being altered by situation. Here also *Silene inflata* may be seen, as in inland fields, without being changed into *Silene maritima:*—see Hooker's *Flora Scotica.* Thomas, of Bex, has sent me German specimens of the Pheasant-eye Pink, as *Dianthus plumarius.* Is he correct? or what is *D. plumarius*?—*D. arenarius* I have from Swartz. Should any plant not as yet published in English Botany fall in my way, fresh specimens shall be sent to Sowerby for the Appendix.—*Rosa dumetorum?* and a few mosses he received from me in the autumn. The Scotch botanists have detected six new species. Did Harriman transmit *Juncus castaneus* to you from Teesdale? I was not aware we possessed this rare plant. Adieu, &c.

N. J. Winch.

Robert Brown, Esq. to Sir J. E. Smith.

My dear Sir, Dean Street, February 2, 1825.

It is very gratifying to me, to find that you are likely to adopt the greater part of the genera I have proposed in *Cruciferæ*. As to M. DeCandolle's labours in this difficult family, I think he has made several improvements in his divisions of genera, and I remember I thought,—but I confess I have not sufficiently studied it,—that his *Diplotaxis* would probably be adopted. I have no hesitation however in saying, that he has pushed his divisions from modifications of form and direction of cotyledons much beyond their value, and, in giving nearly equal importance to all these modifications in every part of the order, has proposed subdivisions or tribes which, though more natural than Linnæus's *Siliculosa* and *Siliquosa*, are still in some degree artificial; one curious proof of this is,—and several others might be mentioned,—that not having thought it necessary to examine the most common plant of the family, namely *Bursa Pastoris*, he has placed it where it really ought to be, but where it cannot remain according to his own system.

Believe me, my dear Friend, most faithfully yours,

R. Brown.

A peculiarity in Sir James's literary compositions may be mentioned; it is, that he seldom wrote any hing more than once, and his manuscript was sent to the press as it came from his hand, without any material correction or interlineation, in a distinct legible character, that appeared more like a corrected copy than a rough draught. This was the case not only in scientific description, but with other compositions, such as his biographical memoirs of botanists printed in Rees's Cyclopædia.

The person who records this was first struck with the circumstance when he wrote the Life of Dombey*, which came

> "Warm from the heart, and faithful to its fires,"

with scarcely an erasure of the pen, and perfect in the minutest particulars of orthography and punctuation.

Another peculiarity, as it appears, was this, that when pressed for time, he frequently wrote most to his own satisfaction. Such was the case with his prefaces and dedications; always delayed till the volume was near its completion, and then hurried by his printer, he generally sat down after tea, and

* "The collections of seeds and botanical collections of this unfortunate and injured man, through the mismanagement of the Spanish court, have been lost to the gardens of Europe.—Among a few of his botanical discoveries which are preserved, are the magnificent *Datura arborea*, the beautiful *Salvia formosa*, and the fragrant *Verbena triphylla*, or, as it ought to have been called, *citrea*[1]; this last will be a *monumentum ære perennius* with those who shall ever know his history."—J. E. S.

---

[1] *V. triphylla*, now *Lippia citriodora*. Kunth.

would fairly write what was wanted, without premeditation or doubt about its plan, as he would have written a letter.

These are proofs of the correct and energetic mind he was endued with;—he loved repose, his bodily constitution required it; but his mental vigour, when occasion called forth exertion, discovered a power beyond what he appeared to possess. It may not become the present writer to speak of Sir James's style as she is inclined to do, but the Prefaces to his works always seemed to her happily expressed, and the Dedications appropriate. That to his Exotic Botany is one of the best among the latter, and shows the truth of the poet's observation,

> "Affection lights a brighter flame
> Than ever blaz'd by art."

She is even tempted to apply to him the eulogy he bestowed on Sir William Jones, "who honoured the science of botany with its cultivation, and like every thing else he touched, refined, elevated and elucidated it. No man was ever more truly sensible of the charms of this innocent and elegant pursuit; and whenever he adverted to it, all the luminous illustrations of learning, and even the magic graces of poetry flowed from his pen."

The following passage from a letter of Mr. Roscoe's, dated the 6th of January 1805, by no means falls short in expression of those sentiments of esteem which are found in the Dedication of this work to that accomplished and learned historian, naturalist, and poet.

"My dear Sir,—It was not till late last night that I had the pleasure of receiving through the hands of my booksellers the first number of Exotic Botany, and of perusing the affectionate, and to me, highly gratifying address which you have done me the great honour to prefix to it. To such parts of it as are commendatory, I can only say, that although it be an arduous task, I will do the best I can to justify you to the world for the favourable opinion which you have ventured to express; and in this respect I feel as if I had been paid beforehand for a work which I have yet to perform: but in your kind and friendly expressions of attachment and esteem I experience the most unalloyed and perfect satisfaction, because I know that affection can only be repaid *in kind*, and that I am rich enough to make you a return. May this public seal of our friendship not only confirm it whilst we live, but long continue to unite our names in future times, as associates in our studies and pursuits, in our dispositions and our hearts."

The Prefaces to the Reflections on the Study of Nature, his Tour on the Continent, his Introduction to Botany, and to the English Flora, (which last is indeed a brief history of the progress of botanical science in this country,) are well worthy of being read for the information and sentiments which they contain.

In all Sir James's literary compositions his only aim was to express his thoughts with clearness and brevity, in the most common words. "Hard words," he observed, " never taught wisdom, nor does truth require them."

He never sacrificed the accuracy of his meaning to a well-sounding sentence, considering it a frequent cause of obscurity in writers, and one form of affectation; and affectation he defined in the words of Lavater, as " the vain and ridiculous attempt of poverty to appear rich."

The facility with which he wrote had its origin in a habit of thinking much on the subject previously to his committing it to paper. Long before he began his English Flora, he occasionally said, " I have it in my mind, and only want time to write it down." This must be understood with some latitude:—as soon as he did begin, it grew beneath his hand, and a thousand ideas, dormant till then, crowded for admission, and new arrangements took place as occasion required; but still, the foundation had been laid, and to its stability the superstructure owed its strength and beauty.

When we consider the variety and number of works that came from his hand, the frequent bodily indispositions that retarded his progress, together with the interruptions occasioned by an extensive correspondence, by the delivery of lectures, by occasional journeys, and the indulgence of social intercourse,—it is remarkable that he accomplished so much; but he had two hands to his work, which experience proves to be effectual in surmounting difficulties;—an unconquerable inclination, and great order and method, without which, however dull and technical it may be esteemed, nothing can be well done. They comprise the secret of much perfection in human operations, and the best œconomy of our most precious and evanescent property, time and temper.

In the common occurrences of life, a promptness of decision was one of the characteristics of Sir James's mind; it implies, indeed, an experienced judgement: but he considered the habit of indecision more wearisome and unprofitable than even a wrong determination; and having chosen for the best, he would not permit himself in vain regrets if he failed in his designs. Upon more important occasions, he usually quoted a rule of his friend the Bishop of Carlisle, "*Let us do the right thing*," and then whatever the result may be, we shall want no consolation. This was a principle upon which he invariably acted; and to those who have felt the peace of mind which accompanies such a motive, it will not require any other recommendation.

The letters of Sir James's inestimable and long-tried friend, who has been just alluded to, will best illustrate the works above enumerated. As a classical scholar Dr. Goodenough's criticism is full of instruction, as a systematic naturalist he was not behind any in precision, nor was it his disposition to compliment at the expense of sincerity; his remarks are therefore valuable, and were continued at pretty regular intervals during an unbroken friendship of more than forty years. The Bishop died in the autumn of 1827, a few months only before his attached friend. His last letter, in the month of May that year, was written to express his regret at being unable to attend the anniversary dinner of the Linnæan Society on the 24th. A similar reason prevented Sir James himself from being present. After repeated attempts to reach the metropolis, increasing infirmity put a stop to an annual visit, which for many returning seasons had been the period of much enjoyment, connected with the celebration of the birth of Linnæus.

CHAPTER VI.

The Bishop of Carlisle endeavours to persuade Sir James not to remove from London.—Motives which induced him to return into Norfolk.—Letter of Mr. Crowe.—Lectures at the Royal Institution.—Flora Græca;—Letters from Sir Joseph Banks;—Sir J. E. Smith;—Thomas Hawkins and Thomas Platt, Esqrs.—Bishop of Carlisle's Correspondence.

THE following correspondence of the Bishop of Carlisle, chiefly critical, and relating to Sir James's literary pursuits, comprises a period of thirty years. With it are mingled a few letters from other learned men, and they are thus kept apart for more easy reference to the foregoing Chapter.

The Rev. Samuel Goodenough, D.D., to J.E. Smith.

Dear Sir, Ealing, Sept. 16, 1793.

I have the pleasure of telling you that I am arrived at Ealing from Hastings, where I have been for a month, safe and sound. Our eyes were blest with the sight of the India, Jamaica, and Oporto fleets, and variety of shipping of all descriptions, beyond what I ever saw before. I was well all the time, met with very agreeable company, and in short *unbent* thoroughly. Having been at Hastings frequently before, there was not so much remaining to be found. However I improved myself somewhat in marine plants. Among other things, the fishermen brought one day to shore the Blue Shark, *Squalus glaucus*. As I observed that Pennant has spoken of this species very slightly, having never seen one, I have taken a very particular description of it, and a young lad of the place has given me a very correct outline drawing of it. I should be glad to know if it would be worth while to make up a paper upon the subject for the Society. Has any one given in to our Herbarium *Asplenium marinum*? because I have a very good specimen to spare. I found two new *Ulvæ* at Hastings, but only single specimens of each. And now give me leave to ask how my paper on the *Carices* goes on: is it all printed yet? I could wish to let Sibthorp see a copy of it before he finishes his *Flora Oxoniensis*.

The Baroness Itzenplitz is brought to bed of a son at Kew. The Baron favoured me with a note announcing the happy tidings.

Pray favour me with a few words upon the Blue Shark, my paper on *Carices*, and the *Asplenium marinum*.

I am, dear Sir, ever yours,

S. GOODENOUGH.

From the same.

My dear Sir, Ealing, March 21, 1796.

The retirement of my situation and the neglect of the newspaper which we read, kept concealed from me till the latter end of last week, that you had completed your happiness. You must now be doubly valuable to me. I most sincerely give you joy, and shall be glad to know when I may tell you so in person. If there be in this world a state in which all is not vanity, it is that where two meet and unite on rational ideas, and on the sacred principle of making each other happy.

Pray hold me still among your friends, and recommend me to Mrs. Smith's good opinion.

I remain ever your most sincere Friend and Servant,

S. GOODENOUGH.

I forgot to say how sincerely Mrs. Goodenough joins me in all I say, and all the girls too.

From the same.

Dear Sir, August 25, 1796.

The same channel of information which informed me of your return from Norfolk, mentioned your intention of quitting Middlesex. I rode over to Portland Place the next day, when lo! (see what it is to live out of the world!) I found you had returned some time, and set out again for Hafod the day before. I found from the ticket at the corner, and from your servant likewise, that all was too true, and that your house was to be let immediately. Directly I began grieving for you and for the Linnæan Society. At the distance of Norwich you will be quite buried alive. I wished to talk over with you once more, how necessary it is, *that he who would reign over many must be perpetually contend-*

ing with many. How many matters do we ascertain, not merely from positive and laborious study, but accidental conversations with persons but meanly informed upon the subjects discoursed of! You will have nobody to question your authority at Norwich; the consequence of this will be, that if you once settle in any error, such is our infirmity, you will continue in it. Then how will you from time to time be aware of the new discoveries which may be made?

If you are not here upon the spot, you will in this respect be greatly behind others. But all these things will keep you out of sight. You ought to be always in the centre, and the ruling and animating power. I hear you propose being three months in town; that is not sufficient;—reverse it;—be nine months here, and three at Norwich. Then again the concerns of the Linnæan Society will require more attendance than three months. The scheme of vice-presidents is proper enough, but a *substitute* (for such after all is a vice-president) cannot give life like the principal. And yet much do we want life; when recourse has been had to several persons in the room, and scarcely one has been prevailed on to take the chair, many have been disgusted and their ardour cooled.

Do think upon all these things before it be too late.

The more the Society flourishes, the more credit to the President: if the Society dwindles away, the President has an empty name.

What credit has the potentate of a barren land,

uninhabited? Better therefore to be much in the busy world, and little out of it; and therefore, as I said before, you had better reverse the matter; be nine months here, and three there.

I want you very much to give me specimens of *Hypericum dubium* and *quadrangulum*, and indeed any other choice specimens which Hafod may afford.

My very particular thanks are due to you for introducing me to Sir Thomas Frankland; I wish this had taken place before we had finished our paper on *Fuci*. I think his correspondence will make it necessary to make an *addendum* to our paper.

I have long since gone through your *Flora*, &c. —I am interrupted.—Always yours affectionately,

S. GOODENOUGH.

These affectionate letters want but little explanation; and it must have been a powerful impulse which made the friend to whom they were addressed resist the strong persuasions of so sincere, so warm and judicious a counsellor.

Sir James had not long before removed from his house in Great Marlborough-street, and taken one near Mr. Lee's nursery-garden at Hammersmith, which was a primary inducement with him to make the change. Domestic convenience, however, influenced him to leave London altogether, and he returned to his native place for a permanent residence in the autumn of 1796. In the city of Norwich he found himself among those who knew and esteemed him; and there, as his medical skill was not unfrequently put in requisition among his intimate acquaintance, Mr. Crowe of Lakenham, one of his earliest and at all times most cordial well-wishers, who had great confidence in his judgement, endeavoured to persuade him to the practice of his profession, and told him he should repeat the advice annually, convinced that he would rise to great eminence in it.

There certainly were moments when Sir James took this advice into consideration, and reflected with some satisfaction upon the friends and confidence which his professional pursuits might procure him. His own experience of the value of a medical friend made him look with complacency upon being useful in a similar way to others: but these were transient thoughts, for he knew that his favourite occupation must frequently be interrupted if he once seriously engaged in practice as a physician, and some apprehension as to his health enduring the requisite fatigue was a bar against a prompt decision to try the experiment.—The writer reflects with as much self-satisfaction as upon almost any determination in her life, that she never encouraged her husband to relinquish his proper pursuit for a new object:—not that she made any sacrifice in this; a comparison of the possible advantages attending such a change weighed but little against the pleasure of seeing him in a much happier situation; in his peaceful library, amidst the *Floras* of Greece and India, of the Alps and England.

To keep up as much as possible a connection with the Linnæan Society and his friends in London, Sir James spent at least two months every spring in town, and for full twenty years he delivered an annual course of lectures on botany at the Royal Institution in Albemarle-street, when his associates there were men of the first talent in their respective departments of science and taste*.

Sir James also gave at different periods botanical lectures at the institutions of Liverpool, Birmingham, Bristol, and London.

* Among these were Sir Humphry Davy, Campbell, Opie, Sidney Smith, and other men of eminent attainments.

The following letters from Sir H. Davy relate to arrangements for these courses of lectures.

Dear Sir, Royal Institution, 1804.

It is in the contemplation of the Royal Institution to extend considerably the subjects of the lectures of the establishment, and amongst other courses for the ensuing season, a short one on botany is proposed. I presume to inquire of you, at the request of one of the most active and enlightened members of the board, whether they may entertain any hopes that you will be disposed to undertake such an object. I trust your goodness will pardon the liberty I am taking. I am sure the committee will be disposed to concede to any terms that you may propose. By your assistance the Institution would be materially benefited, and a new impulse would be given in the metropolis to that science in which you have so long taken the lead, and which you have so much extended by your labours.

I am, dear Sir, with the greatest respect,
Your obedient humble Servant,
H. Davy.

Dear Sir, March 22, 1805.

I am requested by the managers of the Royal Institution to inquire whether it will be agreeable to you to deliver a third course of botany in 1806. I hope and trust you will reply in the affirmative, because I know that your last course gave universal pleasure and satisfaction. I am, dear Sir, yours,
H. Davy.

When his health was good, the occupation was one he enjoyed. He arranged previously the heads of his lecture; but for words he always trusted to the ideas which arose in his mind while he was delivering it, and in general he exceeded the allotted time, and had more to say than could be compressed into the space of an hour. A printed abstract of the subject he intended to discourse upon was not omitted, for the convenience of himself and his auditors; and of these sketches he composed a great variety, as the succession of his courses required. Of one of these Dr. Goodenough, in the year 1795, tells him, "I am quite charmed with your Syllabus. I would advise you, *while you are a lecturer* (do not defer it till you have given up, it will not be half so well done), to draw out all that matter at full length, and publish it as suits you; it would be another *Philosophia Botanica* in a fashionable dress."

Perhaps there is not to be found in the records of any scientific association a more gratifying example of the existence and the expression of kindly feelings, of the absence of all jealousy, and of the most active exertions in the support of science, than has been displayed by the members of the Linnæan Society, not only with a view to maintain the spirit and utility of their institution by individual efforts, unfettered by rivalry or intrigue, but by a mutual feeling of friendship and esteem existing between themselves and him who was so long chosen to preside over them.

The honour which the Society conferred upon its

President, in electing him to that office year after year, was reflected back upon itself by the constant attention which he paid to its interests and by the unwearied devotion to scientific pursuits, which his more retired situation enabled him to exhibit with fewer interruptions than could have been expected had he continued to reside in London: and that his situation at Norwich did not injure him in the estimation of learned men, or hide from him the progress which science was making, the works he composed subsequently to his retreat are acknowledged proofs.

The Rev. Dr. Goodenough to J. E. Smith.

My dear Sir, February 21, 1797.

I rejoice to hear of the comfortable arrangement of your family, and shall never cease regretting that that comfort could not be accomplished in the neighbourhood of London;—you must allow your friends to regret in silence. I have not been at the Society since you went, except once at a meeting of the Fellows. I have introduced a most respectable member, Sir Thomas Frankland.

Salisbury's nomenclature is I think extremely improper, not to say ridiculous. I am sorry that he has persisted in his errors even to printing them. I was present at a very warm dispute between him and Dryander, who in his blunt rough manner finished his argument with "If this is to be the case with every body, what the devil is to become of botany"? I did not interfere then at all, but took a private opportunity of persuading him (but I laboured in vain) to invert the order; and instead of giving new names, and quoting the Linnæan ones under them, to retain the old names, and remark under each how much better any name he thought of would accord.

Mr. Salisbury has a happy firmness, which some people will call obstinacy, which makes him rise superior to every opponent.

You are quite right about *Agrostis littoralis*. Dr. Withering and I corresponded about the arista of the corolla. He had either prepared that page for the press, or actually printed it before this took place. But, as I told him, my specimens which he saw could have convinced him, that both calyx and corolla occasionally are with an arista. I do not scruple to abominate, without the least qualification, the undermining the Linnæan fabric. But Thunberg is answerable for this envious superficial daubing. It would be worthy of you to set the world right in this particular.

Sowerby has failed very much in his figure of *Fucus kaliformis*. My idea when I saw it, was like the story of old churches in briefs,—that it must be wholly taken down and rebuilt. I find we are to have a visit from the French. They will send our specimens flying after Gigot d'Orcey's butterflies, and purchasers after both. If they take it into their heads to come, it will be too serious a matter to joke upon.

Yours most sincerely,

S. Goodenough.

J. E. Smith to the Rev. Dr. Goodenough.

My dear Sir, Hafod, Nov. 9, 1797.

You have perhaps heard of the reason of my coming to Wales,—which is to see my young patient Miss Johnes.

I am very glad you like what I have done about *Orobanches.* I have also sent a paper to our Society upon some foreign ones. The characters I use hold good in them. By looking over Buddle's herbarium, and that of Sherard in my way hither, I have got much light about *Menthæ* and *Bromi.* I have made ample notes, and think I now understand the former genus. The genus *Bromus* is as ill done in all our British authors (and perhaps in Linnæus) as any one in the system. By this time you have got English Botany for November, and my preface. I am very glad *now* Curtis did not accept my offer, though I would then have steadily kept to it. I will always do justice to his botanical abilities on every occasion. I hope our Society will go on very well.

Your ever faithful Friend and Servant,

J. E. SMITH.

James Crowe, Esq. to J. E. Smith.

Dear Sir, Lakenham, Nov. 21, 1797.

I am most exceedingly rejoiced indeed to hear from you that Miss Johnes is so much recovered as to be pronounced so decidedly out of all danger from the most formidable attack that could perhaps be made upon her by nature without depriving her of life. My dear Sir, after your experiencing the pleasure of restoring to her friends and the world so amiable an object, one so likely to diffuse blessings around her, and be in every respect a shining ornament in society, as I have often been informed by your honest and intelligent mind,—I say, after having experienced this pleasure, how can you hesitate to benefit the world with a more extensive practice than the limits of your favoured friends call for! To them I know you can refuse nothing, and I have vanity enough to hope I shall always be on that list; so that it is not from a selfish motive I thus renew a subject I have taken the liberty so often to mention. I feel in the most sensible manner for Mr. Johnes. What a loss would he have had! An only child,—and such a child! His only danger now will be excess of happiness; but you say he has a great mind, and in truth he has had occasion to put it to the test.

I am sorry Mr. Johnes did not receive the white poplars; they were sent according to the directions, and also a bundle of the large sort of white willow. These last should be cut into lengths of twelve or fourteen inches; and both of them put into nursery beds, placed one foot distant from plant to plant. They will soon become great trees, to be removed to such places where they may remain, and add another ornament, I hope, to that now astonishingly ornamented country by art and nature. The white poplar increases rapidly by layers and also by suckers, which are plentifully thrown out by plants a few years old. The willows are propagated by cuttings; and this plant not only will make very large trees, but if used as osiers and so cut every year, are as good a kind as any the basket-makers employ. A moist soil, but not too wet, is the most proper for both; but they do not like a gravel under any circumstances.—I cannot help returning again to the subject of Miss Johnes. I hope she will be prevailed on to come to London, and so be within your reach; for I dread physical ignorance, but much more physical knavery. The first is almost a constant resident in the country, and the latter in London. But in London you could attend her at any time within a few hours notice. If she travels, you might be during the whole journey at her elbow; and for her benefit the enthusiasts in natural history would give up the information to be derived from your intended review of the treasures at Oxford. Indeed I am inclined to believe you have already been too diligent to leave behind much to glean.

Menthæ and *Bromi* will now be traced through all their dark recesses.

From your faithful Friend,

JAMES CROWE.

The Rev. Dr. Goodenough to J. E. Smith.

Dear Sir, Windsor, Nov. 18, 1799.

Your packet came here just as I was setting off

for dear Bulstrode. Upon my return last Tuesday, I set to it, as Shakespear says, incontinently. I have not been able to run through your synonyms of *Carices*.

Your *binervis* is a good species. I believe it grows near Hastings in Sussex. It always struck me as different from *distans*, but I did not think of your excellent mark of difference. *C. Micheliana* is a good species, and entitles you to all praise. Pray give your *vesicaria* a new name. The one I have so called is so named (confounded with yours) in the herbarium. It has always been received abroad as *vesicaria*. I have somewhere among my papers some very cogent remarks upon the propriety of so naming it. Yours is not a vesicarious plant; most probably it is Linnæus's β, or rather what he thought a variety of it. Its fructification is more allied to our *sylvatica*, which I take for granted was Linnæus's *vesicaria* β. Pray do as I say; you will find it will be received as hypercriticism if you do not.

If the world may be suffered after all its bloody struggles, I trust science, and that most natural one, botany in particular, will flourish more and more, and, under your correct auspices, stride on to perfection.

I have used *deficit* with an accusative case. There are good and elegant authorities.

I hope you see in all I say and do, a mind truly attached to you.

S. Goodenough.

From the same.

My dear Sir, Windsor, Feb. 13, 1800.

You must know that riding the other day through water that I had passed a hundred times, all at once my horse plunged into a cavity over head and ears. I fortunately kept my seat. The water, however, at fair standing came over his back. All filled with water, by good luck we scrambled out. My servant could not help me; it was over his head. I then found a little unnoticed bridge had been carried away by the floods just before. All this happened at Old Windsor, in the main road to London. Who could have expected such a thing? I had three miles to ride home dripping wet and chilly. It ended in gout, as I feared.

You are so deep in willows that I cannot come near you. My idea has always been that they should be described in two states,—fructification, and leaf. They are full as distinct in the one as the other. I am glad to agree with you both in *amygdalina* and *triandra* being the same. I have met with it at Bath with an almond leaf as like as *ovum ovo,* and in Battersea fields quite otherwise, but bearing the sign *corticem abjiciens* most remarkably. The *rubra fissa,* Hoffman, I found in a holt, close (on the north side) to the town of Ely, not on Prickwillow Bank, which is six miles from it nearly. I longed to add a habitat of what I always thought *S. arenaria,* but I am not sure of *your* plant. The one I mean is that on the sandy downs about Walmer Castle and Deal, where, when the wind blows fresh, the under side of the leaves turning up gives the whole a silvery appearance. I am glad you throw out *hermaphroditica*.

Again in *Polypodiums* I cannot follow you. I always thought we found *P. Thelypteris* near Bury; but by your list that is *Oreopteris:* it grew only in one marshy place, about six or seven miles from the town. Have you sufficiently attended to *aculeatum* and *lobatum?* As they grew in Sole's garden together, they seemed very different; *lobatum* as smooth again as *aculeatum:* then its pinnæ were adscendent, &c., &c.

I am heartily glad at seeing that so much progress is made. I am getting better every day; so send when you please.

S. G.

From the same.

My dear Sir, Windsor, Oct. 17, 1801.

I wish to say that I am alive, and am disappointed at not hearing from you: I always look up to you for consolations in natural history,—a bit of something new: then especially those new *Carices* which have been discovered since I wrote. Any thing to a poor wretch who has no field for exploration, no companion, but still knows the value and the comforts of the charming science, would be acceptable.

You all seem to be mistaken, I fear, about my

Carex fulva. I left veritable specimens with the Linnæan Society of all stages of its growth. I was the first who ascertained Lightfoot's *C. tomentosa* to be the *filiformis* of Linnæus, which I suppose you did not recollect. I begged the Queen to let me examine her herbarium for this purpose.

I augur from the arrival of peace at last, (Oh that it may last!) that natural history in particular will raise its head. We shall meet folks whom we have not seen for years, and of course shall have much to tell, and so much to ask that we shall never have done. I hope we shall keep the lead in science, as we have undoubtedly in naval glory. I dread, however, the introduction of revolutionary principles; as you have often heard me say, "my nerves will not shake till the peace comes."

Send a body a little natural history news. I have little food here, but I love it as much as ever.

Yours ever,

S. Goodenough.

From the same.

My dear Sir, Deanery, Rochester, Nov. 3, 1802.

I was very sorry to hear, when I was appointed to the Deanery of Rochester, under what a severe affliction you were labouring. To that I attributed my not hearing from you, and to that you must attribute my not writing to you a letter which I was assured you could not read. It would gratify me very much to have a proof that you have shaken off your *holy* adversary St. Anthony.

I have some time since quitted Windsor, and am most comfortably lodged at the Deanery at Rochester, whither you will in future direct to me. I have a charming house, with a very neat and convenient garden, and a little orchard adjoining, and completely secluded from the town of Rochester. All this on the fair side. On the other hand, the neighbourhood for the most part is naval, military, and suited to the various business of a Dock-yard. Unfortunately I have all my life long been accustomed to the conversation of clergymen, men of letters, and liberal pursuits. Our minor canons are very respectable indeed,—beyond most situations of the kind; but I cannot hear of a person who has the least turn for *any* branch of natural history, so that I seem to stand quite alone, a solitary being.

If ever you should come into Kent, I hope you will take your quarters at the Deanery in your way. I will promise you a warm room and warm welcome; and once for all let me say, that I hope you will not think that any elevation of rank in life will make me different from what you saw me at Ealing. I flatter myself I have a mind above such nonsense. I admire the gradual progress of your English Botany. I am told that I am well situated for botany; that must be ascertained as next year (if I live so long) opens. The whole country around me seems chalk, and very thickly clothed with wood. But without a companion one has no heart to move, especially as you everywhere hereabouts see loose fellows from the navy and Dock-yard, with their doxies.

Your *Carex divisa* is a good figure; but had I known of your publishing it, I should have desired you to have noted that strange variation of the culm, from sharp angles in moist rich situations, to almost round in stiff half-moist clay. I found it in great plenty in all its varieties in the Isle of Sheppy. I have a great notion that Hudson, in his first edition, took your *L. chrysophthalmus* for *Juniperinus*, Linn.

Yours,

S. GOODENOUGH.

From the same.

My dear Sir, Bulstrode, April 1, 1804.

I waited till I came to this place, that I might write under the Duke's cover before I thanked you for your very kind attention to me in sending me two copies of the third volume of your most excellent *Flora Britannica*. I see every thing you do with most partial eyes. I thought that I had long since done growing, but really I fancy myself some inches taller from the deference you have paid to my humble attempts at the genus *Carex*. I have just heard that there is a lady botanist at Rochester. I shall endeavour to find her out, and hope that she is a proficient. It is really hard that during my whole residence at Windsor, as well as Rochester, I have not had a naturalist within my reach. The moment I can find one I shall run over my stores with him, and communicate with pleasure.

But this horrid war turns all men's minds to drums, trumpets, and arms.

You may depend upon my noting down any errors which I may perceive; but from what I have read, I should think there are none, for all seems quite correct.

How admirably well you support English Botany. The numbers improve as they go, I think. I wish to recommend the variety δ of *Hedypnois autumnale* to your notice, as I have found it several times just as Petiver's figure represents it. I cannot but think it a distinct species: the outermost lobe of the lower leaves always large.

There is a plant of which I have not a correct notion; viz. the *Picris hieracioides.* I always took a dwarfish plant, about one foot high, of a hard roughish tendency, to be that; but the Eton botanists assured me, that a smooth plant which grew just over the ferry lane at Datchet, about three feet high, was *P. hieracioides.* I do not find my dwarf plant in English Botany. How will you contrive about the *Fungi?*

An acquaintance of mine, whom you formerly saw at my house, a good entomologist, and a friend of Dr. Sibthorp's, Miss Hill, has lately turned her very acute mind to the study of marine plants. I have encouraged her to proceed, and I think she will produce some valuable observations. Time will show. An old friend of yours at Leyden and Edinburgh, Dr. Vaughan, is stationed as a physician, and he has good practice, at Rochester. He threatens that he will take to botany this summer. I know if while you are in town you would take a run down to us that he would be very glad to renew his acquaintance with you. I have a good large house, and you will always find a bed for you at your leisure.

Your faithful

S. G.

From the same.

My dear Sir, Deanery, Rochester, Nov. 21, 1804.

You cannot think how glad I am to receive any letter from you, particularly when it calls me to literary investigation. Having always been a student in Divinity, as my first duty, that bore a very principal part in my pursuits even when at Ealing, when interest ought to have kept me entirely to Classics. Notwithstanding, however, that I was so wholly engrossed by those two profound studies, Natural History, as you always saw, was a study quite congenial to my feelings, and continually sweetened the toils of the other two. Happily released from Classical interest, Divinity now remains the great object of my life. You may judge then how delightfully a little botanical, or rather I should say Natural History literature, relieves a long and laborious application to that ponderous employment. I need not say more to persuade you to think how gladly I always turn aside to your call,—never leading me into any thing uninteresting or cold or trifling, but into rich and most gratifying conceptions.—*Verbum sat.*

My reason for preferring *archetyp*us in my former letter, was because the noun in Greek is τυπος (*typus*), not τυπον (*typum*). It is true that Aristotle himself uses adjectively αρχετυπον (*archetypum*); but I think he uses it in the abstract sense, το αρχετυπον, *a thing that represents the original form*, meaning something rather conceived than real. But you apply it to a thing at hand, *the original* itself present. I did not recollect that you used *archetyp*um in the nominative case in *Flora Britannica*. If you did use it in the accusative, the argument will not apply. I still think *archetypus* better than *archetypum*, for 'specimen,' especially where *specimen* is contrasted with *icon, figura;* for I think αρχετυπος means *forma originalis præsens*. I do not think your generic name *Anotion* right. When a vox hiulca occurred, I observe the Greeks often inserted a consonant to please the ear, *euphoniæ causâ:* but in this word they could not, for by inserting the ν (*n*), they introduced a new word of different meaning. *Anotion*, written in Greek will stand either ανοτιον or ανωτιον. The first word signifying *non humidus*, without moisture—if any thing; the second, *non dorsalis*, without a *back*. Now it happens that Athenæus has an expression, κεραμιον αωτον, *vas non aures habens*, in which he makes αωτος, which is naturally a noun, declined (as I observed on αρχετυπος) like an adjective. This was no uncommon thing with the Greeks themselves, when they wanted to express a new idea. We have the same in our own language. We say *starlight*, making *star* an adjective for *starry*, &c. Now the original word is *αωτος*, and when declined adjectively has only two terminations (like *tristis, triste*), *ὁ και ἡ αωτος και το αωτον*. But as in botany the word *βοτανη* or *ποα* (*planta*) is usually understood, the generic name therefore should be *Aotus*, and all the specific names should be of the feminine gender: and this observation will account to you why Linnæus was advised to apply feminine specific names to so many generic nouns apparently not feminine: the feminine word *ποα* or *βοτανη* was understood. You are very happy in *Eriocalia:* the name is after the true Greek etymology, and applies with great exactness. I have some doubts about *stylidium;* not but that it is framed after the Greek analogy: the terminati on *ιδιον* (*idium*) being used by the Greeks to express a diminutive. But it happens that the original word *στυλος* (*stylus, columna*) has its diminutive already formed by the Greeks themselves. They called it *στυλις* (*stylis*) and *στυλισκος* (*styliscus*), both signifying *columella*, not *στυλιδιον*. I am of opinion, therefore, that you had better use either *columella*, the Latin word, or *stylis*, which makes *stylidis* in the genitive; which last remark I mention in case you should wish to use the word in any of its oblique cases. *Linguiforme* is undoubtedly after the true analogy of Latin words.

When you come to town, you and Mrs. Smith both must come and stay with us at the Deanery. You may work here in my study below; I have a few good books and specimens: and she may work above with my females; and I promise her she shall be taken care of. Whoever comes here, uses the privilege *uti possidetis,*—there is no restraint.

I remember very well making a hasty sketch of the Air-plant. I remember also my ideas of the physiology of the plant, which I gathered from Kæmpfer, and the directions which I gave to the duchess's gardener, under whose management the plant flowered for the only time, I believe, in this kingdom. I wish you are not running into too many species of Lichens of the *Crustacei,* and *Liprosi* divisions. Have they been watched from year to year in their changes?

I am encouraging a beautiful charming young student in botany in general, particularly in *Fuci,* &c., Lady Mary Thynne. I intend, as soon as I have a little leisure, to write to Mr. Turner for the favour of a few specimens both for myself and her.

Your faithful and affectionate

S. Goodenough.

Sir J. Banks to J. E. Smith.

My dear Doctor, Soho Square, Feb. 22, 1805.

After maturely considering the question of the title-page to Sibthorp's book, submitted to me by you at the desire of the trustees, and consulting with your friends about it, I am clearly of opinion that it is much too long, and that it will be better to divide the matters intended to be expressed in it into two parts: the one, that is the tribute to Sibthorp's memory, will be a very proper subject for the title of a book of such extreme cost at his expense; the other, the justice which ought strictly to be done to your talents and judgement by separating your part of the work distinctly from Sibthorp's part, is, I think, a much fitter subject for an introduction; because everything may there be specially stated and the reasons given, which surely ought not to be omitted, why Sibthorp neglected to do the things which you are now employed in. Thus will due justice be done to both his and your claims in a plain and intelligent manner. I therefore submit to your censure the following title, for which we are obliged to your friend Dryander, a man whose correct ideas of justice, as well in matters of property as of literature, I have not yet seen excelled by those of any other of my friends.

"Johannis Sibthorp M.D. S.S. Reg. et Linn. Lond. Socii Bot. Prof. Reg. in Academia Oxoniensi Flora Græca Edidit Jacobus Smith" (Here enter all your literary titles).

Sibthorp's intentions of making the descriptions he had at first omitted in Greece, and his journey there the second time for that purpose; the ill health which prevented him from executing his most laudable intentions; his death immediately after his return home; his will, with an eulogium on the sacrifices he made to science, not only by submitting to personal privations, hazards, difficulties, &c., but also by his magnificent pecuniary provision for this work,—will make excellent themes for your Introduction. You may in it do full justice to the memory of your friend, which a title-page can never do, and at the same time be a perfect and complete declaration of the share of the work you have had, which will come admirably well in that place, under the plea of an honourable care that no fault committed by you should by any misconception be laid to the charge of your friend.

If this plan meets your ideas, the business will be settled I think to the perfect satisfaction of the trustees, and I have no doubt of yours also. If you see any objections to it, do me the favour of stating them, and I will take them into immediate consideration. Being, my dear Sir, with real esteem and regard, very faithfully yours,

JOSEPH BANKS.

J. E. Smith to Sir J. Banks.

My dear Sir, Norwich, Feb. 26, 1805.

I take the earliest opportunity in my power to thank you for the kind attention you have given to the subject of my last, and for your excellent letter received on Sunday afternoon. It has always been my intention to prefix a preface to the *Flora Græca* upon the plan you so ably suggest. In that everything will be fully explained. The exact nature of the materials Sibthorp left, the great pains and expense he bestowed, and his zeal and knowledge, shall have as full justice done them as possible. His manuscripts, when deposited at Oxford, will vouch for all this. I shall also as simply and plainly as possible state what part I have had in the work, and the assistance I have received from you or other friends. On these heads we are perfectly agreed. I only thought it necessary, for the honour of the work and its original author, to express in the title, that it was not strictly confined to Greek plants, lest some carping critic, finding a plate of *Saccharum Teneriffæ* from Sicily, a new tulip from Florence, and a few other such extraneous matters, should object to the plan or to the title. I also thought as we are all agreed that it should somewhere be fully explained what part I have in the work, that such information ought to appear in the title.

I think, however, yours and Mr. Dryander's simplification of the title a great improvement, except that I am in doubt about the word *edidit;* for though I am aware of its true Latin meaning being nearly what I wish, yet an English, or superficial reader will, by casting his eye on it, suppose me to be little more than the publisher of a book written by Dr. Sibthorp. For this reason I strongly objected to saying in the title, that it was "composed from his manuscripts." I also doubt the propriety of the title standing as it does, without an accusative case connected with *edidit.* I shall, however, think about it; and as the title will not be wanted yet for some time, we can finally decide upon it when I go to town in April; and perhaps you or I may hit on some other word, which does not now suggest itself. I would by all means keep on the modest side in my part of the title; and I know I may trust my fame implicitly in the hands of you and my friend Dryander, who will both be ready enough to assert my claim, in case I should be underrated. I know also that my advantages would be tenfold, on the score of my ambition, by such a procedure; but about this I am really very easy, for I love the science chiefly for its own sake: I would only not put an expression into the title that should give a false idea, whether for or against me.

I am now hard at work on the *Prodromus Floræ Græcæ* (having got the great work sufficiently forward), which would be easy enough if Sibthorp had referred from his MS. habitats and catalogues to either specimens or drawings; but as he has not, it is a series of laborious criticism and investigation.

Believe me, dear Sir,
Your ever obliged and faithful Servant,
J. E. Smith.

The Rev. Dr. Goodenough to J. E. Smith.

My dear Sir, Deanery, Rochester, May 7, 1805.

I am quite rejoiced to hear that you are in town, and that you are disposed to pay me, what in my idea you have long long owed me, a visit at Rochester. I have a charming retirement here, a good house, delightful garden, quite out of the tumult of the town, and what I call a very respectable library; so that my wants are very few. I really feel but one at present, and that is the want of literary society. Had I a naturalist within my reach, or a good classical scholar, I should feel myself in delights. But we cannot have everything. I have lately exchanged my Oxfordshire preferment for a living in Kent; Boxley, about seven miles from this place, which makes a very convenient *arrondissement* of my worldly possessions.

If I live, I shall attend you on the anniversary the 24th, from which these few last years I have been necessarily absent.

Nil mihi rescribas, attamen ipse veni.

More of the title when we meet. *Edidit* in that form of words does not necessarily require an accusative after it, the word *opus* or some such thing being always understood.

You certainly are more than an editor; I would not put myself so low as that, though there are some would have you considered in no other light.

The sooner you come and the longer you stay the happier you will make me and Mrs. G., and perhaps ultimately the better it may be for you.

Ever yours most faithfully,

S. GOODENOUGH.

The Rev. Dr. Goodenough to J. E. Smith.

My dear Sir, April 4, 1806.

To speak *en masse*, I cannot object to one of your proposed alterations. All is right in point of Latinity, and seems to express what was wanted more minutely than what was written before. I understood from your copy that modern Greece and ignorance were nearly synonymous terms. When I added "*&c.*" after Homero, Aristoteli, Theophrasto, I did it to save the trouble of enumerating every author of authority who might be classed with them. You see that Pliny refers to multitudes who wrote before him. Every one will be aware that there are many who have touched upon the plants of Greece, though they are not commonly mentioned. Their memory is almost defaced, as Horace observes, *Quia carent vate sacro*.

Your title, as you send it now, will stand very well: only I would suggest, that perhaps *inven*it would be better than *invene*rat—"which *he found*," not "*had found*." Perhaps you wrote *invenera*t for *inveneri*t*, which possibly may be best of all, as it implies something indefinite, "which *he found occasionally*."

In the title of *Prodromus* the word "*omnium*" seems necessary: for the *Flora Græca* contains Sibthorp's thousand plants, which he wished to be engraved, &c.; the *Prodromus* gives an account of *all* which he found in his course of travel.—*Flora Græca Sibthorpiana* cannot be improper.

I wish you were settled at or near London again.

Yours ever,

S. Goodenough.

J. E. Smith to the Rev. Dr. Goodenough.

My dear Sir, Lowestoft, Sept. 2, 1806.

My long stay in London this year made me so far behindhand with my periodical publications, that I have been occupied with them ever since my

* *Invenerit* is the word used in the title-page.

return. We are now here for a while with my wife's family, unbending; yet I long to be at work again. I take advantage of my present leisure to chat a little with you, my good friend; as I could not enjoy your conversation in person, on account of our several engagements.

If you see Exotic Botany, you will observe in the Number for August what I have done about *Globba marantina* and other *Scitamineæ*. I flatter myself I have done some good in this Order, and shall do more. Mr. Roscoe, in a most excellent paper to the Linnæan Society, first led the way to a true knowledge of the genera of these difficult plants, founded on the form of the filament.

I have lately had a most bountiful present from my old friend Dr. Buchanan, long resident in India, of his whole herbarium of Nepal and Mysore plants; about 1500 species, with all his MS. descriptions, and near 200 fine drawings. He has given me them entirely, for publication in Exotic Botany.

I mean to go home the beginning of next week, and, except one or two little excursions, I shall be stationary here through all the winter. My work laid out for the season is to finish *Flora Britannica;* to write a popular Introduction to Botany; to finish and publish Linnæus's Lapland Tour, besides going on with *Flora Græca,* and my two periodical works. I like a variety of employment; and we lead a quiet domestic life that allows me to be very much master of my time.

The first fasciculus of *Flora Græca* is printed, but waits, I believe, for the coloured frontispiece.

The first fasciculus of the *Prodromus Floræ Græcæ* is not yet done. I shall beg your acceptance of the only copy I can give away to anybody. I hope it will be ready by the time the *Flora* is. It will contain the first five Classes. The *Umbellatæ* have given me much trouble, but I think I have learned to know them tolerably. The tribe of *Silene, Dianthus*, &c. is also a very numerous and difficult one in this work.

How glad I should be if anything could tempt you to visit this coast! My wife begs to join me in best respects to all your family; she would be as happy as myself in entertaining them here or at Norwich. Do you know our new bishop? I once thought nothing but your coming to the see of Norwich could console us for the departure of the late bishop; but really the present is a very amiable benignant character, and not without taste for botany, at least he esteems it much. Pardon me for intruding on you with this long scrawl; and believe me, dear Sir, with the warmest affection and respect, ever yours,

J. E. SMITH.

Dr. Goodenough to J. E. Smith.

Boxley, near Maidstone, Sept. 9, 1806.

My dear Sir,

Your letter of the 2nd instant followed me to this my new charming retirement. The house itself is the only spot in the parish which can be called dull; add to this, in very wet weather it is rather damp. You now know all the worst: but the whole country round is beautiful in the extreme, —such hills, such rich valleys, such extensive views, such rides, such roads, as are not to be met with scarcely anywhere. An excellent neighbourhood, and within a mile and a half of the market town, Maidstone, where we get tolerably well supplied with everything. I shall long for you to see it. But this must, I fear, be a distant pleasure.

I should have been glad to have had a quiet talk by ourselves concerning Salisbury's strange conduct. I never admired him; and from a warm conversation I once had with him about changing of the nomenclature of the genus *Erica*, I thought him too wild to take a lead. In that same conversation Dryander took part with me, and dressed him in his roughest manner. When you have to consider that you and Sir Joseph Banks divide the empire of the botanical world, you have a clue to all the awkward oddities to which such a circumstance necessarily exposes you. Well managed, these whimsicalities are mere ephemera: in angry hands the possession of a butterfly might disturb the universe: nothing is so prolific as strife.

I have seen here *Hieracium umbellatum*, *Chrysosplenium oppositifolium*, *Hedypnois hieracioides*, *Crepis biennis*, with many others fond of a chalky soil. I have looked in vain through the neighbourhood of Rochester for *Viola hirta*. *Orchis militaris* β is surely no variety.

S. Goodenough.

Dr. Goodenough to J. E. Smith.

My dear Sir, Rochester, Dec. 11, 1806.

I have just received your kind present of the first part of the Prodromus of Flora Græca. I like the idea of one good figure to every plant. As this first volume reaches only to *Pentandria,* I should apprehend that there will be two or three volumes more. How superb must *Flora Græca* itself be! I see a new work cut out for you: you only can execute it. To some species is added *icon nulla*. Then there should be a figure given, and you may count upon the number of purchasers beforehand by the sale of the Prodromus. No purchaser of the Prodromus will be without the figures of these new and rare plants. Indeed I shall look upon this work as one of the prides of my library. Would the times were more tranquil, that the work might make its way upon the continent, where it must be very interesting; but Bonaparte's rage against us increases hourly with his wonderful success. He seems to say, as Achilles did over the body of Hector, "I wish I could eat you raw." I protest now, I do not see how peace can take place, for nothing can satisfy his ambition. He has already all the land, and now he talks how hard it is that he cannot be upon equality with us by sea:—the English of which is, that he must have free access to us, *i. e.* invade us, and then—the whole world is at his feet.

It will be a pleasure to me to meet you in town

in the spring, and talk over all our pleasurable conversations; and a greater to see you here.

Yours,

S. G.

J. E. Smith to Thomas Platt, Esq.

Dear Sir, Jan. 1, 1807.

In reply to your favour of Dec. 29, it is impossible not to see the justice and candour of what you say on the subject of my engagement with you; and I feel and acknowledge that you have shown confidence in me, and that our dealings together have been more like old friends than persons hitherto unknown to each other. But I wish you and all the world to know that the protracted publication of the work (except what arose from my being almost blind for some months, and therefore unable to work at it) has been owing to the confused nature of the state in which our deceased friend left the materials, and which no one could have suspected beforehand. For instance, there being no names to either specimens or drawings, except a few; which has occasioned me infinite trouble, and (if I may say so) required eminent botanical knowledge in order to combine the materials together. This you will readily perceive would not have been the case if the same names which are in Dr. Sibthorp's journals and catalogues, my only guides to the places of growth, had been written on either specimens or drawings; because then a person of moderate knowledge might have turned at once from one to the other. Now it has cost me many weary months, and will cost many more, to surmount this unexpected and unnecessary difficulty. For this reason I conceive you must be justified in paying me that additional compliment of presenting me with the works, even in the eyes of the most rigid judges. But you well know you are amenable to nobody, and that it is left to your discretion how to use the funds in the best manner for the publication of the work. Nevertheless, I would have the whole transacted as if the world saw it, and so I am sure would you. I therefore accept the books on this ground from you and Mr. Hawkins, as Dr. Sibthorp's executors, but with even more satisfaction as a pledge of your favour and friendship for me, and of your approbation of my labours. The work is my pleasure, and I always long to be at it.

Wishing you and Mrs. Platt, and all dear to you, many happy years,

I remain, dear Sir,

Your faithful and obliged Servant,

J. E. Smith.

Dr. Goodenough to J. E. Smith.

Boxley, near Maidstone, Sept. 21, 1807.

My dear Sir,

I am always so glad to see your handwriting, ever touching upon matters of usefulness and entertainment, that I could wish you to be beset with difficulties every day, if those difficulties would induce you to apply to me. My search after truth is so strong, that I like the discussion of difficulties; for I have always found that if I have not been able to solve the difficulty itself, I have always gained knowledge of matters connected with it more than sufficient to repay my labour. In the beginning of my study of botany, I remember being employed for three days in the investigation of *Hypericum procumbens*. (You will say, how stupid I was!) But I believe I learned more from ascertaining *what it was not,* than from any other discovery which I ever made.

From all this preface, one would think that you had now thrown down some arduous matter in my way. That is not the case; but it is a pleasure to talk with you.

I am of the same opinion now with respect to your Syllabus as I ever was, and have still no doubt but that in your plain illustration and management it would become a very popular *Philosophia Botanica.* I am a great advocate for throwing out criticisms and hints of improvement upon the Linnæan system, as it is the surest mode of strengthening and establishing it. What a great advantage would it have been to the world, had Aristotle begun where Linnæus did! All the subjects of natural history would have been so well known, that we should now have been advanced to the philosophical parts of the study,—the gradations of nature in structure and œconomy; and, by having the whole before us to compare together, perhaps been enabled to find out the true uses of everything. As yet, we are not got beyond the elements; but it appears to me that we,—that is, you who are the enlightened professors of natural history,—build surely as you go, and that you have laid a foundation for solid and progressive improvement.

How I admire your English Botany!—the first thing of the kind. You must not talk of a close yet, when so many things remain to be figured.

I long for a completion of *Flora Britannica*. I shall look over all English plants again. Shall you be able to maintain all your new Cryptogamics? Do set somebody upon *Ulvæ* and *Confervæ*. Oh that I may see them illustrated before I die!

Ever yours,

S. Goodenough.

Dr. Goodenough to J. E. Smith.

Dear Sir, Rochester, Dec. 20, 1807.

Many thanks for your kind letter; but I am grieved to think by the expressions of it, that a letter from me to you the moment I received intelligence of my advancement, has not reached you. You know me very well,—what I have been, that I hope to continue; neither will one old friend have ever to say that I have forgotten him.

Rose Castle is distant, but it opens out a new country. The labours of my mind will be relieved there, as elsewhere, by the delights of natural history. I anticipate the great pleasure I shall have in seeing you and Mrs. Smith there.

My friends are too partial to me: I can never hope to answer their expectations; but it is flattering to find the King taking the lead in bearing testimony to me, and all the friends of my intimacy joining in the same strain.

Adieu, dear friend; and always be assured that I am, and shall be, as truly yours as ever.

I have not time to go on with your book yet. When I can frank, you shall hear from me.

Yours ever,

S. Goodenough.

J. E. Smith to the Bishop of Carlisle.

My dear Lord, Norwich, March 28, 1808.

I have waited, and hoped, and despaired, and hoped again (but now I despair), that you would perform your very kind promise, "*as soon as you could frank*," of letting me hear from you.

In despair then of the speedy performance of this promise, I venture to take up my pen, trusting that this will find your Lordship somewhere or somehow, and that I may express how ardently I wish to pay my congratulations in person. How happy should I be to see Rose Castle and Carlisle, where I have been so many years since! I hope to visit my sister Martyn at Liverpool this summer. I must quote a passage in one of her letters: "If the new Bishop of Carlisle or any of his family should pass this way, I should be happy to spread my cloak in their path, as Sir Walter Raleigh did for Queen Elizabeth."

My dear Lord, I presume to hope you will make my congratulations acceptable to Mrs. Goodenough and all your amiable family. Long may you be honours and blessings to each other!—so sincerely prays your ever devoted

J. E. SMITH.

From the same.

My dear Lord, Norwich, April 3, 1808.

My friend Dawson Turner has at length informed me how I may direct a letter to you. If I could be jealous of so good a friend, I should grudge Turner this letter, as I have so long languished and hoped for, and been promised one, "*as soon as I can frank.*" But I will not indulge mean passions; so I do most heartily rejoice that the above-mentioned excellent and amiable friend has found favour in the sight of one among the very few whom I have known and loved more and longer. Lambert too writes that "our friend the Bishop of Carlisle is to dine with him on Monday, and is to be at the meeting of the Linnæan Society on Tuesday." This last piece of intelligence, my good Lord, is what makes me trouble you now. You would by law, as the oldest member amongst the Vice Presidents, be in the Chair. If this be your intention, and I hope and trust there is no objection to it, it would give me peculiar pleasure; because a paper of mine on a new genus of mosses is to be read, and the Latin characters and descriptions would certainly not come so well from any other mouth. Another reason is, that I call this genus *Hookeria*, after a friend here, F.L.S.*, who is one of the best of our cryptogamic botanists, and a very promising young man,—who devotes an independent fortune to literary and other commendable objects.

Your Lordship's most faithful
J. E. SMITH.

The Bishop of Carlisle to J. E. Smith.

My dear Sir, April 5, 1808.

I have begun my career in the House of Lords. You cannot think how my time is cut up by it. Fifty-eight hours confinement there (as my predecessor's fate was) in one week is rather too much We have to thank All the Talents, or, as they are humorously styled, *all the papers*. I do not think it possible that I can be at the Linnæan Society to-morrow, for I was told today that there will be a late debate.

We had a very agreeable day at Lambert's. Sir A. Hume was there, Admiral and Mrs. Essington, General and Mrs. Grose, MacLeay and Marsham, and Lord Seaforth! He contrived with all his deafness to hold conversations and to play at cards.

Your Piquoté (I never knew before how to spell that word, neither do I know its etymology now) pink is a curious plant. I never saw a yellow one

* Dr. W. J. Hooker, now Regius Professor of Botany in the University of Glasgow.

before. Your yellow carnation surely is something extraordinary. I wish you would dispatch one to me in a letter; the sight of the flower will be very gratifying.

I long for your coming up. You will find me at No. 14 Berners Street, just as you found me at Ealing of old, unchanged in all but in name.

Yours, most truly and faithfully,

SAMUEL CARLISLE.

J. E. Smith to the Bishop of Carlisle.

Norwich, April 7, 1808.

How very unfortunate am I, my dear Lord, about your letters! and those the most interesting to my vanity and improvement: as to my heart, I had rather read that "you are unchanged to me in all but in name", much rather, than even your instructive remarks; yet never did I suspect the contrary. I remember a long and critical letter about my Tour, franked by the Duke of Portland, never came to my hands; neither most assuredly did this about my Introduction! But I shall hear your remarks to double advantage in conversation, and I shall bring my MSS. and Prodromus with me. I fear my letters and MSS. will prove as great a *bore* as "all the papers"; but I must be allowed to vent my chit-chat to my long alued correspondent, about botany or the Linnæan Society, or some such old favourite subject, now I feel that I can do it without taxing him further than for a few moments' patience to read (for I will not be so unreasonable as to expect him always to answer) what I may chance to scribble. Sir Thomas Frankland, when in parliament, used to write to me every now and then the most delightful *epistolia* of the moment; perhaps four or five lines, about some plant found, or some bird seen or heard, &c. &c. I do not despair of having some such produced by the beauties or rarities about Rose Castle. The very name of the spot enchants me. May all its present occupiers ever find it (as I am confident I shall when I visit it) "*rosa sine spinis!*"

I recollect a most charming day spent in the woods of Corby Castle near Carlisle, with Yeates and Broussonet, where I caught *Elater cupreus*, and a new *Scarabæus*, which I gave Marsham. I wonder he has not mentioned me for the former; he is very deficient in habitats for that genus. The latter is his *Sc. arvicola*, but Corby Castle is not in "Yorkshire."

I found *Elater pectinicornis* on Cromford Moor, near Matlock.

As I am on the subject of beetles, I beg leave to observe, that one of the commentators in Johnson's Shakespeare explains the "shard-born beetle"* of

* Macbeth, Act iii. sc. 2. [Shakespeare also uses the word in the following passages which elucidate his meaning. They have however been overlooked by entomologists: See Kirby and Spence, i. 392.—*R. T.*

.... Often, to our comfort shall we find
The sharded beetle in a safer hold,
Than is the full-wing'd eagle. *Cymbeline*, Act iii. sc. 2.
They are his shards and he their beetle.
Ant. and Cleop. Act iii. sc. 2.]

the poet, as born or produced among broken stones or pots, *shards*. Surely it means *borne* or flying about on *shards* or *shells* (*testæ*)? I stumbled on this the other day, and was surprised at it much*.

Yours,

J. E. Smith.

J. Hawkins, Esq. to J. E. Smith.

Bignor Park near Petworth,
April 16, 1808.

Dear Sir,

Your letter reached me by a very circuitous route and after much delay, or it would have been answered sooner. I take for granted that you are now in town, so I shall address this to you at Sir Joseph's, and shall beg you to recollect in future that I am settled here, where I shall not only be happy to receive your letters, but yourself too, whenever business or inclination prompts you to visit this part of the kingdom. My present residence has some claims to your notice, having been that of Charlotte Smith, whose numerous little poems on subjects of natural history must have engaged your attention, and from whose sister Mrs. Dorset, who is equally eminent in the same line, I purchased it three years ago.

I took for granted you would apply to me whenever you found it expedient in the progress of the work; and I am happy to learn that a second part of

* In reply the Bishop says, "I think that you must be right about *shard-born*: quandoque bonus dormitat Homerus.—We have in Scripture *born of four*; i. e. carried by four people;—so it must mean here."

the Prodromus is nearly ready for publication. I have received the second part of the first volume of the *Flora Græca*, and am much pleased with every part of it; as I was with the other, a few trifling errors excepted, which I ventured to point out, and from which this latter part seems to be free. It had escaped my memory that I communicated anything on the subject of *Briza elatior*.

Sommavera will be of great use, provided you follow him with caution, in explaining the fanciful modern Greek names.

Both he and my friend the apothecary of Zante have given a number of the ancient names of plants, which in reality are not in vulgar use. In fact, the modern Greek names of plants are generally so barbarous, that people of any education among the natives are ashamed to make use of them, and gladly avail themselves of those which they have culled from Matthiolus and other popular authors, which they do not hesitate often to put off upon travellers as the real vulgar names of the plants. Many therefore of Sommavera's names of plants must be rejected as not strictly neo-grecian, and for the same reason many of the Zantiote names.

Mr. Platt spoke with much satisfaction of his tour to Norwich.

I remain, dear Sir,

Your faithful and obedient Servant,

J. Hawkins.

Αζόγερα Argol.—Αζώγερας Zacynth.—*Anagyris fœ-*

tida. It will be proper to retain both names, as the etymology is uncertain.

Κουτζοπιὰ μουκλιὰ,—*Cercis Siliquastrum*; perfectly barbarous and unintelligible, and therefore I would recommend its being omitted. I suspect it to be Albanian.

Κουκάκη Zacynth.—better κουκάκι; means dwarf bean, if anything.

Αγριοςαφήδα Zacynth. should be Αγριοςαφίδα, from its resemblance to the currant grape.

Βεκα λουνγκα Zacynth. *Lythr. hyss.*—A lame attempt of the apothecary to give a scientifical name to a plant. The right way of spelling Beccabunga in modern Greek is μπεκκαμπουνγα.

Μεσσαδρουλα, *Reseda undata*, should be μεσαδρουλα. Etymology uncertain.

Αγγίστρα, 'Οχητρα, *Reseda alba.*—Αγγείστρα is I believe correcter orthography, but its meaning I cannot explain. "Οχεντρα is 'serpent'.

Κουκουλοφάνια Lacon.—*Euphorbia spinosa*; right accent. Etymology inexplicable.

Αγουλνάρια Lemnos,—*Euphorbia Gerardiana*; either ι ο or α seems to be wanting between λ and ν. Etymology too difficult.

The Bishop of Carlisle to J. E. Smith.

My dear Sir, Rose Castle, Oct. 1, 1808.

First let me say that I have forwarded your letter to Mr. Don. I did not know that *Limosella* was not to be found in Scotland It grows on Hounslow Heath. I cultivated it for years in my garden at Ealing.

I am vastly disappointed at not finding a single Naturalist in this part of the world. One clergyman formerly gave himself up to the study, but he seems to have neglected it totally of late years.

As I ride out I always keep my eyes upon the watch. I have found here upon very elevated moist situations, that variety of *Plantago maritima* which Hudson took to be *Lœflingii:* it seems to me different from *P. maritima* in many respects. It is much smaller, and its leaves have not the smallest tendency to be *toothed.* Next year, if I live, I hope to be able to find it in proper state for examination in its several stages of growth. Are you aware of its being found so far from the sea, and in such situations?

We have in all our moist hedges, *Salix pentandra.* Having been accustomed to see it in its *tame* state in Kew Garden, I protest I did not know it in this large free luxuriance of growth.

Just by Rose Castle I observe *Campanula latifolia.* I have seen it also in several shady lanes.

You mention *Fumaria capreolata* as a plant of the South. I have found it here in several of the corn-fields, particularly among the turnips.

Senecio saracenicus is here, but I fear that it has escaped from some garden. I find a favourite Oxfordshire plant, *Sanguisorba officinalis.* I have found also *Serapias latifolia, Salix repens,* and a tree which they call here Bird-cherry or Heckberry. By the description of the flower (it was out of flower before I came into the North, neither can I discover any fruit) I should suppose it to be *Prunus Padus*; *Rubus corylifolius*, if I discriminate the plant rightly, is by no means uncommon. Many of our hills and wastes seem to be in the same wild state in which they were left in Noah's time: I have traversed several of them without finding any thing at all uncommon. *Erica vulgaris*, *Juncus squarrosus*, *Tormentilla reptans*, *Lichen rangiferinus*, and such things, cover the whole.

I am glad to hear Dryander is so profitably employed, and particularly yourself, who I know are never idle. I have nothing ready to offer upon your Introduction; indeed you are too deep for me to presume to inform.

I was quite surprised the other day to find in the grass by the road side one single and large specimen of *Gentiana Amarella*; I could not by any research discover a second.

I long to see the fourth volume of *Flora Britannica*; and in that, nothing whatever will more move my curiosity than what you will give us upon *Confervæ*. Do figure them away in English Botany, that I may have some chance of knowing a little of them before I die.

Rose Castle is very prettily situated, and the air seems remarkably salubrious and vivifying. The wind being north, we have already cold weather: the thermometer in the house, half way between Temperate and Freezing point.

As yet we are not got into the way of supplying ourselves with necessaries readily, being six miles from the market-town, which is at best but sorrily supplied.

Your philosophical mind would not dislike the spot.

We are all of one mind about Sir Hew and Sir Harry and the Portugal Convention,—shocking!

Yours ever,

SAMUEL CARLISLE.

From the same.

My dear Sir, Rose Castle, Oct. 24, 1808.

If *Hypericum* be not already named, I would beg to submit whether *ciliatum* or *crinitum* be not the proper specific name.

Don seems to me one of your most valuable correspondents. What a number of new things has he occasionally discovered! * * * * * should not disgust him; most particularly he should avoid putting such a man to needless expense. The old Duchess of Portland used to say, talking of her assistants, "I do not mind what their bodies suffer, for they do not mind it themselves; but I always take care that their *pockets* shall not."

I am not sorry to see all ranks of people eager for an investigation of the Portugal business: Ministers are as desirous as they. How can they control the blunders of officers at such a distance? You will find that Ministers could have nothing to do with it.

I am all anxiety for the poor patriots of Spain. What shadow of right has Buonaparte to interfere, invade, and call them rebels? Are they not true to their country? Never was there in the history of man a more barefaced invasion of natural right and justice.

I shall be glad to hear from you, if it be but the chit-chat of a moment. In what state of forwardness is the fourth volume of *Flora Britannica?* How I long for it!

S. CARLISLE.

From the same.

My dear Sir, Rose Castle, Nov. 12, 1808.

Natural History totters as it were under its own weight; every one publishing, quite to the nausea of purchasers, the commonest things. When Shaw published the *Cock Sparrow* and the *Common Snail,* I thought it high time to discontinue the Naturalist's Miscellany. I thought Don's discoveries very valuable; the *Hypericum* and *Equisetum* were quite new to me. How is it that such large plants have lain hid from ages? I fear much that roguish tricks have been played by more persons than your humble servant, who often attempted to naturalize foreigners. Witness the *Sisymbrium polyceratium* in the streets of Bury, which poor Laurents and I sowed there. I cannot help thinking his *Lamium* to be a distinct species. If the leaves are constantly petiolated, surely it ought to be so named. The corolla appears also to me to be different from *amplexicaule*.

Hugh Davies sent me, while at Rochester, his

three species of articulated *Juncus;* but one of them had no leaves to it, so that I could judge of its distinction. I remember *Sempervivum sediforme* at Kew many years ago: I always contended with old Aiton that it was a *Sedum;* but I think Solander had laid down the law to him that it was a *Sempervivum.* I am speaking of a plant so named; perhaps it might be that which you now mention.

I am glad to see you write and quote Greek so fluently, and talk so big of Dioscorides. I would advise you to get another contemporary of his, Theophrastus: they would mutually explain each other; and Bodæus's notes are very copious and very explanatory.

I am sorry to see that envious creature Salisbury again barking at you;—I should rather say, *hear,* for I never see any thing of his. I shun all fellows who have a twist in their head. Never mind him; you are confessedly the king of natural history:—I beg your pardon, I should rather say, as Buonaparte, villain! does, *emperor.* All kings and emperors are abused by one party or the other, but it is beneath them to notice such nibbling.

The great Philip of Macedon used to say that he rejoiced at being abused; "For," said he, "I mend myself if this abuse be well-founded, and thus they are mortified at doing me good:—as to falsehoods, they prevail only for the moment."

This place, quite calculated for philosophical retirement, with its pure air, free from all nauseous effluvias, the very opposite to the perfume of crowded towns, continues to agree with us exceedingly. I shall be truly glad whenever you and Mrs. Smith will direct your steps to see us here.

I am glad to find *Prodromus Floræ Græcæ* is in such progress. How I long to see *Flora Britannica* complete!

Your very affectionate
S. CARLISLE.

J. E. Smith to the Bishop of Carlisle.

My dear Lord, Norwich, Nov. 17, 1808.

If you are so kind as to write me often such amusing and instructive letters, I shall be a troublesome correspondent.

I have Theophrastus by Bodæus a Stapel, and often consult it; but as Sibthorp's professed object was to illustrate Dioscorides, it is not my business to go further, at least in the *Flora Græca* or *Prodromus.* Now I want your Lordship's kind advice, both critical and botanical.—Mr. Salisbury (whom I wish in this case to consider as an indifferent person) makes *Nymphæa alba* a distinct genus from *lutea,* and I think rightly: I have said so in my Introduction, page 385. He calls it *Castalia,* from *casta* (chaste), because the petals "chastely fold over and cover the organs of impregnation";—such is his idea. I should like it better if it could allude to the Castalian fountain. Now I am about to insert these plants in my *Prodromus Floræ Græcæ.* If you will be so good as to turn to Dioscorides, you will find *N. alba* is his true Νυμφαια, which therefore ought to have been retained as *Nymphæa,* especially as there are many species of it, some of them known in our gardens,—as *N. cærulea, rubra, odorata*, &c. I wish therefore to keep this name as it is; and as Dioscorides says that the flower of his Νυμφαια αλλη, which is our *N. lutea*, is called βλεφαρα, I think this will be an excellent generic name, expressive of the closed position of the calyx, which *shuts up* over the petals entirely, and partly over the internal organs. *Blephara* I presume is feminine, is it not? The generic characters will be something like these.

NYMPHÆA. *Cal.* 4—5-phyllus. *Petala* plurima, germini inserta. *Stigma* radiatum, sessile, medio nectariferum. *Bacca* supera, multilocularis, polysperma.

BLEPHARA. *Cal.* 5—6-phyllus. *Petala* plurima, receptaculo inserta, dorso nectarifera. *Stigma* radiato-sulcatum, sessile. *Bacca* &c.

The main point on which I wish for your Lordship's opinion is the generic name, and whether I shall follow Salisbury or not.

The characters are all my own, except what regard the nectaria (which are his, and excellent), and are pointed out in English Botany, t. 159, 160. It is curious that the *N. alba* is strictly gynandrous, according to the more correct character of *Gynandria*. I think I have quoted Dr. Hull oftener than I have commended him;—he is a mere compiler.

I hope I shall never forget Philip of Macedon: but I think in the above case he would have profited by his enemy;—if so, I am right.

The Bishop of Norwich is always gratified to hear of your Lordship's remembrance, and charges me with his best compliments. He is quite recovered; though he had a fall lately at Holkham, opening a door by mistake, which led down a flight of cellar steps. He told me "it would have made a better story, if he had been a jolly wine-bibbing bishop;" but he drinks scarcely more than I do.

Lady Amelia Hume's yellow *Chrysanthemum indicum* is now in flower, *right glorious* at my elbow: it looks by candle-light of a rose-colour.

I will send a bit of my *Sedum ochroleucum* in a frank. It will grow. Your devoted

J. E. SMITH.

The Bishop of Carlisle to J. E. Smith.

My dear Sir, Rose Castle, Nov. 23, 1808.

I had the pleasure of receiving yours of the 17th instant yesterday. I am glad now, as I have been at all times heretofore, to receive communications of your literary difficulties. To begin, then—in Greek, for I must talk to you as a Grecian, this would be Και δη λεγω σοι. Much as I wish for peace and forbearance and *condescension to men of low estate*, (and in point of scholarship thus must I style Salisbury,—of *very* low estate,) I must hold up both my hands against allowing Salisbury to desecrate the name *Castalia*. To make the name of the nymph of the fountain where Apollo and all the Muses drank the purest lymph, serve for the denomination of a plant inhabiting foul, stagnating, fœtid water, and that too in a *Flora Græca*, which is to preserve the memorial of all Grecian excellence in the natural world, will be an offence of the grossest sort: *Religio vetuit.* A bad name, Linnæus says, had better be retained, than that a change should be made. But really there is reason in roasting of eggs. You cannot be bound down to a name that is execrable, and which must excite in all minds ideas of execration. Besides, how contrary is it to all rules of analogy, to make a noun *Castalia*, from *Castus -a -um?* Is there any other word of the form in all Latin? Would *latus* make *latalia?* or *bonus, bonalia?* Really if such things, so very gross, are to be allowed, natural history Latin must soon come to be a language fit for barbarians. Dryander himself cannot plead for such coarse liberties. Still further, *Castalia* is a Greek word, and has nothing to do with Latin etymology or meaning. In point of priority of language, *Castus* might have been formed (but who can say that it was?) from *Castalia, the chaste Greek virgin* (it would have been *castalus,* not *castus*); but it was *impossible* that *Castalia* could be formed from *castus*:—so much for that *absolutely inadmissible* word. Philip of Macedon would not have attempted to profit from such conceit and folly: he would have placed such a correction amongst his *contemnenda,* *καταφρονηνεα,* to speak as a Grecian.

Βλεφαρα is the plural of the neuter noun βλεφαρον. I suppose Dioscorides meant that the petals, particularly the outward ones, or our calyx, closed *curvingly,* if I may so say, as the eyelids do: and if you observe, he calls the *flower* βλεφαρα, not the *plant* itself; so that it would be false unity to make

Βλεφαρα a generic name. If you look into your Dioscorides, you will see a marginal reading taken from another copy, Νουφαρ, *Nuphar*. That would come nearer to a true generic name. If you wish to examine more deeply into this name, turn to your Theophrastus and read the explanation. This is amply detailed by Bodæus, p. 1103, and continues to half of the second column on p. 1104. See particularly the bottom of p. 1103, where he tells you that *νουφαρ* is a Mauritanic, or modern Greek word. Νουφαρ makes a medicine, *νουφαρον*. So that that very circumstance favours the idea of making *Nuphar* a generic name. But go back to Theophrastus himself, and you will see that he brings forward another name for it, *μαδωνια*. Perhaps such a termination of a word may seem suitable to the purpose; see p. 1093 (last line but two), where he says it is a Bœotian name. As you profess to be bound to Dioscorides, I declare I should prefer *Nuphar*, and should suppose it a feminine noun, and make the species ranged under it, *lutea*, &c., &c. *Madonia* is formed from *μαδων*, *calvus*, bald, having no hairs upon it. Observe towards the end of p. 1103, how it is said, *μαδων ποα και λειον*, *a glabritie caulium et foliorum*, &c.—*λειος* is *lævis*. Either of these names *Nuphar* or *Madonia* is highly classical. I scarcely know which is best: perhaps *Madonia* is; but *Nuphar* is Dioscorides's. You *must*, and you do reject Salisbury's *Castalia* here upon irrefragable grounds. In your Introduction, you have pledged yourself, not to the name *Castalia*, but merely to the separation from *Nymphæa*.

In your generic description of *Blephara*, should it not run, "Stigma radiato-sulcatum, sessile, *immune*?" *immune* opposed to *medio nectariferum* of *Nymphæa*.

As soon as Christmas is over, we shall begin to pack up for London for the winter. How I dread the angry countenances, tedious speeches, and late nights of this next session of parliament!

I am always, my dear Sir,

Most sincerely yours,

SAMUEL CARLISLE.

Pray always say kind things for me to my good Brother of Norwich.

From the same.

My dear Sir, December 14, 1808.

Flora Græca surely may raise its voice occasionally on classical grounds. On this idea I propose the three following emendations of your note of observation at the end.

α. ex fructu antidotum νουφαρον parabant veteres.

β. ascitis insuper nectariis.

γ. at minus bene Nymphæam antiquorum veram, nomine Castalia, ad novam et plane abnormem etymologiam formato, distinxit.

Nomen Castaliæ virginis illius prænobilis Græcæ, unde Castalius fons ex quo Apollo atque omnes Musæ lymphas hauriebant purissimas, ad plantam in palustribus sæpe fœtidis nascentem dignandam, desecrare, religio vetuit.

As nothing *could* be more ignorant than Salisbury's derivation and use of the word *Castalia*, I think you are fully entitled to check him, and give a classical reason.—Would it not be better to say Νυμφαια αλλη, s. Νουφαρ Diosc.?

I did not remember *estur* used in *classical* writing. But if it have obtained in medicinal Latin you are free to use it.

Sedum ochroleucum thrives bonnily.

If the Spaniards are foolish enough to fancy themselves warriors and to fight pitched battles, there will be a quick end put to all their hopes.

Yours ever,

S. CARLISLE.

From the same.

My dear Sir, Rose Castle, Aug. 25, 1809.

I wish I could give you any information about *Ethulia*. As Linnæus was no scholar, he adopted names, and formed names which are passing all powers of accounting for. Had he written it *Esthulia*, I should, from combining it with Vaillant's name *Sparganophorus*, have supposed it derived from εσθης, *indumentum*, and ουλος, *mollis*; alluding to its contributing somehow to soft clothing or swaddling. I do not know the plant. Can it be supposed to be derived from αιθω, *uro*, and ουλος, *perniciosus*? Is there anything hot or poisonous in it?

I know of no analogy or authority which will justify your elegant reference to Thule in Seneca: Linnæus never had such a classical thought in his head. When no one offers an etymology, all I can say is, that every one is at liberty to *guess* at it. If you choose to refer to Thule, who has a right to condemn you, unless he will give you a true origin? Where have we in compounded words, *e* for *extra* in the local sense of *beyond*?

Euclea was once a name for *Diana*. (See Constantini Lexicon in vocem Ευκλεια.) Was her name derived from her celebrity or her chastity? In either case it is derived from the *idea* of celebrity. For κλειω, to shut up, would never form a derivative κλεια: it is against rule.

If Linnæus did really form it from κλειω, *claudo*, he acted, as he often has done, from not knowing better. The *manifold coverings* evidently justify the idea that he did so: but it is contrary to rule; nouns are usually formed from the Aoristus, or Perfectum of verbs.

I am quite sorry that you do not come here this year, as well as next. So many things happen *inter poculum et labra*, that I always dread delays. However, if you will keep your word, and come next year with Mrs. Smith, we will endeavour to put up with the disappointment.—I fear that the death of Lady Amelia Hume will take Sir Abraham off from natural history.

If you should find any clue to *Euclea* and *Ethulia*, let me know, and I will hunt again. I remember of old that I could not do anything with *Ethulia*.

S. CARLISLE.

From the same.

My dear Sir, Rose Castle, Oct. 30, 1809.

You need not fear burthening me or the Carlisle post with your correspondence. To me, I may say, *Labor ipse voluptas.* The postboys must travel the road whether you or your friends write or not; and as for the poor horses, the only animals to be considered compassionately, they love the contents of your correspondence so well, they would eat them if they could.

Quæ regio in terris nostri non plena laboris?

I was sorry to see our Carlisle choristers' impudence blazoned forth in the public papers: it is making such *chits* of wonderful consequence. But this is the age of servants and children.

You once gave me a foreign specimen of *Ophrys Corallorhiza.* As this is now found so near Edinburgh, I wish you would bring with you next spring a good native specimen or two. How I dread the being in London! Independent of eternal noise, smoke and dirt, what a fiery session have we to expect! God guard our dear King's life yet awhile. When a change comes, Buonaparte will not be inert. I wish our troops were all well at home from Spain and Walcheren. These expeditions and the Catholic question will produce such a flame within a fortnight after the meeting of parliament, that I fear nothing will be able to extinguish. Just so Carthage fell,—the enemy at the gates, and the leading citizens immersed in party-animosities. Just so, you say, the French revolution began. The danger of these things may perhaps awaken us to a sense of union and moderation: but I fear that nothing but sufferings will correct us.

Yours ever,

S. Carlisle.

From the same.

My dear Sir, Rose Castle, Nov. 12, 1809.

I write you this amidst the hurry of packing for town, for which noisy, smoky, foggy place we are tomorrow morning to exchange the pure serenity of the unparalleled air of Rose Castle. My comforts also there will be sorely abridged by the death of my most steady friend and counsellor and patron the Duke of Portland. His excellence was but imperfectly known: never was a sounder understanding. I am sure that old as he was, and distracted by his sufferings, this Administration *must* miss him. I could write volumes in his praise.

How we are to be extricated from our difficulties I know not. I hope for good, and pray to God to grant it.

In great haste,

Yours ever,

S. Carlisle.

From the same.

My dear Sir, Nov. 24, 1809.

I was delighted upon my return to town, with a

sight of the second volume of the Prodromus of Flora Græca. I prize that work and its authors so much, that I could not refrain from writing expressly to assure you, with what sincerity I thank you for this choice present. My friendship for you sets me always a nibbling. Upon opening the book, I found repeatedly this expression, (Hab[t].) *in Delphi*, *Atho et Olympo Bithyno montibus.* Now the word *Delphi* is plural only. The Greeks called it Δελφοι, and from them the Latins called it *Delphi*, and declined it plurally only, *Delphi, Delphorum, Delphis, Delphos*, &c.

Schoolboys have been often taught, but very improperly, to say in English, the Oracle at Delph*os:* it should be the Oracle at Delph*i;* and thus in your work it should stand in all such descriptions as above. *In Delph*is, *Atho et Olympo Bithyno montibus.* All this is upon the supposition that you mean the place where the Oracle at Delphi is situated. Perhaps something else may be designed by coupling the above with the expression, *In Delphi monte Eubœæ*, which occurs several times. Is Delphi here a modern name? or is it the ancient one? If so, I should apprehend that still it ought to follow the original declination, *Delphi -orum* &c.

Sir J. Banks having a slight touch of the gout, I had the honour of sitting in the Chair at the council, and at the meeting of the Royal Society in the evening. A paper of Davy's was read, full of acute remarks, observations, and experiments upon hydrogen and oxygen gas, and to which unfortunately I have had no opportunity of turning my mind.

It is said that Lord Wellesley joins the Ministry. People seem confident that a great naval victory has been gained;—time will show.

Yours,

S. Carlisle.

From the same.

My dear Sir, Dec. 26, 1809.

Fucus discors and *abrotanifolius* are certainly one and the same. Miss Hill and Mrs. Griffiths can give you the details of proof more minutely than I can. If you choose to consult them, send their letters to me, and I will forward them. Unlike as *F. discors* and *abrotanifolius* may seem, the difference is only in the breadth of the lower leaves, which are rather rudiments of new branches: this difference is occasioned by the season of the year, depth of water, &c.; as those scientific ladies, who have watched them thoroughly, can demonstrate to you. I duly forwarded your letters to Mrs. Griffiths, a most intelligent investigator.

I took your Chair at the Linnæan Society at the last meeting. How shocked was I to see Salisbury's surreptitious anticipation of Brown's paper on the New-Holland plants, under the name and disguise of Mr. Hibbert's gardener! Oh, it is too bad! I think Salisbury is got just where Catiline was when Cicero attacked him: viz. to that point of shameful doing when no good man could be found to defend him. I would not speak to him at the anniversary of the Royal Society; I often caught his eye upon "the Bishop's wig."

Yours,

SAMUEL CARLISLE.

From the same.

My dear Sir, Rose Castle, June 20, 1810.

I had a brother who formerly lived at Hurley, close by Sir William East's. I have been hospitably received there, as I believe everybody is. I have the pleasure to tell you that the young folks after whom you make inquiry (I mean the Carnations) are alive and merry: I managed them myself, keeping them in their original ball of earth wrapped up in newspapers, and keeping the whole constantly moistened with water. I remember once keeping roots of *Ligusticum cornubiense* in the same manner for a month before I could get home and plant them: they lived for years; and perhaps they may be in existence now at Bulstrode, where I left them in the poor Duchess's time, and saw them afterwards growing year after year. How the name of Bulstrode agitates my mind!

Monotropa hypopithys certainly used to grow in Bisham woods. The Rev. Sir Henry Parker sent it to me from thence many years ago, when I was a young botanist. I was down in Cumberland time enough this year to see *Primula farinosa* in great abundance, lining our road-sides, and in our meadows. I have brought some roots into my garden.

I shall long to show you Rose Castle. Whether it be a paradise or not, every one's own mind must determine ;—Satan could not find a paradise in the garden of Eden.

Yours most sincerely,

S. CARLISLE.

From the same.

My dear Sir, Rose Castle, Sept. 14, 1810.

I am always glad to have a little botanical nibbling with you. All I give you is my opinion, whether right or wrong, and you always take it in good part. This encourages me to speak with greater freedom. Your German botanist Tlüggé is I think quite right about *Paspal*us (Gr. Πασπαλος) instead of *Paspal*um. There can be no dependence in general upon Linnæus for learned criticism. I believe he took all upon trust, accordingly as he consulted books or men. I could easily believe that he met with the word *paspalum* as the accusative of *paspalus*, and supposed it to be the nominative of the right name: or he might have seen the accusative in Greek, πασπαλον, and have thought it a nominative: or he might have thought that, like many words ending in ος, πασπαλος was read also πασπαλον in the nominative. Many have done it before Linnæus. The old Latins observing in Greek κασσιδα the accusative of κασσις, 'a helmet', taking it for a nominative, made their noun *cassis*, or *cassida*. Monstrous! So also κρατηρ, 'a bowl', made κρατηρα in the accusative. The Latins ignorantly adopted them both indifferently as nominatives. The Greeks themselves dealt with Hebrew words in the same way, of which the Greek Septuagint will give you hundreds of instances. *Paspal*us is undoubtedly the true reading. I once before let off a squib upon this very word; I thought it had been to you.

I went upon a botanical expedition the other day to our high mountains;—a great undertaking for me,—ten or twelve miles out. No sooner had I reached the spot, than clouds began to gather, and tremendous rain ensued. My girls were of the party: they must needs climb the highest pinnacle, their brother Samuel being with them, and got back without one dry thread: providentially no one took cold. I got back to the carriage without being wet through. I had just time to find three species of *Lycopodium,—Selago, clavatum,* and *alpinum; Empetrum nigrum, Pteris crispa, Aspidium thelypteris;* I believe *Parnassia palustris,* and a few Lichens; when I was fairly driven off the field by wind and rain. We set out immediately after breakfast, and could not get back till six o'clock in the evening. There's for you! Should there be fine weather I shall be for trying again.

S. CARLISLE.

This place gains upon me most exceedingly. I view it with daily delight.

——*mens expleri nequit, ardescitque tuendo.*

From the same.

My dear Sir, Rose Castle, Nov. 9, 1810.

I had a great deal to do in the former edition of *Hortus Kewensis*, and cannot but think the work highly useful and most highly honourable to Great Britain.

What a dreadful loss will the death of poor Dryander be! I do not think that he is to be replaced, in all his bearings. Possibly he was a dull plodding genius as to brilliant and classical effusions: but that said genius fitted him for every other situation which he filled. Plodding is the first quality of a librarian. None but he could have worked up the *Bibliotheca Banksiana*. Who so fit to investigate *dried* plants, and trace out synonyms in the musty journals of foreign Academies. Then he was a walking Dictionary, or rather Repertory, for all inquiries into natural history. Then also his usefulness as a patient drudge in all matters which were proposed to him, was an excellent quality. His bluntness had its great effect with innovators, impertinents, and popinjays. He had a consciousness of his real worth also, which made him a very independent character. I really am quite sorry that we have lost him. What will Sir Joseph do?

As soon as I come to town I shall have a long arrears of English Botany to run over, for I take in that work in town.

Ever yours,

SAMUEL CARLISLE.

J. E. Smith to the Bishop of Carlisle.

My dear Lord, Norwich, Dec. 3, 1810.

As to poor Dryander's loss, and his character, I can only mournfully assent to what you have so well expressed! The immediate object of my present writing is to consult with your Lordship about supplying his place, as far as we can, in our own scientific circle. I find it not easy, independent of his vast attainments and peculiar usefulness, to find any one to put into his place of Vice President.

We are lamenting the state of poor Lady East, and soon expect to hear of her death. She is one of the best and most amiable of human beings; a most delightful companion, as to taste, sense, cheerfulness, and everything that makes society charming.

I have seen much of our excellent Bishop lately, and cannot but profit by his society in many respects. He is much pleased by your Lordship's mention of him.

I find Mr. Brown is in Dryander's place at Soho Square;—his manner will be more *suaviter*, but not less *fortiter* with coxcombs and blockheads.

I hope *Hortus Kewensis* will go on.

Your Lordship's ever faithful

J. E. SMITH.

The Bishop of Carlisle to J. E. Smith.

My dear Sir, Dec. 5, 1810.

I received your obliging favour of December 3rd,

and answer it without delay. Dryander's place in the natural history department is not to be filled up. The man is not to be found in the whole world;—Brown will fill admirably well the Botanical line at Sir Joseph's; a Librarian for the Royal Society may be found;—a Vice President for the Linnæan;—a Medallist for Mrs. Banks. But his biblical knowledge, his local acquaintance with Sir Joseph's prodigious collections, is not to be had elsewhere. I attended at the Linnæan Society last night, and took the Chair, when I announced his death and the vacancy of the V. P., and enlarged upon what I have said above.

I gave a little sketch of the origin of the Society; how anxious we were to get scientific men, and them only, wherever they could be found. That no man's *rank* in life should exclude, *knowledge* being the only object in electing members. Immorality alone (and ignorance at first) was to be a bar to our choice. That our Society owed its stability to our never admitting Society-hunters, but only practical men who would have an interest in its welfare,—"*cum multis aliis hujuscemodi*". Observing as I spoke that I made great impression, I went no further for fear of disturbance.

S. Carlisle.

From the same.

My dear Sir, London, May 25, 1812.

Everything has gone off today beyond my utmost expectations. In the morning I first took the precaution of consulting every leading man in the room respecting the choice of the Bishop of Durham for our honorary member, and at length proposed him. He was elected unanimously. I immediately wrote him word of it, dating it *from the Society's rooms*. He returned me an immediate answer expressive of his thanks for the honour done him, but sent it to my house in Berners Street;—the consequence was, that I did not receive it till my return home; so that I could not deliver his thanks. There were nearly thirty members present. We sat down to dinner about fifty-three: among others were present the Bishop of Winchester, Lord Stanley, Sir Nash Grose, Sir George Staunton, Sir Thomas Frankland, Mr. Poulter, Mr. Woodward, D. Turner, Davies of Trinity College, &c. &c. I was sorry to see a falling-off of some of our old members,—Dickson and Fairbairn and Francillon,—none of whom were present in the morning or at dinner. Hoy was there. We drank the usual toasts,—yours of course; when I took an opportunity of repeating what I had before said in the morning respecting your utter inability to attend, your great regret at this hindrance, which now occurred for the *first time* since the formation of the Society.

The Vice Presidents were drank of course.—After I had returned thanks in a short speech, Sir Nash Grose would give my health individually, as President for the day. It was rather hard to return thanks *twice;* however, I somehow contrived it. I returned Sir Nash's compliment by giving, "The Laws of Old England, and our very able Ministers of those Laws."

This induced Sir Nash to make a speech of thanks. As it happened, all these little things (together with Marsham's, MacLeay's, and Brown's thanks) enlivened the day not a little, and sent every one home expressing what a pleasant day we had had.

I must say that I never knew a day go off more harmoniously.

Marsham and MacLeay will give you all the detail of the finance matters and other particulars more immediately within their provinces. Dillwyn was with us; Mr. Forster, who has lately, he told me, lost his father; Dr. Maton, Mr. Lambert, Mr. Symmons, Mr. Rackett. Dr. Rees was at the rooms in the morning. I pressed him to dine with us, that I might have an opportunity of improving my acquaintance with him. He excused himself from dining with us, pleading that he was not well.

Mr. Hodgson, Mr. Knight, Mr. Davies Giddy, were with us in the morning, but were engaged to dine at other places.

Sir A. Hume and Lord Valentia, the Bishop of Salisbury and others, had promised to dine with us, but did not come. However, "All's well that ends well." We must not expect at any time to have everything our own way.

This will relieve Mrs. Smith from no small anxiety, and your many other friends.

It will be a great pleasure to me to hear you are better.

I am, dear Sir,
Your very faithful humble Servant,
S. Carlisle.

From the same.

My dear Sir, London, Feb. 25, 1813.

You are very good in turning your thoughts towards us and our distresses. Indeed we have sustained a great loss: in one sense, perhaps, I the greatest; for my poor daughter was possessed of a mind and understanding all usefulness and resource: I had depended upon her to arrange many matters in case of my death, having no idea but that she would have survived me. Nothing was too difficult for her comprehension; nothing happened cross and untoward but for which the resources of her mind could find some remedy or alleviation. But so was the will of God; and we have nothing to do but to accept his pleasure, and be thankful that no worse thing befell us. In all distress it is fruitless to look back and regret; we should only still press forward, and look to the comforts which still remain. It is both our duty and our happiness so to do. What a gap is made in our little domestic society! What a permanent loss in every respect!

I am sorry to tell you that Sir J. Banks is very ill. I called there today, when the servant, with most mournful countenance, said he had been better yesterday, but was worse today. Sir Everard Home said this morning that he was rather alarmed about him. He is one that cannot be well replaced. I shall still hope that his strong constitution, as of iron, will carry him through. His servant described him to me as being in great pain. If this proceeds from gout, it is surely no bad symptom. Moderate quantities of *l'eau medicinale* failed in relieving him this time: he took an immoderate quantity, which produced immoderate effects; hence malady and weakness ensued.

Mrs. Goodenough is much affected, but bears all with Christian patience.

I am, dear Sir,

Your most truly affectionate

S. CARLISLE.

From the same.

My dear Sir, Rose Castle, Oct. 25, 1814.

I received your very agreeable letter of the 18th. Both you and the Bishop of Norwich must have had a rich treat in the library at Holkham. You took your lines very appropriately; you the History of Italy, and he the MSS. of the Greek Testament.

If you can remember to answer such a trifle, I wish you would in your next tell me if the spirits (asper ʽ and lenis ʼ) were marked in those MSS.

That missal so full of miniatures must be a great curiosity. Do the Roman Catholics attach any *religious* worth to the possession of such costly missals? Is the possession of them deemed doing God service, or a proof of piety? I have heard these points argued, but I own never satisfactorily.

I never see the Monthly Review now. When old Griffiths died, the managers of that work took such liberties with our church establishment, and the doctrines of our church, (not in argument, for that would not be unfair, but in invective and sarcasm,) that I left off taking it in. When I return to town, I will endeavour to get a sight of it.

Yours,

S. Carlisle.

From the same.

My dear Sir James, Rose Castle, July 26, 1816.

I am always happy when I receive a letter from you: it is amusing to compare the different accounts of things from different persons in different situations; they usually vary very much in those accounts. But there is one particular now in which all accounts from all quarters agree, viz. about the weather: we have had scarcely a day without rain since we came into the North.

I admire some of those agricultural meetings, particularly when they are purely agricultural, like Mr. Coke's, and not as they have been in this county, mixed up with politics.

I am glad to hear that the House of Buckingham has so fair a promise in Lord Nugent. The world never will move better and in more natural order than when the greatest men in it are the wisest and the best. A *novus homo* has a deal to struggle through before he can get into a commanding situation. A man of rank, and of real knowledge at the same time, comes into it without an effort, and thus a great deal of time and contention is saved.

I do not like this Indian nor this Algerine war. Wars are prolific monsters, and soon people the world with those of their own kind. Would that war was completely banished from mankind!

My son Edmund has been at Malvern. He was puzzled there with that variety of *Ervum tetraspermum* which has footstalks with one flower only. I do not recollect ever to have seen your variety of *I. Xiphium.*

S. Carlisle.

From the same.

Dear Sir James, Rose Castle, Sept. 15, 1819.

I will begin answering the last matter which you mention,—your new edition of the Flora: I hope that you will not abandon that idea. Had you not so very lately given a new edition of your Compendium complete, I should have advised an edition of it, and also have wished that you had announced your intentions of publishing your edition for the ladies, and stating what you had got already for the Flora.

As that edition of the Compendium cannot be nearly taken off yet, I fear you must risk your edition at once. It must be a most respectable work, far beyond what has been produced in any country under the sun. I cannot say anything to the English Flora, not knowing your plan. The old plan of your Flora Britannica was, I always thought, extremely good, and such as needed no mending.

I have never seen Fonthill, but at a great distance *en passant*. What an amazing inclosure! fourteen miles of wall!

Yours most sincerely,

SAMUEL CARLISLE.

From the same.

Dear Sir, Berners Street, Nov. 29, 1819.

The famous American serpent is at length ascertained to be no fiction. It seems that there has always been a *rumour* of this animal: Aldrovandus mentions it among others. However it has never been caught and described. It has now been seen by three hundred people at once, and hopes are entertained that ere long this will be taken. It is of immense size and length.

This feverish anxiety cannot last long, as things must come to a crisis soon. And Cobbett's landing with Tom Paine's bones will be adding oil to the flame. All this is frightful.

I am, dear Sir James, yours,

S. CARLISLE.

From the same.

Dear Sir James, Rose Castle, Sept. 28, 1821.

I had been expecting a letter from you according to annual custom, I own; but that did not stop me from writing to you, but an uncommon variety of business, and a greater variety of domestic troubles than usually falls to one's lot at one time.

We were hindered coming down hither at first by the two servants whom we leave here in our absence being seized with typhus fever: they could not join us for a considerable time after our arrival. This confused the regularity of our house, and gave us no small trouble, and some expense. As soon as we were all set to rights in this respect, my housekeeper's husband, a healthy young active man, was suddenly seized with a swelling in his thigh and leg, so that his clothes were obliged to be cut away to free him from them: this makes her very uneasy, and paralyses her services. My coachman's wife, who keeps our house in Berners Street, and has been expecting to be confined all this month; a person has come down from London and told him she is very ill, so that this keeps him upon the fret. Then the other night, Mrs. Goodenough's maid, a nice delicate young woman, fell down suddenly in a fit, and has not been thoroughly herself or useful since. One of my labourers has just lost his cow, and he has a son in a rapid decline. To add to all, our letter-carrier is taken suddenly ill; so that I have nothing to hear but complaints, and nothing to see but moping figures. A pretty picture this of domestic inconveniences!

Lady Melville the other day gave me a very fine pot of *Gloxinia speciosa*. How like the flower was to *Digitalis*, at first sight. The cold begins to pinch it. I am very much afraid that when I go away, it will go the way of all flesh, *i. e.* die. It is beautiful in flower and health just now, and has a great many flowers.

Yours always,

SAMUEL CARLISLE

Sir J. E. Smith to the Bishop of Carlisle.

My dear Lord, Norwich, March 3, 1822.

I am steadily at work at my English Flora, and hope to get it into the press when I go to town, which I cannot spare time to do before the beginning of May. I find so much to do in this work that it will be quite an original Flora, the whole subject being revised after so many years experience; and nothing, or worse than nothing, done by most writers since I began. To try to fix the language, revise all generic and specific characters, enrich and correct synonyms, add remarks on natural affinities, clear away the mischief done by compilers,—all this is an Herculean labour. The object of the book is *botanical determination.*

I am now finally disposing of the grasses, and hope I have done some good, as also in the *Triandria Monogynia.* Brown is very great in these.

My present employment makes me a very bad correspondent.

I remain

Your Lordship's devoted and most faithful

J. E. SMITH.

The Bishop of Carlisle to Sir J. E. Smith.

Dear Sir James, Berners Street, March 8, 1822.

You are entered indeed upon a Herculean task, if your English Flora be intended to embrace what you propose; to fix the language, revise generic and specific characters, enrich and correct synonyms, add remarks on natural affinities, and clear away the mischief done by compilers,—and to correct habitats. One would think that this is sufficient for a first edition. As the object of your work is *botanical determination*, it certainly might fairly include the accenting of the generic and specific names throughout. But have you time for all this before the beginning of May? I shall very gladly lend you all the assistance which I can give. When I turned my thoughts to this subject, I adopted two characters to mark the quantity of the penultimate and antepenultimate syllables. Where the syllable was pronounced long, as in *pu̍rus, edu̍lis, Polypo̍dium,* I put one straight stroke (̍); where it was pronounced short, as *ma̎gnus, palu̎stris, Chrysa̎nthemum,* I put two straight strokes (̎). *I* preferred this to the more learned characters of ¯ and ˘, because we accent a living language according to the *use,* not the *nature* of the syllable. Thus there are some refined critics who would pronounce *polypo̍dium polypo̎dium,* the *o* in *podium* being an *omicron,* and thus by nature short. But this in the present state of things would be perplexing, because great tenderness and attention should, I think, be paid to practical men, and ladies and gardeners, who have been accustomed to pronounce so. I give you my opinion and my reasons; and therefore, if you enter upon this part of your work, you will judge for yourself.

I do not see that you are called upon to give the etymology of the generic and specific names: it would open into dispute, and take off the mind from your main object, *botanical determination*. These things must, I think, be left generally to scholars.

Yours,

SAMUEL CARLISLE.

Sir J. E. Smith to the Bishop of Carlisle.

My dear Lord, Norwich, March 13, 1822.

I derive, as usual, great assistance and encouragement from your Lordship's instructive letter. I think I may be spared explanations of generic names, except such as are new to British readers.

As to accents, your Lordship's plan is excellent: I well remember your telling me about it,—but perhaps it is more than I want. Common English readers, not wanting to make Latin verses, would never understand *ma̍gnus*, and *palu̎stris*, &c. It will be enough, I apprehend, to put a over the syllable on which the accent is to be laid. This I can do easily in most instances; and when I am puzzled, I know whom to consult.

DeCandolle has introduced a new term for the

single grains or parts of compound fruits, as in *Uvaria*:—*carpella* (plural?); but should it not be *carpiola*? as the original word is Greek. I do not however admit it at all, any more than his *sepala* for segments or leaves of a calyx. I must adopt Schreber's genus *Spartina*, for *Dactylis stricta*: it is very distinct. Is not *Spartina* a good name, alluding to the broom-like, or cord-like stems and leaves? The French have subsequently called it *Limnetis*,—a good name; but it ought not surely to supersede the older one.

Mr. Coke is too good and too prosperous to escape envy. I regret that party should make such men as him enemies. As to the bulk of each party, it matters little. I esteem him one of the best of human beings. I allude not to politics, which I often regret should disturb so fine and happy a mind as his.

Mr. Roscoe is preparing for our Society a very excellent distribution of the species of *Canna*, now very numerous.

I am ever, most respectfully,

Your Lordship's,

J. E. Smith.

The Bishop of Carlisle to Sir J. E. Smith.

Dear Sir, March 14, 1822.

When I put my two strokes over *mägnus*, &c., it was more to show you my plan, than any wish that you should mark words, dissyllables or others, which the common genius of our own language would so readily point out. Whether the single stroke will sufficiently teach the pronunciation of words of three or more syllables, a little use and experience will convince you. If, for instance, the word *Ricinus* occur, will the same mark ′ indicate that the *i* be pronounced short, as if two consonants followed it? I only throw out this for your consideration.

DeCandolle was quite mistaken about his term *carpella:* it is evidently taken from the Greek word καρπος, and is a diminutive of it, and so far is of classical analogy. But καρπος is a masculine noun, and so must its diminutives be; and therefore the word should be *carpelli* not *carpella*. But why is not Linnæus's word satisfactory? Thus in *Rubus,* he says, "Bacca composita *acinis* monospermis." Who is at a loss?

I can make neither head nor tail of *Sepala.* I should think with you that the term is not called for.

As *Dactylis stricta* is so distinct a genus, I certainly should prefer the older appellation of *Spartina* (although it does not quite suit my ear and taste) to the more modern one *Limnetis.* I call *Limnetis* no name at all; it is derived from the Greek word λιμνη, *palus*—i. q. *palustris.* But there may be others found perhaps growing in marshes also;—in this case, as in many of the Linnæan genera, the term *palustris* does not assist us.

I am, dear Sir James, yours,

Samuel Carlisle.

From the same.

My dear Sir James, Rose Castle, Sept. 6, 1822.

I observe one odd thing this year. Although last year we had such a multitude of wasps that they ate up all our fruit, this year not one has been seen yet. Another odd thing took place:—the swift left us in the beginning of August; and during last week, when there were violent rains, storm and tempest, there was not a swallow or martin to be seen; but this week they have appeared again. Was it that the wind and storms destroyed their food? and now the serenity of the atmosphere has brought them all back again,—I cannot say in such numbers as before. Indeed this year they came some weeks later, and were by far less numerous than usual.

I envy you at the sea side.

Yours,

SAMUEL CARLISLE.

The Bishop of Carlisle to Sir J. E. Smith.

My dear Sir James, Berners Street, Jan. 25, 1823.

I am not at all surprised at Cobbett's manœuvres; I only wonder that such things have not taken place before. For when parties meet, such a multitude of inflammatory speeches are made, that the passions of the ignorant must be set to work. The principal people concerned, think that the rabble are too far off to presume to take a lead; but that distinction is set aside, more or less, at every meeting, and the consequence is to be dreaded. Now at the York meeting, it was fine talking to say that our Constitution is made up of three distinct *limited* powers. But what is to keep them limited? Mr. Wortley could not be heard, so that here *in limine* there was one unlimited power that *overruled.* They are three limited powers, I grant you; but the argument is too fine drawn for the lower classes to be held by it. There is nothing new in all this; it is thoroughly well understood, and has practically been observed as far as circumstances will permit: but occasionally one part or other has outrun its limits, and evil has inevitably ensued. I own I dread seeing things pushed to these nice extremities: I say this as a true Whig.

I am glad to see that you have taken the *Umbellatæ* in hand. I have been more puzzled in that class than in any other. The *Involucra* and *Involucella* are by far too unstable to form decisive distinctions. If the fruit will furnish distinctions broad enough, you will have done a great work. Will your system hold good through all the foreign genera?—I can write no more than to assure you and Lady Smith how truly I am

Yours,

S. Carlisle.

From the same.

Berners Street, Dec. 11, 1823.

My dear Sir James,

What a *winter summer* we have had! The weather at Rose Castle was all the time as cold as at Christmas: we had fires in all our rooms from morning till night, just as we have in the depth of winter.

Your accurate distinction between *Geum* and *Dryas* is very satisfactory. How many times I have looked at both those genera, without having an idea or a suspicion of your nice distinction!—but I have often observed that old men go on in the errors which they imbibed inadvertently in their early age. It would be well if they could begin again according to the old adage—*Disce docendus adhæc.*

The singular generic character of *Icosandria* is so very striking, that I believe it gives one a careless way of examining its plants. Hence our errors and our difficulties so late to be discovered.

The produce of our gardens was sadly damaged this year. We had no apples: our pears did not ripen: strawberries and raspberries were in plenty, but they had not by any means their usual flavour. Yet, what was remarkable, we had mulberries, of which we have had none but twice since I have been at Rose Castle. Our corn was very late. When I left Cumberland, the 3rd of November, there were multitudes of acres not reaped, all of them almost as green as grass. We have *one peach-tree* from which we gathered ripe and well-flavoured fruit the *last* week in October.

Sincerely yours,

SAMUEL CARLISLE.

The correspondence between these friends continued to the end of 1826 with its original spirit; but from that time the declining health of the Bishop became visible in his handwriting, and his subsequent letters were chiefly upon the progress of infirmity. Like most admirers of Nature, he possessed warm domestic and social affections, and great love for the enjoyments of home. Speaking in one of his earlier letters of a family possessing the same qualities, who had been bereaved of a tender parent, "The description," he says, "of such an harmonious family, is as delightful as it is serviceable. When one wants nothing from without, how sure and constant our comforts are! and that is the case with a family at unity in itself. Oh, how good and joyful a thing is it to dwell together in unity! It is a pleasure even for strangers to look upon.

"Parting is a hard task. I do not wish for stoicism enough to set me above such fine feelings. In the end they are a happiness for us, ever recalling the connexion while it subsisted, teaching us the value of everything of the kind, and inviting us to look forward to other modes of it."

END OF THE FIRST VOLUME.

PRINTED BY RICHARD TAYLOR,
RED LION COURT, FLEET STREET.

Printed in the United States
By Bookmasters